Our Common Illiteracy  KW-266-314

90 0859983 1

Environmental Education, Communication and Sustainability

edited by Walter Leal Filho

Vol. 10

PETER LANG

Frankfurt am Main · Berlin · Bern · Bruxelles · New York · Oxford · Wien

Rolf Jucker

Our Common Illiteracy

Education as if the Earth
and People Mattered

PETER LANG
Europäischer Verlag der Wissenschaften

Die Deutsche Bibliothek - CIP-Einheitsaufnahme

Jucker, Rolf:

Our common illiteracy : education as if the earth and people
mattered / Rolf Jucker. - Frankfurt am Main ; Berlin ; Bern ;
Bruxelles ; New York ; Oxford ; Wien : Lang, 2002
　　(Environmental education, communication and sustainability.
　　Vol. 10)
　　ISBN 3-631-39117-X

The author would like to express his gratitude
to the Leverhulme Trust
for supporting the research for this book
with a Study Abroad Fellowship.

ISSN 1434-3819
ISBN 3-631-39117-X
US-ISBN 0-8204-5483-4

© Peter Lang GmbH
Europäischer Verlag der Wissenschaften
Frankfurt am Main 2002
All rights reserved.

Printed in Germany 1 2 3 4 6 7

www.peterlang.de

We have to look after sustainability ourselves. Because nature won't force us. According to its rules "fools" who jeopardise their vital long-term interests are simply dismissed from biological evolution. Unfortunately, this doesn't work on an individual basis, but collectively, so that the reasonable, if they want to prevent this, have to find ways to stop the foolishness of the unreasonable. (Dürr, 2000, 9)[1]

As *Homo sapiens*'s entry in any intergalactic design competition, industrial civilization would be tossed out at the qualifying round. It doesn't fit. It won't last. The scale is wrong. And even its apologists admit that it is not very pretty. The design failures of industrially/technologically driven societies are manifest in the loss of diversity of all kinds, destabilization of the earth's biogeochemical cycles, pollution, soil erosion, ugliness, poverty, injustice, social decay, and economic instability. (Orr, 1994, 104)

All human institutions, programs, and activities must now be judged primarily by the extent to which they inhibit, ignore or foster a mutually enhancing human-earth relationship. (Thomas Berry, quoted in: O'Sullivan, 1999, 43)

While it is unwise to believe in any one environmental projection of the future, it is important to bear in mind that nature bats last *and* owns the stadium. (*Natural Capitalism*, 2000, 316)

The question is whether any civilization can wage relentless war on life without destroying itself, and without losing the right to be called civilized. (Carson, 1991, 99)

Civilized people depend too much on man-made printed pages. I turn to the Great Spirit's book which is the whole of his creation. (Native American Indian, quoted in Zinn, 1996, 514)

Modern man has failed in his effort to be god. (Gustavo Esteva)[2]

The real alternative [to the dominant Euro-American society] is the self-governing of society through the "evil horizontal" of a comprehensive soviet system; and such self-government is the opposite not just of the state, but also of the market. (Kurz, 1999, 789)

The real historical challenge (...) must be addressed in something other than ecocratic terms: how is it possible to build ecological societies with less government and less professional dominance? (Wolfgang Sachs)[3]

To talk about the future is useful only if it leads to action *now*. (Schumacher, 1993, 9)

Illusions mistaken for truth are the pavement under our feet. They are what we call civilization. (Barbara Kingsolver)[4]

---

[1] All translations, where a professional English translation of material originally published in other languages couldn't be found, are mine.
[2] Esteva, 'Development', in *The Development Dictionary*, 1992, 6-25, here 23.
[3] Sachs, 'Environment', in *The Development Dictionary*, 1992, 26-37, here 35.
[4] Barbara Kingsolver (1999), *The Poisonwood Bible* (New York, HarperPerennial), 532.

Preface

As is customary in places like this I would like to pay tribute to the numerous people without whose help this book could not have been written. First and foremost I need to mention my children, Meret and Mathis, and my partner Susanne Schüeli, who made sure that I kept on living in the real world while writing and researching, which is not so easy in the world of academics (as we shall see). Next I was greatly helped, again in much more than merely intellectual ways, by my friends: Paul Thomas, my Buddhist who showed me that down-shifting can even work in a metropolitan context; Antje Ricken whose feminist perspective and wit have prevented me from descending into cynicism; Heinz Lippuner, once my thesis supervisor, long converted into a regular discussion partner, equipped with an unusually sharp eye for our contemporary follies, trained in many decades of exposure to indigenous, especially Native American peoples and thoughts; Vrena Sulser, always reminding me of the perspective of current students; Christine Tresch, for her almost unconditional, sympathetic support, yet precise listening and criticism skills; Stefan Schütz, for being the original inspiration, through his literature and our discussions, for this project; he, together with Volker Braun, through their sharp and comprehensive analysis of contemporary capitalist society *through the medium of literature*, are responsible for my continued infatuation with good literature (not that there is too much of it around); Volker Braun also for writing the most beautiful poems in German today; and last, but not least, Chet Bowers, whose insistance on eco-justice and non-commodified ways of learning has been very productive for me.

There are others who have contributed directly or indirectly. My colleagues in Swansea, particularly Tom Cheesman, but also Terry Holmes, Beth Linklater and Marie Gillespie deserve special mention, but also my head of department, Rhys Williams, who has always generously supported me, even if he privately might have disapproved of what I was doing. During my research year in Switzerland, which was generously supported by a Study Abroad Fellowship of the Leverhulme Trust and a sabbatical granted by the University of Wales Swansea, I was particularly encouraged by the crew at the Collegium Helveticum in Zurich. All of them, but in particular Christian Pohl, Michi Guggenheim and Regula Schälchli, have shown an openness, generosity and readiness to help, to an extent which I have never before experienced in the context of a university institute.

There is another group of people I want to pay special tribute to. I have, in most cases, never met these people, never spoken to them, yet still they have exerted the most decisive influence on the formulation of my ideas during the last decade. I am talking of Noam Chomsky, Wolfgang Sachs, Douglas C. Lummis, David W. Orr, Vandana Shiva, Fritz Schumacher, Ivan Illich, Neil Postman,

Mathis Wackernagel and the people from the Corporate Europe Observatory (CEO).

Lastly I would also like to thank my parents, my brother and my sister for their continued support, and for making time to discuss ideas.

Swansea, January 2002

Contents

Introduction

Rational individual action leads to irrational collective results.
(Huib Ernste)[5]

The comedy of humankind is that it lives below its knowledge.
(Volker Braun, 1998, 111)

One of the most ironic features of our times is that, despite our unrelenting search for new knowledge, we do know quite enough already. As I will attempt to show in chapter 1, we do in fact know, in a sufficiently robust manner, the ecological, economic, social and political parameters which we need to observe in order to live sustainably within the carrying capacity of our life-support system, planet Earth.

But if I say 'we', that is not quite correct.[6] There are societies on Earth that embody this knowledge implicitly and live accordingly within the limits of nature. As for the rest of us, the situation is rather more embarrassing, since it amounts to a comprehensive illiteracy. We are or have become thoroughly unfamiliar with the *real* world, with both the physical conditions of our planet as well as the social, economic and political conditions which exist today and which would need radical alteration to bring about such things as equity, social justice and self-determination (see Jacobs, 1996).

To a large degree this illiteracy is self-inflicted. The relevant knowledge is available, accessible for most of us in the so-called developed countries – and we, after all, would easily win any competition for unsustainable behaviour. And if we take the notion of responsibility for our own lives seriously, we cannot but acknowledge, as Ehrenfeld once put it, that we have "no right (...) to be ignorant" about sustainability (quoted in Orr, 1996, x).

Yet, as I shall outline in chapter 2.b, things are slightly more complex. There are massive vested interests, there is a "deep cultural pathology" of denial, as Thomas Berry has called it (quoted in Orr, 1999a, 221), there is a huge worldwide propaganda industry at work, virtually all our educational systems are geared to unlearn whatever we would need to know, and the commodification of

---

[5] Ernste, 'Kommunikative Rationalität und umweltverantwortliches Handeln', in *Umweltproblem Mensch*, 1996, 197-216, here 212.

[6] I will nevertheless carry on using 'we', fully aware that I am writing from a Euro-American perspective. When I say 'we', I mean the affluent North and the elites of the South and do *not* intend to implicate indigenous or poor people wherever they may live. I am doing this because I believe – as the book should make clear – that 'we', as defined here, have to face up to our collective responsibility: 'we' are primarily responsible for the destruction of the earth and therefore primarily called upon to act. To quote Sartre: "With us [= the Euro-Americans], to be a man is to be an accomplice of colonialism, since all of us without exception have profited by colonial exploitation." (Sartre, 'Preface', in Fanon, 1967, 21-22)

information creates new barriers of access,[7] so that it is difficult to lay the blame for this illiteracy squarely at the individual person's door.

This raises important questions for a book like this. It would usually be assumed that to talk about (university) education would be to do just that. But, as I will attempt to show in chapter 3, this would perpetuate the illiteracy rather than counteract it. As long ago as 1971, Ivan Illich observed with his customary sharpness: "Any attempt to reform the university without attending to the system of which it is an integral part is like trying to do urban renewal in New York City from the twelfth story up." (Illich, 1971, 38)

In essence, this means that we have to take the notions of transdisciplinarity,[8] intellectual honesty and responsibility seriously. We cannot carry on praising and demanding interdisciplinarity in the mission statements of all our research funding institutions and in our educational declarations of intent, only then to marginalize and discourage it in our actual endeavours. We have to start practising what we preach. We need to overcome specialisation – which, in our context, more often than not actually means blindness to the contexts and consequences of our specialist 'solutions' – and establish what Wolfgang Sachs has called for: new notions of excellence within our professions.[9] We have to venture into all aspects and areas where these are relevant to the topic. I know that this is not easy, because we are constantly taught that only the experts know. I shall have more to say about the state of affairs this expert advice has landed us in (chapter 2.a), but the mere observation that it was not the illiterate peasant farmer from Thailand, but the highly educated experts from the best universities in the world who have wreaked industrial destruction on Earth should suffice to indicate that we must begin to develop a more accurate vision. Undoubtedly, we will make mistakes; we will find it difficult, because we are thoroughly unaccustomed to thinking in systems, in adequately complex form. But it is not impossible, and we shouldn't be afraid of making fools of ourselves, because it is an absolute necessity that we get better at looking at the whole story: our survival will depend on it. And if we don't try, we will never get better at it.

I can give you an example from my own discipline, German Studies, of the unwillingness of university people, working in specialist disciplines, to expand their horizon and look beyond their narrow view on issues. But before I do that, let me just quickly justify why I, a German Studies scholar, should dare to write a book about sustainability in general and university education in particular. According to the traditional view, I would be entirely disqualified from taking part in

---

[7] See Noam Chomsky, 'Equality. Language Development, Human Intelligence, and Social Organization (1976)', in Chomsky, 1992, 183-202, here 189.

[8] A good definition seems to me to be Dieter Steiner's: "*Transdisciplinarity = interdisciplinarity + participation* [= to plan and carry out projects whenever possible in suitable cooperation with affected people and users]" (Steiner, 1998, 308; see also Nowotny, 2000, 2).

[9] In a talk at the Schumacher Lectures in Bristol, 30.10.1999; see below p. 250.

such a discourse because I am not a specialist: I have no formal training in ecology, the natural sciences, or pedagogy, for that matter. But, as I will attempt to show in chapter 3, precisely this way of thinking lies at the heart of our troubles. If we do not accept our responsibility (as far as it goes) for the status quo and take seriously the claim routinely made in international documents and manifestos of political parties that sustainability is of concern to every one of us, we will never be part of the solution. Usually one faces two objections at this point. Firstly, that somebody is not entitled to take part in a discussion because of a lack of formal training, and secondly – the other side of the coin – that one cannot know anything about the issue in question, because one is not a specialist. More often than not, these are just convenient excuses for not facing up to the reality around us or for justifying the status quo. And this is precisely the reason why I intend not only to talk about university education, but also about education in general. Furthermore, EfS (Education for Sustainability) itself cannot be achieved without both a clear notion of where we should be heading (chapter 1) and what is unsustainable in our current world (chapter 2). I also decided to attempt this broader view, since on the one hand, most people I talked to during my research had not even a vague notion of sustainability, and on the other hand, academics talking about EfS themselves rarely spend much time on the context and instead instantly dive into specific pedagogical questions. I shall argue that this is part of the problem, not the solution.

But let me proceed to the promised example: a few years back I was editing a volume for what is considered a prestigious international series in German Studies. The call for papers asked for contributions on utopian concepts in contemporary literature *and* society. The intention was to look at interesting, mutually productive ways in which literature could inform society about its future developments and how an understanding of contemporary issues can enrich our readings of literature. But the volume – I have to admit it, even though I was responsible for it – was a failure. Not even a third of the contributors took on the challenge of trying to bring the two fields of literary and social study together or to transcend their disciplinary vision in new ways. The others, in varying degrees, just carried on doing what they had been doing in literary studies throughout their careers: dissecting texts as if they existed in a void.

This is the attitude we will have to overcome. We will have to be prepared to work from real problems towards real solutions, and to face up to the complexity that this entails. We should stop reducing issues so that they fit the artificial boundaries of subject disciplines.

But in order to succeed with this task we will need a willingness to critically reassess almost all our deeply held assumptions about 'the market', 'freedom', 'democracy', 'education' and much else. We need a radical openness of thought, a commitment to precise thinking, a readiness to double check utterances which we have become accustomed to believe, but also, as will become clear, a dedication to action rather than mere words. And above all else we need cooperation and dialogue. Rather than just borrowing 'ideas' from other disciplines, we need to integrate them into a fuller picture and we shouldn't then be afraid to be enlightened by other people, be they specialists or practitioners. Modesty will emerge as one of the most important values for a sustainable society (see chapter 1.c and Haraway, 1997), and that is what we need in discussions as well. This will prevent us from blunders of the sort the postmodern French philosophers (and their followers) committed when they dabbled in chaos science, mathematics and physics without understanding much, but in an attempt to lend authority to their impenetrable, preposterous and often false or banal arguments, as Sokal and Bricmont have shown (1998). We have no other hope than a "rational vision of the world" for our endeavour. This by no means excludes feelings and passion, but it does exclude "superstitions, obscurantism and nationalist and religious fanaticism" (Sokal/Bricmont, 1998, 195), in particular the twin quasi-religious fanaticisms that have long obscured our vision, namely 'scientific progress' and 'neoliberalism'.[10]

But it also requires that we be honest about our findings. If, as chapter 1.c will show, a sustainable world is a necessity, and if that logically entails certain paradigms (for example, no growth, accepting limits, fair distribution, co-operation instead of competition), then we should have the courage to stick to these findings and apply them to all relevant areas, even if that requires a substantial reorientation. A very good example of what I mean can be found routinely in such newspapers as *The Guardian* in the U.K. Every Wednesday it carries a supplement called *Society* which deals in a thorough and approachable manner with the most pressing social, economic and ecological problems which beset our world. But if you care to read the main sections of the same broadsheet it is as if all these insights (for example that GNP is a thoroughly inadequate, positively misleading measure of human endeavours, see below p. 136) are non-existent. There, day in, day out, events are interpreted with the same old patterns of thought which have been shown to be wrong and inadequate (more about this in chapter 2.b). We have to start to face the consequences of our knowledge and *act on it*.

The above insights have also forced me to adopt a particular form, potentially alien to many readers. I am quoting, and often at length, from a great many sources. This might be misinterpreted as an attempt to intimidate the reader. In fact, it is an effort to counteract two shortcomings of standard academic writing. Firstly, academics often fill their texts with references in the form of 'see Suchandsuch, p. 101'. This leaves the readers with no chance to check for themselves whether the reference given is appropriate. If the reference passage, how-

---

[10] For an elaboration of this position, which argues that we cannot afford to throw away the only tools for understanding we have, namely our reason, if we want to "uncover the structure of reality with enough approximation to understand how it works", see Postman, 1999, 61ff.

ever, is quoted, it allows the readers to judge for themselves, thus increasing transparency. Similarly, the footnotes are intended as a working tool for those who might be further interested in a particular point. Just imagine the opposite scenario: I would claim and propose whatever and just expect you to believe me without explanation, reference or indication where the idea originated from and where you might look for further information, discussion and elaboration. Unless you are looking for a high priest telling you the gospel so that you can stop thinking, would you be satisfied? Secondly, we academics are trained not to accept a correct insight or remark by somebody else as such; our individually centred intelligence and our professional ethos almost force us to dissect and criticise and interpret everyone else. I deliberately do not do that. I cannot see any sensible reason why I should try to reformulate and reinterpret a particular statement by somebody else who has said what needs to be said aptly and precisely, and better than I ever could. This means accepting that we might not have the last word and might not be the most valuable contributors to the debate. I therefore understand my text as a multi-voiced conversation about the issues at stake, without the grotesque pretension that I have invented it all by myself. And lastly, the fact that I dissect the arguments of 'opponents', but not of 'friends' has the simple reason that the voice of the other side is so much more amplified today, that I don't feel I have to dissect arguments I share just for the sake of an old academic custom.

In order to show that sustainability, despite being abused in many quarters as a fig leaf to hide the continuation of business as usual, is not just a useful concept, but in fact deserves our most serious attention, I will demonstrate in chapter 1 three things: a) every genuine indicator or study which has appeared in recent times shows clearly that we are accelerating the degradation of the biosphere on which we depend for all of our activities. This also means that we are eating into our natural capital, which implies in turn that we are leaving less for generations to come; b) I will then proceed to document that this knowledge has, in fact, filtered through into international, national and regional documents. Both sustainability and education for sustainability are now enshrined in various ways in government strategies, constitutions etc.; in c) the task will be to put some flesh on the bone: what *is* sustainability, beyond lip service in the chat rooms of international diplomacy and business?

Having shown both the international intentions to create a sustainable world and the parameters such a world would have to satisfy, the next question to be answered is: how does our current global society shape up to the demands thus established (chapter 2.a)? And, after demonstrating that unsustainability is too weak a word to describe the real state of play, we have to explain the gap between these international declarations of intent and the often very active refusal to do anything which might move us in the declared direction. This means that we have to look at those factors which legitimise the unsustainable status quo, such as the media and current educational practice at all levels (chapter 2.b).

Only against this background can we then focus on education. In chapter 3 I will establish what education for sustainability should entail, with a special focus on the university level. This section will be complemented by examples of best practice – not only from the university sector – which should help us to envisage the necessary change and give us tools and ideas with which we might accomplish it.

1. Sustainability as Intention and Paradigm

a. State of the World

It is disturbing to notice that at the same time as the state of the environment seems to be retreating from the consciousness of people in the richer nations,[11] there is no overall improvement on the ground, quite the contrary. Every major study, every indicator, every attempted measurement of the state of the world's ecosystems points in the same direction: downwards. To quote just one of the more recent reports jointly issued by the United Nations Development Programme (UNDP), the United Nations Environment Programme (UNEP), the World Bank, and the World Resources Institute (WRI):

> Earth's ecosystems and its peoples are bound together in a grand but tenuous symbiosis. We depend on ecosystems to sustain us, but the continued health of ecosystems depends, in turn, on our care. Ecosystems are the productive engines of the planet, providing us with everything from the water we drink to the food we eat and the fibre we use for clothing, paper, or lumber. Yet nearly every measure we use to assess the health of ecosystems tells us we are drawing on them more than ever and degrading them at an accelerating pace. (*A Guide to World Resources 2000-2001*, 2000, 1)

An assessment of the ecological footprint of our lifestyles (Wackernagel/Rees, 1996; *Sharing Nature's Interest*, 2000), the yearly *State of the World* report by the World Watch Institute,[12] an estimation of the total material throughput of our economy (MIPS, Schmidt-Bleek, 1997),[13] the *Living Planet Report 2000*[14] or the most authoritative study on the ecological impact of Germany (*Zukunftsfähiges Deutschland*, 1996),[15] all arrive at the same conclusion: we cannot go on like this, we are seriously degrading the resource base upon which all life on Earth, including our own, depends:

> Humans are facing an unprecedented challenge: there is wide agreement that the Earth's ecosystems cannot sustain current levels of economic activity and material consumption, let alone increased levels. (Wackernagel/Rees, 1996, 1)

---

[11] For an explanation of this phenomenon see chapter 2.b.

[12] Latest accessed version: Lester R. Brown, Janet N. Abramovitz, Linda Starke, Christopher Flavin, Hilary French (2001), *State of the World 2001* (New York, Norton).

[13] See also: MIPSonline [http://www.wupperinst.org/Projekte/mipsonline/index.html].

[14] Issued jointly by WWF, UNEP World Conservation Monitoring Centre, Redefining Progress and the Centre for Sustainability Studies, see http://panda.org/livingplanet/lpr00/.

[15] An abridged translation into English, interestingly enough omitting the discussion on necessary political changes, appeared under the title *Greening the North: A Post-industrial Blueprint for Ecology and Equity* (1998).

> The human world is beyond its limits. The present way of doing things is unsustainable. The future, to be viable at all, must be one of drawing back, easing down, healing. (Meadows/Meadows/Randers, 1992, xv)

> The gap between what we need to do to reverse the degradation of the planet and what we are doing widens with each passing year. (*State of the World*, 1998, 183)

> It is thus obvious that the "advanced" societies are no model; rather they are most likely to be seen in the end as an aberration in the course of history.[16]

I don't want to bore the knowledgeable amongst you with already well-known facts, but just a short list of data drives this overall assessment home with all the necessary clarity. It also makes a mockery of the widely-held belief that we wouldn't be left with any environmental problems if only we could prove that climate change is neither happening nor our fault.

- If one starts from the assumption that every human being has a right to the same amount of natural resources – and it is hard *not* to argue that without falling into fascistic notions of 'my race is better than yours, therefore...' – this means that currently every person can make use of an average ecological footprint (total ecological productive area on Earth divided by number of inhabitants) of roughly 2 hectares (including sea space).[17] Quite apart from the fact that this figure is steadily shrinking due to the loss of productive areas and the increase in population, it means that:
  - o 20 per cent of the world's population occupy around 70 per cent of the global footprint.[18] Because the total global footprint is, according to a conservative estimate, 37 per cent larger than all the ecologically productive areas combined, the wealthiest 20 per cent alone occupy a footprint as big as the planet's total carrying capacity: 70 per cent of 137 per cent is nearly 100 per cent.
  - o Since the average American has a footprint of 9.6 hectares, we would need an additional four planet Earths to provide the resources, should, as current ideology has it, the entire global population want to lead a similar lifestyle.[19]
- *World Population* has doubled since 1960 and is expected to reach at least 10 billion by 2050.[20]

---

[16] Wolfgang Sachs, 'Introduction', in *The Development Dictionary*, 1992, 1-5, here 2.

[17] For an introduction to the concept of ecological footprints see Wackernagel/Rees, 1996 and *Sharing Nature's Interest*, 2000.

[18] They consume 40 per cent of the meat, 60 per cent of the energy, 85 per cent of the paper and own 85 per cent of all vehicles ('Sustainability: the facts', in *New Internationalist*, No. 329, November 2000, 19).

[19] For the newest footprint data see 'The Ecological Footprint' page at *Redefining Progress* [http://www.rprogress.org/programs/sustainability/ef/] and the 'Footprints of Nations Study' [http://www.rprogress.org/programs/sustainability/ef/ef_nations.html].

[20] United Nations Environment Programme (1999), *Global Environment Outlook (GEO) 2000: Overview* (Nairobi, UNEP), 2 [http://www.unep.org/geo2000/].

- The *World Economy* has tripled in size since 1980, and is predicted to quintuple in the next 50 years (*A Guide to World Resources 2000-2001*, 2000, 6).
- *Biodiversity:* Per year, up to 17,000 species become extinct (Wackernagel/Rees, 1996, 31); more than 50 per cent of the original mangrove area in many countries is gone; the wetland area has shrunk by about half; and grasslands have been reduced by more than 90 per cent in some areas (*A Guide to World Resources 2000-2001*, 2000, 9).
- *Global Warming*: In the last 100 years the CO_2-concentration in the atmosphere has risen by 30 per cent (Wackernagel/Rees, 1996, 31) and is now higher than at any time in the last 160,000 years;[21] global emissions of carbon from the burning of fossil fuels have increased nearly fourfold since 1950;[22] the chief culprit in recent emission growth is the transportation sector, the fastest growing source during the last 20 years; here as elsewhere growth in demand has outstripped any efficiency gains that might have been achieved (*State of the World 1998*, 1998, 113-115); a worldwide CO_2 reduction of 50-70 per cent is necessary to slow climate change, yet world energy demand is forecast to grow by 65 per cent by 2020.[23]
- *Agricultural Land:* Three-quarters of the agricultural land worldwide has poor soil fertility and about two-thirds of agricultural land has been degraded in the past 50 years by erosion, salinisation, compaction, nutrient depletion, biological degradation, or pollution; about 40 per cent of agricultural land has been strongly or very strongly degraded (*A Guide to World Resources 2000-2001*, 2000, 10).[24]
- *Food Availability*: Since the 1950s the grain area per person has fallen continually, but cropland productivity has dramatically increased, though this trend has reversed since the mid-1990s (*State of the World 1998*, 1998, 79-81); productivity increases were also bought at a tremendous environmental price, because, quite apart from the pollution from fertilizers and pesticides, the overall energy input per kilogram of yield has risen dramatically, making modern methods far less efficient overall than traditional ones (Shiva, 1991, 196); "mono-cropping and dependence on a handful of basic crop varieties is endangering genetic diversity. Just 20 plants now supply 80 per cent of humanity's food and most farmers grow identical types of wheat, rice and potatoes";[25] in fact, modern farming methods seem to be directly related to a loss of biodiversity.[26]

---

[21] Or even longer: "Scientific examinations of the Vostok ice core in the Antarctica found CO_2 concentrations in the atmosphere at their highest levels in 420,000 years" ('Sustainability: the facts', in *New Internationalist*, No. 329, November 2000, 18; *Natural Capitalism*, 2000, 316).

[22] Marco Morosini, 'Ein tot geborener Traum. Die Concorde ist nicht nur technisch gescheitert', in *Die WochenZeitung*, No. 40, 5.10.2000, 24.

[23] 'Sustainability: the facts', in *New Internationalist*, No. 329, November 2000, 18-19.

[24] For further details see also: United Nations Secretariat of the Convention to Combat Desertification (UNCCD) [http://www.unccd.int/main.php].

[25] 'Sustainability: the facts', in *New Internationalist*, No. 329, November 2000, 18.

[26] "The Food and Agriculture Organisation says that around three-quarters of the world's plant species have been lost this century. It has warned that the large-scale loss of plant genetic resources, vital for agriculture and food security, is a matter of major concern. The warning is based on a survey of 154 countries, carried out in 1995. Over 80 countries reported that the spread of

- One third of German consumption of agricultural goods is dependent on farming areas in other countries (*Greening the North*, 1998, 64).
- *Hunger*: "More than enough food is already being produced to provide everyone in the world with a nutritious and adequate diet (on average, 350 kg of cereal per person) – according to the United Nations' World Food Programme, one-and-a-half times the amount required. Yet at least one-seventh of the world's people – some 800 million people – go hungry. About one-quarter of these are children. They starve because they do not have access to land on which to grow food, or do not have the money to buy food, or do not live in a country with a state welfare system. (…) Hunger has seldom been the result of an aggregate shortage of food; instead it has consistently been one of inequalities in political and economic power. Hunger, as economist Amartya Sen points out, is the inevitable outcome of the normal workings of a market economy. (...) Food goes to those who have the money to buy it. (...) In Costa Rica, for instance, while beef production doubled between 1959 and 1972, per capita beef consumption in the country fell from 30 pounds to less than 19. The reason? US consumers could pay higher prices for the meat than Costa Ricans. It is this market logic that explains (…) why Ethiopia was using some of its prime agricultural land at the height of the 1984 famine in the Horn of Africa to produce linseed cake, cottonseed cake and rapeseed meal for export to Britain and other European nations as feed for livestock." (The Cornerhouse, 1998, 3ff.); "80% of all malnourished children in the developing world in the early 1990s lived in countries with food surpluses; the WHO estimates that roughly half the global population suffers from poor nutrition – of that 50% eat too little and 50% eat too much; obesity is the second-biggest killer of Americans after nicotine, claiming at least 250,000 lives a year; liposuction (an operation to reduce fat) is the leading form of cosmetic surgery in the US with over 400,000 operations performed a year".[27]
- *Forests*: The expansive growth in consumption and trade of forest products means that since 1950 the demand for wood has doubled, the demand for paper has risen five times; half of the forest cover once on Earth has now gone and each year 16 million hectares disappear [= 7.9x size of Wales]; between 1960-90 1/5 of tropical forest cover was lost, but 90 per cent of the loss happens not in tropical, but temperate and boreal forests; the results are soil erosion, more extreme weather conditions, CO_2 fixation lost and global warming (*State of the World 1998*, 1998, 21-23).
- *Fisheries*: Between 1970 and 1989 there has been a 25 per cent decline in the yield of high-value species worldwide; 11 of the world's 15 major fishing areas and 69 per cent of the world's major fish species are in decline; by 1992 catches of Atlantic cod were down by 69 per cent since their peak in 1968; West Atlantic tuna stocks were down more than 80 per cent between 1970 and 1993; modern fishing fleets are so efficient that they can wipe out entire fish species, then moving on to less valuable ones; due to overfishing ever smaller fish are caught, leaving populations ever younger and declining; because the marine environment is increasingly polluted

---

modern, commercial agriculture and the introduction of new varieties of crops was the main cause of the loss of plant genetic resources." (Madeley, 1999, 28-29)

[27] 'Sustainability: the facts', in *New Internationalist*, No. 329, November 2000, 19.

(algal bloom due to waste run-off etc.), fishermen are forced further off-shore, re-enforcing the pressure on existing stocks (*State of the World 1998*, 1998, 60-67).[28]

- *Coastal Ecosystems* have already lost much of their capacity to produce fish because of overfishing, destructive trawling techniques and the destruction of nursery habitats (*A Guide to World Resources 2000-2001*, 2000, 12).

- *Traffic*: The total health effects of traffic pollution in Britain are estimated to cost $24 billion per year, while road congestion imposes further costs of $26 billion per year; road accidents cost Britain over $16 billion per year, arriving at a conservative grand total of $66 billion per year (Myers, 1998, 89); the number of motor vehicles world-wide has grown from 391.1 million in 1980 to 676.2 million in 1996.[29]

- *Ozone Hole:* In 2001 the hole in the ozone layer above Antarctica was "significantly larger" than that of 2000.[30]

- *Water*: Freshwater withdrawal has almost doubled since 1960;[31] water demands are so high that a number of large rivers decrease in volume as they flow downstream, with the result that downstream users face shortages, and ecosystems suffer, both in the rivers and in adjacent coastal areas; many underground water resources, known collectively as groundwater, are being drained faster than nature can replenish them; high withdrawals of water, and heavy pollution loads have already caused widespread harm to a number of ecosystems;[32] "nearly half the world's major rivers are going dry or are badly polluted; in China 80% of the major rivers are so degraded they no longer support fish life".[33]

This list, unfortunately, could go on and on. It is also true, of course, that there are some positive developments in isolated areas, such as cleaner air and water and a reversed trend in deforestation in parts of the US and Europe. But we shouldn't fool ourselves into believing that just because the rich Euro-American[34] countries have the means to improve certain aspects of environmental degradation, the gen-

---

[28] Fisheries are a very good example of the unsustainability of many high-tech tools. Modern fishing boats are such potent catchers that they literally destroy the fisheries they are designed to depend on (see Bowers, 2000, 50-51).

[29] United Nations Environment Programme (1999), *Global Environment Outlook (GEO) 2000: Overview* (Nairobi, UNEP), 3 [http://www.unep.org/geo2000/].

[30] J. D. Shanklin (2001), *British Antarctic Survey (BAS) Ozone Bulletin 01/01* [http://www.antarctica.ac.uk/met/jds/ozone/bulletins/bas0101.htm].

[31] See WWF, 1998 [http://www.panda.org/livingplanet/lpr/water_graph.html].

[32] United Nations Sustainable Development (1999), *Comprehensive Assessment of the Freshwater Resources of the World* [http://www.un.org/esa/sustdev/freshwat.htm].

[33] 'Sustainability: the facts', in *New Internationalist*, No. 329, November 2000, 18.

[34] Throughout the text I use the term Euro-American. In a geographical sense, this is very imprecise, because the lifestyles, ideology, political, social and economic system meant by the term are not confined to Europe and America any more, but have spread to Japan, Southeast Asia and the ruling elites and wealthy classes all over the globe. Yet, as opposed to the more frequently used terms Northern or Western, it denotes clearly the historical responsibility of the Europeans who have brought us the twin brothers of Industrial Revolution and Colonialism and of the United States who is at the forefront of perpetuating and spreading this unsustainable legacy everywhere.

eral global trend towards more destruction is halted: "Our current affluence [in the so-called developed countries] is deceptive since it is based on consumption of resources at the expense of ecological stability, global justice, and generations to come." (*Greening the North*, 1998, 4) All we have done in the majority of cases is export the polluting industries to countries of the South: "Today Germany imports virtually all its oil, three-quarters of its natural gas, and most unprocessed minerals." (*Greening the North*, 1998, 5; for the US see Zinn, 1996, 556-557) This statement means in reality that all the destructive work to get these raw materials out of the ground is done in other countries, i.e. is *externalised* from the German point of view.[35] That this is a common pattern was shown by a United Nations survey which

> found that transnational corporations regularly "adopted lower environmental standards in their operations in developing countries than those in developed countries, thus exposing workers and communities in developing countries to dangers that would not be accepted in developed countries." The report attributed the ongoing prevalence of double standards to corporations' "greater concern over maximizing profitability ... rather than ensuring maximum safety of their operations". (Karliner, 1997, 149) [See box 1 for an example of this environmental colonialism.]

Box 1: Incineration

Incineration is a case in point. With a saturated market and after more than a decade of popular protest against this technology's social and ecological impacts, the hazardous waste incineration business has stagnated in the industrialized North. Meanwhile, exports to the Third World and Eastern Europe are on the rise. For instance, just after the Clinton Administration announced a moratorium on new incinerators in 1993, U.S. consultants were pushing the technology in Malaysia at what the Third World Network's Sobhana William describes as "rock bottom prices". Taiwan is building more than twenty waste-to-energy incinerators. And in Argentina, despite a dearth of public information, Greenpeace documented more than eighteen incinerator proposals in 1995.

Joshua Karliner (1997), *The Corporate Planet. Ecology and Politics in the Age of Globalization* (San Francisco, Sierra Club Books), p. 164.

The other sinister aspect of this trend is that the existence of lower environmental standards and "pollution havens in the South gives corporations leverage in their efforts to bargain down standards in the North" (Karliner, 1997, 154; see there for concrete examples).

The most important thing, therefore, is to remember that behind the above mentioned global data there are individual human beings on the ground whose

---

[35] This is true more generally: the wealthy minority everywhere are living at the expense of the poor: "It suddenly became evident [in the 80s] that women in the industrialised countries and middle-class women in the South are not only victims but also beneficiaries of the international exploitative system." (Bennholdt-Thomsen/Mies, 1999, 13)

livelihood is often directly affected by the deterioration of the ecosystems (see boxes 2 to 4).[36]

Box 2: Shrimps for life

Ibrahim, 90, is one of the oldest residents of Badarkhali village, and in his lifetime he has seen a drastic transformation of his home from dense mangrove forest to barren, unproductive shrimp fields. In a matter of 15 years, shrimp aquaculture has become a US$9 billion industry, active in over 50 countries. Commercially-grown shrimp is extremely lucrative, extremely risky and extremely destructive. It has been vigorously promoted since the early 1980s by multilateral development banks, UN bodies such as the FAO and UNDP, governments and commercial interests around the world. (...) Pollution, environmental destruction, coastal deforestation, soil erosion, the collapse of fisheries, and social conflict over land are some of the common consequences of commercial aquaculture. In the late 1980s, widespread disease wiped out most of the farms in Taiwan, forcing the world's leading exporter of shrimp to take a closer look at the long-term consequences of commercial aquaculture. (...) In shrimp aquaculture, as with all risky ventures, certain 'externalities' had to be managed. Among them were the impact of intensive industrial activity on the coastal community; the management of large quantities of polluting effluents and saline water; the acquisition of coastal land normally used for agriculture and often highly populated; and the existence of delicate coastal ecosystems, mangrove forests and fisheries, a critical source of livelihoods for the villages along the coast. (...) For many coastal dwellers it has become a question of life and death: in Bangladesh, for example, over 100 villagers have been killed in conflicts related to land acquisition for shrimp culture. Less than a decade after its introduction to Asia, it is clear that the aquaculture industry cannot manage these externalities, nor do they enter into its short-term, profit-centred calculations. The industry operates on a hit-and-run basis, typically on a five-year horizon, sufficient to get a return on investment, and then move on. It is also clear that commercial shrimp aquaculture is one of the most destructive cash crops ever – a cash crop that has endangered the lives and livelihoods of millions of coastal people around the world.

Faris Ahmed, (1998), *In Defence of Land and Livelihood: Coastal Communities and the Shrimp Industry in Asia* (Penang, CUSO; Canada, Inter Pares and the Sierra Club)

Box 3: Shiny gold

For the second time in less than two months, an Australian gold mining company has been responsible for a cyanide spill that has endangered lives and the environment in another country. The lethal chemical washed into a river system last week when a freight helicopter chartered by Dome Resources dropped a one-tonne box of sodium cyanide pellets about 80 kilometres north of the Papua New Guinea (PNG) capital of Port Moresby. In the previous incident, on February 1, a cyanide overflow from Esmeralda Exploration's gold mine dam in Romania caused massive environmental damage in Eastern Europe. It threatened the water supplies of 2.5 million people, and destroyed fish and wildlife in Romania's River Dsomes, Hungary's River Tisza and Yugoslavia's Danube.

Mike Head, 'Cyanide spill endangers villagers in Papua New Guinea' (World Socialist Web Site, 28 March 2000) [http://www.wsws.org/articles/2000/mar2000/png-m28.shtml]

---

[36] For more stories of indigenous people directly affected by the gas, oil and nuclear industry, see the Indigenous Solidarity Statement, issued for Earth Day 2001, calling for a transition to clean, renewable energy [http://www.earthday.net/goals/humanrightsstatement.stm].

Box 4: Living Earth pulled away under indigenous people's feet
The Papua New Guinea island of Bougainville has been seriously affected by years of civil war, provoked by operations at the Panguna copper-gold mine. As many as 10 per cent of the island's population may have died in the war, most of them innocent civilians. The mine, which is 53.6 per cent owned by Rio Tinto, has "devastated the rainforest, wiped out all life from the Jaba river and silted the Empress Augusta bay to a depth of 30 metres", alleges Roger Moody. The strength of feeling was summed up by Perpetua Serero, one of the leading women campaigners: "We don't grow healthy crops any more, our traditional customs and values have been disrupted and we have become mere spectators as our earth is being dug up, taken away and sold for millions."

John Madeley (1999), *Big Business, Poor Peoples* (London, Zed Books), p. 94.

We would do well to heed Wolfgang Sachs' advice to try to sympathise with and visualise these directly affected people. Otherwise, their exploitation will never stop and even well-meaning attempts might backfire if they turn into global environmental management and become as brutal and insensitive as any global measure which, through its distancing effect, cannot cope with real people and real conditions on the ground any more:

> As soon as worldwide strategies are launched to prevent the boat from capsizing, things like political autonomy or cultural diversity will appear as the luxuries of yesteryear. In the face of the overriding imperative to "secure the survival of the planet", autonomy easily becomes an anti-social value, and diversity turns into an obstacle to collective action. Can one imagine a more powerful motive for forcing the world into line than that of saving the planet? Eco-colonialism constitutes a new danger for the tapestry of cultures on the globe. (Sachs, 1999, 100)

Nevertheless, irrespective of the way we are dealing with the environmental problems (for more about that see 1.c below), it seems quite clear by now that only fools could deny that they are a reality which must force humankind to act decisively.

b. The International Consensus

Since even the world's leaders cannot entirely ignore hard evidence, the acknowledgement that we are in trouble and urgently need to do something has penetrated the national and international political discourse. 'Sustainability' as a term[37] – returning more hits from internet search engines than 'sex', as a friend half-jokingly remarked – has had a remarkable career, quite unlike its counterpart, namely sustainable action (see chapter 2.a).

Everybody pays lip service to it: Alex Krauer, the CEO of UBS, the biggest Swiss bank, explained in a recent interview that most businesses now are firmly operating within a sustainability framework. The message seems to be that we are there already, which consequently leads Krauer to oppose any measures to speed up the transition to a future powered by renewable energy.[38] The World Economic Forum (WEF) has done much to propagate this sort of message globally, through its "Environmental Sustainability Index (ESI)", which, using dubious, misleading and biased data, "represents some of the worst eco-villains as the world's good guys" and tries to sell the clearly identifiable lie that "the most 'eco-friendly' nations were the world's most industrialised" while "the 'eco-offenders' were the poor".[39] Even more blatant is the greenwash promoted by multinational companies in advertisements. The October 2000 issue of *National Geographic*, for example, featured at the very beginning three multi-page advertisements by transnational corporations (TNCs), at least two of which have an abysmal environmental record (Shell and BP). But if you read the ads you would be forgiven for thinking that if anybody can save the environment and is doing something about it, against the resistance of almost everybody else, then it is these companies. Just a few quotes: "Can 100,000 people in 100 countries come together to solve the paradox of a world that wants both mobility and clean air? We [BP] think so"; i.e., we, BP, are these innovative people and we are saving the world. Or: "Time and again at Shell we're discovering the rewards of respecting the environment when doing business." Presumably in Nigeria?! And Canon portrays itself in a six page "special advertising supplement" (for entirely altruistic motives, I presume....) as the quintessential agency for saving endangered species. It is *so* blunt, so blatantly a lie, it makes you sick, and yet how can the general public know unless it has a very good knowledge of history as well as access

---

[37] Recently traced back to the German forestry specialist Hans Carl von Carlowitz who used the German equivalent "nachhaltig" in the full sense we use it today in his work *Grundsätze der Forst-Oeconomie* in 1757 (see Ulrich Grober, 'Der Erfinder der Nachhaltigkeit', in *Die Zeit*, No. 48, 25.11.1999, 98).

[38] '"Ins kalte Wasser springen". Interview mit UBS-Verwaltungsratspräsident Alex Krauer', in *Tages-Anzeiger*, 28.8.2000, 25.

[39] The Ecologist and Friends of the Earth, 'keeping score', in *The Ecologist*, 31 (2001) 3, 44-47, here 44.

to independent information about the atrocities committed by these and similar companies?

Nevertheless, the message is clear. Both Krauer and these ads admit implicitly that we *do have* serious environmental problems, only to say that the public needn't worry since big business has it all under control (see, for more on this, chapter 2.b). Important for us here is this implicit admission by the corporate sector that sustainability has moved centre-stage.

This, of course, is also true of the political sphere. The 1992 United Nations Conference on Environment and Development (UNCED) in Rio, better known as the Earth Summit, was, despite all its shortcomings, a very powerful signal to the public that virtually all nations of the world (108 heads of states were present) considered the issues surrounding ecology and development, which are now normally labelled with the 's'-word, worthy of top priority. And despite the almost unanimous admission at the Rio+5 conference in New York (23-27 June 1997) that the world has not progressed towards sustainability, the main document from UNCED, *Agenda 21* (1993), "signed by world leaders representing 98% of the global population" (Nebel/Wright, 2000, 20), motivated 150 countries to set up national advisory councils on sustainable development policies and continues to have considerable impact, particularly on the local and regional level: over 1,800 cities and towns worldwide have since created their own "local Agenda 21" (see especially the ICLEI initiatives).[40] In fact, the ratification of *Agenda 21* means that, at least formally, the question of whether we should embark on sustainable development has been decided; the signatories have committed themselves to it. So the question is not so much *whether*, but *how*.

As a result the sustainability discourse has penetrated the official documents of many countries. The European Union entitled its fifth environmental action programme (1992) *"Towards Sustainability": A European Community Programme of Policy and Action in Relation to the Environment and Sustainable Development*;[41] since 1992 the Federal Republic of Germany officially supports sustainability;[42] the British Government adopted a *Sustainable Development Strategy* in 1994;[43] and Switzerland has enshrined the principle of sustainability into its revised constitution, valid since 1 January 2000.[44]

---

[40] International Council for Local Environmental Initiatives (ICLEI) [http://www.iclei.org]. Source for figures: 'UN Conferences: What they have accomplished?' [http://www.un.org/News/facts/confercs.htm].

[41] COM(92) 23 final, 27.3.92 [http://europa.eu.int/comm/environment/env-act5/5eap.pdf]; see review (2000) [http://europa.eu.int/comm/environment/newprg/99543_en.pdf].

[42] Edgar Kreilkamp, 'Sustainable development als Chance für die Lehre', in *campus courier. Zeitung für das Projekt "Agenda 21 – Universität Lüneburg"*, (Sommersemester 1999) 1, 2.

[43] See the first report of the UK Government Panel on Sustainable Development [http://www.sd-commission.gov.uk/panel-sd/panel1/].

[44] See Preamble, Art. 2, especially Art. 73, but also Art. 104; adopted 18.4.1999, valid from 1.1.2000 [http://www.admin.ch/ch/d/sr/1/101.de.pdf].

Indeed, all these documents, treaties and trends seem to indicate a considerable international consensus that humankind, if it wants to secure its survival, has to embrace sustainable development leading to a sustainable society.[45] And whenever the discussion focuses on sustainable development, there also seems to be unanimous agreement that education plays a crucial part in achieving this transition. In fact it seems, and is written into international, European and national treaties, that education for sustainability[46] (subsequently referred to as EfS)[47] is urgently needed, on all levels (primary, secondary, tertiary education). These documents note for example that

- education in general, but tertiary education as the "nursery of tomorrow's leaders" in particular, is of crucial importance in any potential progress towards sustainability;[48]

---

[45] These documents, and also many education for sustainability guides, share another feature. They are laden with unexamined assumptions about 'development' (growth is good), 'education' (is always, by definition, good), positive connotations of industrial agricultural and other industrial productivity; they promote top-down approaches (ministries, international bodies etc. have to achieve/implement sustainability). All these concepts are used in such a generalized fashion that they become unhelpful, pulling a smokescreen over the real issues of power, democracy and ecology. They also tend to fall into the trap of using business lingo, which actually is detrimental to sustainability (like 'wise use', focusing on end-of-pipe and technological solutions). See as two examples from politics and education: The National Assembly of Wales (2000), *Creating an Environmental Vision: Progressing the Environment Agency's Contribution to Sustainable Development by Way of a Better Environment in England and Wales* [http://www.environment-agency. gov.uk]; Rosalyn McKeown (1999).

[46] I use this term because I prefer it (see Huckle/Sterling, 1996, xiii; in the strong sense, see Plant, 1998, 40, 44) and because it avoids the inherent contradictions in the term "sustainable development" (see Sachs, 1999, 34). However, I am fully aware that 'Education for Sustainable Development' is more often used, but maybe precisely for the reason that it is more malleable to one's own (unsustainable) agenda. Yet I'm not too bothered about the term. I agree with the conclusion that the fundamental question is "what constitutes good education (...) regardless of its name or label" (*ESDebate*, 2000, 50).

[47] I quote a recent attempt at a definition: EfS "enables people to develop the knowledge, values and skills to participate in decisions about the way we do things individually and collectively, both locally and globally, that will improve the quality of life now without damaging the planet for the future." (Sustainable Development Education Panel, 1999, 30)

[48] "Education is critical for promoting sustainable development and improving the capacity of people to address sustainable development issues. (...) Countries should stimulate educational establishments in all sectors, *especially the tertiary sector*, to contribute more to awareness building." (*Agenda 21*, 1993, Chapter 36) [http://www.un.org/esa/sustdev/agenda21text.htm]; my emphasis). See also: European Union, *Resolution of the Council and the Ministers of Education Meeting within the Council on Environmental Education* of 24 May 1988 (88/C 177/03) [http:// europa.eu.int/comm/environment/eet/res88177.htm]; *Conclusions of the Council and the Ministers of Education meeting within the Council* of 1 June 1992 on the development of environmental education (92/C 151/ 02) [http://europa.eu.int/comm/environment/eet/cc92c15102.htm]; Resolution of the European Parliament on environmental education of 17.12.93. OJ C 20/94, 24.1.94 [http://europa.eu.int/comm/environment/eet/parres.htm].

- all countries should "reshape education so as to promote attitudes and behaviour conducive to a culture of sustainability" (UNESCO, 1997, 1);
- EfS is life-long learning and should therefore be available throughout life and to everybody,[49] because "sustainable development is the responsibility of everyone" and "education for sustainable development needs to pervade every aspect of life" (Sustainable Development Education Panel, 1999, 3);
- there is a vital long-term economic, social and political interest in EfS because "prosperity in the long term depends on our capacity to learn about sustainable development" (Sustainable Development Education Panel, 1999, 3).[50]

Now, before we get too excited about this seemingly overwhelming thrust for sustainability, we have to clarify a few questions:
- What is sustainability? What does a sustainable society look like? (chapter 1.c)
- Is our current globalised society sustainable? (chapter 2.a)
- If not, why not? (chapter 2.b)
- What bearing do the answers to these questions have on EfS, which, as we have seen, is supposedly responsible for our transition to sustainability? (chapter 3)

---

[49] The first report of the UK Government Panel on Sustainable Development stressed that EfS is "available throughout life to enable citizens to see for themselves the need for sustainability and to help convey the necessary sense of individual responsibility for a healthy environment". [http://www.sd-commission.gov.uk/panel-sd/panel1/5.htm#environmental education and tra]

[50] See also, with reference to the German situation, *Zukunftsfähiges Deutschland*, 1996, 363-377.

c.   *Sustainability as a concrete paradigm*

> Our challenge is to create a global system that is biased towards the small, the local, the cooperative, the resource-conserving and the long-term. (David Korten, quoted in Madeley, 1999, 178)

> The main problem is not knowing what to do, it is mustering the desire to do it. (Jonathan Rowe)[51]

> We think a transition to a sustainable world is technically and economically possible, maybe even easy, but we also know it is psychologically and politically daunting. (Meadows/ Meadows/ Randers, 1992, xvi)

> There is no template for the transition to sustainability (...) but there is a direction and there are principles. (Tim O'Riordan, quoted in *ESDebate*, 2000, 52)

> Moving towards the goal of sustainability requires fundamental changes in human attitudes and behaviour. (UNESCO, 1997, 1)

If it seems clear that everybody wants sustainability and that governments, businesses and NGOs are falling over backwards to achieve it, it might be worth pondering for a moment what a sustainable society by necessity entails. Very often it is said that there are as many definitions of sustainability or sustainable development as there are people trying to define it (a recent survey has found more than 200 different definitions [see Porritt/Wilsdon, 1998]). But whilst this is certainly true, I believe that the reason for it has little to do with the anticipated goal, but much to do with the different vested interests trying to shape their definition of sustainability so that it fits a 'business as usual' approach which doesn't force anybody to change their ways (see Karliner, 1997, 41). For contrary to all the talk about the impossibility of defining sustainability – doubtless another application of the well-known PR strategy to foster doubt and disorientate the public and thereby ensure that it won't act, as so brilliantly deployed by corporate PR strategists in the case of global warming[52] –, it is relatively easy to define the parameters within which a sustainable society has to be developed.[53]

[51] Rowe, 'Eat, sleep, buy, die', in *New Internationalist*, No. 329, November 2000, 24-25, here 25.

[52] Sharon Beder, 'Corporate Hijacking of the Greenhouse Debate', in *The Ecologist Special Issue: Climate Crisis*, 29 (1999) 2, 119-122, here 119.

[53] I am writing this partly in a provocative manner. Even though it is true that many aspects of the issue are exceedingly complex and that the definition of sustainability as a "regulative idea" – in other words an ideal towards which we strive but which we will only be able to implement to a certain degree, therefore necessitating ever new efforts (Hirsch, 1999, 270) – seems accurate, I would nevertheless polemicise against a dominant tendency in sustainability studies which stresses the almost insurmountable complexity, the impossibility of even defining the problems etc. (see Kaufmann-Hayoz in *Umweltproblem Mensch*, 1996, 7-9 as an example). This, then, is very often accompanied by an emphasis on institutional, economic and financial barriers to sustainable ac-

It was at one point customary to specify these parameters within four areas, the 4 Es: ecology, empowerment/democracy/political sphere, equity/social dimension and economy.[54] Recently there has been a move to collapse these into just three aspects: ecology, economy, and the social sphere.[55] My suspicion is that this has to do with the overall de-politicisation of the sustainability discourse: we do not need to talk about one of the most important areas, namely power in the political sphere,[56] because – so the assumption goes, in line with Fukuyama's theory that we have reached the "end of history"[57] – we in Europe and the US live in the paradise of political systems, i.e. 'liberal democracy', already.

I believe that this reductionism is literally throwing out the baby with the bathwater. I would even argue that the original 4 Es are in a sense misleading, since they seem to imply that the four aspects are equally important, which is as far from the truth as you can get. Therefore, and since I feel that one aspect, which for the last fifty years has become more and more dominant in our lives, has been

---

tion, which apparently prevent people from acting sensibly. Whilst, as chapter 2.a will show, there is no reason to underestimate the real opposition of the powers-that-be to sustainability, the argument nevertheless lacks a healthy dose of existentialism: as it is not very convincing if a forty year old still blames his or her upbringing for every mishap in his/her life, it is simply too convenient to blame exclusively the power structures for one's powerlessness, especially if other members of the same society take their fate into their own hands and *act*.

[54] See Alicia Bárcena; Diomar Silveria, 'Envisioning Sustainable Alternatives within the Framework of the UNCED Process', in *Getting Down to Earth*, 1996, 455-463, 455 and 460. As categories the 4 Es apparently date back to the Chicago School of Sociology (see Dieter Steiner, 'Humanökologie – Überlebenswissenschaft für die Zukunft', in *Neue Zürcher Zeitung*, No. 221, 22.9.2000, 89).

[55] See for example HE21 [Higher Education for the 21st Century] (1999), *Sustainability Indicators for HE* (London, Forum for the Future), [3] or Bolscho/ Seybold, 1996, 73.

[56] People in the South instantly recognise that there can be no sustainability without empowerment: "I think this Commission should give attention on [sic] how to look into the question of more participation for those people who are the object of development. Their basic needs include the right to preserve their cultural identity, and their right not to be alienated from their own society, and their own community. So the point I want to make is that we cannot discuss environment or development without discussing political development. And you cannot eradicate poverty, at least not only by redistributing wealth or income, but there must be more redistribution of power." (Aristides Katoppo, Publisher, WCED Public Hearing, Jakarta, 26 March 1985, quoted in *Our Common Future*, 1987, 31)

[57] In his infamous, and by now distinctly dated, original article Fukuyama claimed that liberal democracy was the "end point of mankind's ideological evolution" and the "final form of government" ('The End of History?', in *The National Interest*, (1989) 16, 3-8, here 4). At the end of the book version, where he compares history with a train drawing into a city (presumably supposed to stand for liberal democracy and the market economy), it is claimed "that any reasonable person looking at the situation would be forced to agree that there had been only one journey and one destination" (Francis Fukuyama (1992), *The End of History and the Last Man* (New York, Free Press), 339). Yet for the context of how the industry-sponsored Fukuyama received such worldwide attention for his dubious claims, see Susan George, 'Eine kurze Geschichte des Einheitsdenkens', in *Le Monde diplomatique*, 2 (1996) 8, 10-11.

left out, I propose another structure (see Figure 1): Because our biosphere, the "thermodynamically closed and non-materially-growing" life-support-system Earth,[58] is the sphere on which everything else depends, it is the most fundamental aspect, defines the possibilities and limits, and encompasses all the other spheres. In Shiva's words: "The real meaning [of sustainability] refers to nature's and people's sustainability. It involves a recovery of the recognition that nature supports our lives and livelihoods and is the primary source of sustenance." (Shiva, 1992, 192) Contained within nature, therefore, we find empowerment, equity, economy, and equipment – standing for science/technology – as the fourth element.[59]

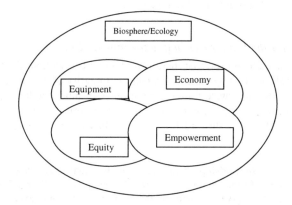

Figure 1

However, I am only separating out these aspects for analytical clarity. In reality, the hardest challenge for sustainability will be how successfully we learn to think systemically, in a fashion appropriate to the complexity of the real world. All these spheres are inextricably intertwined; they form a complex, holistic entity which could be likened to what is called a whole-organism person in biology.[60] Any solutions will be sustainable ones only if they live up to that complexity and escape the currently predominant reductionism – particularly in the natural sciences (just think of the idea that genetically modified crops will feed the hungry [The Cornerhouse, 1998] or of the ultra-darwinism of sociobiology [Rose, 1998]) and in economics (just think of the GNP[61] or the neoliberal dogmas[62]). We now

---

[58] Robert Costanza, Olman Segura and Juan Martinez-Alier, 'Integrated Envisioning, Analysis, and Implementation of a Sustainable and Desirable Society', in *Getting Down to Earth*, 1996, 1-13, here 2.

[59] For a similar structuring, based on the biogeochemical limits, but then emphasising equity, the precautionary principle and proactivity, see Robert Costanza *et al.*, 'Integrated Envisioning...', in *Getting Down to Earth*, 1996, 1-13, here 2; see also the 'Russian doll's model' in *Sharing Nature's Interest*, 2000, 7.

[60] Tim Ingold, 'Evolving Skills', in *Alas, Poor Darwin*, 2000, 225-246, here 238.

[61] For a full critique and alternatives see *Taking Nature into Account* (1995) and below p. 136.

need to apply strategies akin to Life Cycle Analysis, used in environmental science, to all our problems, or we will multiply rather than solve them.[63]

We have to recognise, therefore, that in each of these five spheres, which I will analyse in more detail below, we have to apply a set of guiding principles or thoughts which are, as it were, non-negotiable. Ignoring them very quickly leads to simplified pseudo-solutions. These principles more or less follow if we define sustainability as the achievement of self-determined freedom, with full appreciation of the limits to that freedom imposed by respect for all the other, including future, inhabitants of the planet and within the boundaries of the life-support-system Earth. These guiding principles are:

Planning for the future/ Expanding the time horizon

Once we have recognised that there is no human life without the biosphere to support it, it follows that we need to protect the integrity of the earth, i.e. conserve the natural capital. Otherwise future generations will inherit a depleted resource base. This would mean that our children would have diminished chances to lead a dignified life. The term stewardship, as described by Nebel/Wright, expresses this well:

> Modern-day stewardship, therefore, is an ethic that provides a guide to actions taken to benefit the natural world and other people. (...) Stewards recognize that a trust has been given to them and that they are responsible to care for something that is not theirs – whether it be elements of the natural world or of human culture – which they will pass on to the next generation. (2000, 11)

The consequence for all our actions is that we have to expand the timescale within which we evaluate them. Nothing can be an end in itself any more; it has to be reconnected to the survival of the overall system, the biosphere (cf. Vester, 1997, 31-32, 47). As a result, sustainability

> involves a sense of time that does not begin and end with the expectations of the individual. Rather, it places the person within a continuum where an awareness of

---

[62] For recent accounts of the destructive capacity (such as dearer essentials, worse poverty and shorter lives for the people affected) of these dogmas (free market, deregulation, privatisation), see Gregory Pallast, 'IMF's shock cures are killing off the patients', in *The Guardian Weekly*, 12-18 October 2000, 14 and *Globalising Poverty*, 2000.

[63] Vester, for example, stresses the overarching importance of a complex assessment of a problem (in his example, traffic): "Energy consumption, materials usage, space demand, ecological and human ecological considerations and again and again the re-examination, whether the transfer in question could not be substituted by different means, all this eventually has to be privileged over technical and economic issues, and not vice versa." (Vester, 1997, 132)

past and future generations helps to define what constitutes meaningful knowledge, values, and sense of responsibility. (Bowers, 1995, 39)

This, of course, is in sharp contrast both to the current economic paradigm which by now bases its decisions on quarterly reports, rather than long-term viability strategies, and to the individualistic consumer society focused on instant gratification (see chapter 2.a).

Yet we also need to re-appreciate "that basic social change may require the labours and dedication of a lifetime" (Bookchin, 1980, 20). In other words, commitment to sustainability needs to be a long-term dedication, not just the newest political fad which fades away again with the next fashion.

Complexity

Quite apart from the fact that no reductionism will ever get to grips with the irreproducibility, historical contingency and irreversibility of complex living processes (see Rose, 1998), this means that whenever we evaluate an action, an institution, a purchase or whatever, we have to attempt a complex assessment of all factors, over the entire lifespan of a product or the duration of the consequences flowing from our actions. It is therefore *not* complex thinking if we compare renewable energy and nuclear reactors and exclude the waste question, or if we introduce high yield crops without ever considering the overall resource consumption as related to the output, or if we rejoice about the cheap price of our new Nike trainers and completely forget about the human costs accumulated during their production.[64]

Diversity

Diversity is crucial to human survival, on all levels: biodiversity, cultural and social diversity, diversity in terms of political and economic systems, technological diversity. It is crucial for two reasons: firstly, it is more productive than its opposite number, monoculture: "Biodiversity-based productivity is higher than monoculture productivity. I call this blindness to the high productivity of diversity a 'Monoculture of the Mind', which creates monocultures in our fields." (Shiva, 2000, 15; see also Sachs, 1999, 93-94)

---

[64] At the moment, simple solutions are still preferred and more effective. The more or less successful international action against CFCs in the wake of the Montreal Protocol was possible due to the fact that there were relatively few substances from relatively few producers involved for which substitutes existed. Climate change, however, even though potentially far more dangerous, is far less effectively tackled since the causes, products and processes involved are far more numerous and complex, *and* closer to home in terms of required lifestyle changes (Hirsch, personal conversation, Zurich, 23.10.2000).

Secondly, both our survival and indeed our human capacities, as they developed through evolution, are dependent on it:

> The ecological crisis, in short, is about what it means to be human. And if natural diversity is the wellspring of human intelligence, then the systematic destruction of nature inherent in contemporary technology and economics is a war against the very sources of mind. We have good reason to believe that human intelligence could not have evolved in a lunar landscape, devoid of biological diversity. (Orr, 1994, 140)

It is fairly obvious that this applies to the cultural sphere as well. Just imagine the opposite: the world would *only* know Hollywood, McDonald's and Coca Cola. Even the most fanatic defenders of globalisation would agree that this would amount to an incredible impoverishment.[65]

Acceptance of limits[66]

The biosphere sets us limits which are non-negotiable. Jonathon Porritt has described this very poignantly:

> People somehow assume that petrol disappears when it is burned, or that rubbish no longer exists when it's incinerated. But neither is the case. Waste can change its form, but it cannot be thrown away because the Earth is a closed system with respect to matter. There is no "away". Politicians and regulators have connived in that illusion – and continue to do so. Scientists and engineers have buttoned their lip, and crossed their fingers that the reckoning wouldn't come in their own day. For they know with absolute certainty that one day it will come. That's what the laws of thermodynamics tell us. Everything has to go somewhere. Nothing disappears. These are not just interesting hypotheses or bright ideas. They are laws – like the law of gravity – which cannot be ignored or avoided. (Porritt, 2000, 97)

This is hard for us to accept, because we have been brought up believing in an ideology which stresses unlimited linear expansion. But accepting limits is not a negative concept, on the contrary:

---

[65] The German writer Volker Braun reports in an interview on an international writers' gathering in Sanaa where the question was discussed whether globalisation would lead to one world language for poetry. The writers present were unanimous in their view that the differences and the specifics of every regional culture and language are the most valuable assets when it comes to literature (see Volker Braun, 'Wir sind geschichtsgeschädigt', zur Verleihung des Büchner-Preises interviewt von Lothar Baier, in *Die WochenZeitung*, No. 42, 19.10.2000, 21-22, here 22).

[66] For general background on why it is important for societies to accept limits, see Illich, 1973.

Limits have a double nature, being both restraining and facilitating: they act as constraints only with respect to one particular order of things, but open up possibilities with respect to another order of things. On the one hand, the confines of the canvas restrict the surface available to the painter. On the other, however, they determine the base upon which a sophisticated creation can rise. (...) Limits may set free action, because they foreclose options. As they concentrate attention, they help to mobilize energies and stimulate excellence. (Sachs, 1999, 179-180)

Or, as Paul Hawken asks: "Why should the bio-physical limits to which progress is subject be any more constraining than, say, the confines of a canvas for Cézanne or the number of holes in a flute for Jean-Pierre Rampal?" (quoted in *Greening the North*, 1998, 84) This positive aspect of accepting limits can be exemplified with the related ideas of frugality and the ascetic. With regard to the former, Norberg-Hodge says:

In the West, *frugality* conjures up images of old aunts and padlocked pantries. But the frugality you find in Ladakh [India], which is fundamental to the people's prosperity, is something quite different. Using limited resources in a careful way has nothing to do with miserliness; this is frugality in its original meaning of "fruitfulness": getting more out of little. (2000, 25)

Similar things can be said about the ascetic: quite contrary to the current negative usage, which essentially goes back to a Christian vilification of the term and its equation with renunciation (particularly of things physical), an ascetic way of life means complete self-determination, the absence of outside control over your life, your emotions, your thoughts. It means that you train yourself into using only what you need. With it comes the realisation that "restraining a need results in much stronger satisfaction even with a smaller dosage. This applies equally to eating, drinking and sex. The only condition is not to be enslaved by lust, but train oneself ascetically in its autonomous use." (Schmid, 2000, 10; see Hösle, 1994, 78-80) Thus we should shape our needs to fulfil us according to our demands, rather than being helplessly at their mercy, as is the case with the current consumer society which promises instant fulfilment of every wish. Yet because this is impossible, consumer capitalism has to rely on our forgetfulness, our preparedness to try again and again. Rather than on fulfilment it relies on constant frustration. This is the only trigger to endless (and useless) consumption. What we might, in fact, need is to revitalise the medieval concept of "*mâze*" which was called "the mother of all virtues". Its aim was "to strike the right balance in all things and to pursue the golden mean."[67]

---

[67] Joachim Bumke (1986), *Höfische Kultur. Literatur und Gesellschaft im hohen Mittelalter*, 2 Vols. (Munich, dtv), Vol. 2, 416-430, here 419.

Slowness

"Generally speaking, the ecological crisis can be read as a clash of different time-scales: the timescale of modernity collides with the timescales that govern life and the earth." (Sachs, 1999, 189) Since the first one is adaptable, whereas the latter isn't, it might be sensible to adopt the latter as a measure for our actions. This, invariably, means that we have to slow down. The cycles of innovation are so fast by now – usually for no other reason than stimulating production and consumption – that they cannot be sensibly sustained any more. For that reason, if we want to gain the upper hand again in terms of control over our future, speed reduction, not just on German motorways, is a necessity. Furthermore and most ironically, those societies with the greatest abundance of timesaving devices have ever less time at their hands:

> Across the board, from mobility to communication, from production to entertainment, time saved has been turned into more distance, more output, more appointments, more activities. The hours saved are eaten up by new growth. And, after a while, the expansion of activities generates new pressure for time-saving devices – starting the cycle all over again. Time gains offer only transitory relief, because they encourage further growth of all kinds. In this way, the Utopia of affluence undercuts the Utopia of liberation. Acceleration is therefore the surest way to the next congestion. (Sachs, 1999, 193)

There is a further aspect, which will occupy us in chapter 3. Understanding and learning do not work at great speed; they need, if they are to be thorough and yield real benefits, a lot of time. And all the increasing data mountains and the fastest computers in the world won't change that: data, which isn't thought through, evaluated and contextualised, is worthless (see below, p. 205). This is why the German physicist Hans-Peter Dürr calls for a general "deceleration" of society.[68]

---

[68] "Learning processes, if anything really new is learnt by them, need adequate time. The natural sciences tell us that, due to the Second Law of Thermodynamics, generation and differentiation cannot run at any speed, as opposed to degeneration, destruction, reproduction and copying. Increased speed therefore leads relatively quickly to a weakening, eventually stalling of creativity and intelligence. This is why we and our society need 'deceleration' [Entschleunigung], and this cannot happen sufficiently under the conditions of destructive competition." (Dürr, 2000, 123)

Impact/ Small is *beautiful/ Accepting our responsibilities*

> The Indians, however, did possess one inextinguishable source
> of wealth: the modesty of their desires. (McKibben, 1997, 207)

An insight, which has to permeate profoundly all our thoughts and actions, is that whatever we do has an impact on the environment, on the social fabric and on the economy. We have to learn to integrate this impact into our assessment of a course of action – whether long-term or short-term, whether immediately visible or delayed through feedback mechanisms. This is as true for our decisions in the ballot box, as it is for professional actions in our careers, consumption patterns and so on.[69] There is, in this respect, no division between our private, public or professional lives; and there is also no profession (Berry in O'Sullivan, 1999, xiv) or academic discipline which can exempt itself from a responsibility for its impact (see Hirsch, 1995, 305).

It is well known that the effects of our actions adhere to the following formula: "Impact (on the earth) = Consumption x Technology x Population [I = C x T x P]" (*Lugano Report*, 1999, 32).[70] If we remind ourselves of the finding in chapter 1.a that humankind as a whole is already consuming in excess of the carrying capacity of the planet, it is clear that we have to reduce impact by lowering all three factors influencing it. But 'we' obviously does not mean every one of the world population, it means the wealthy countries and elites of the world who consume over and above their resource allocation based on equitable sharing (see below).

What follows from an analysis of the five spheres, is that small *is* beautiful. This is why E.F. Schumacher's book with the same title (published in 1973) is still so astonishingly up-to-date: Schumacher very clearly saw that as soon as you start to take the complexities of life and the limits of the biosphere into account, only small, transparent, controllable set-ups prove to be sustainable (in fact, if not the one to coin the phrase, he was certainly one of the first Euro-Americans to think within a sustainability framework). The North American Indians always had a very clear awareness of the fact that sustainable communities and a working democracy were only possible within small structures. Rarihokwats, the editor of the famous *Akwesasne Notes*, says in an interview:

> What has to happen is a return to tribal structures, a re-tribalisation and a revitali
> sation not only of the ancient peoples. Those people who need tribal structures
> most urgently in North America are the white Americans. They should form

---

[69] See Vester, 1997, 64-65 and Angayuqaq Oscar Kawagley and Ray Barnhardt, 'Education Indigenous to Place. Western Science Meets Native Reality', in *Ecological Education in Action*, 1999, 117-140, here 136.

[70] This is based on Paul Ehrlich's IPAT-formula (see Paul Ehrlich (1995), *The Population Bomb* [New York, Buccaneer Books]).

small, functioning living units in order to regain their human identity. (quoted in Reichert, 1974, 87; see also Vine Deloria, 1970)

The American Indian people, particularly the Hopi, had a very acute awareness that meaningful communities, and democracy in particular, can only work with small numbers. They found from long experience that communities above around 3,000 people become unmanageable.[71] This is clearly borne out by other indigenous cultures. Smallness can have a positive impact on responsibility, accountability and self-determination, as Norberg-Hodge has observed in Ladakh:

> Since villages are rarely larger than a hundred houses, the scale of life is such that people can directly experience their mutual interdependence. They have an overview and can comprehend the structure and networks of which they are a part, seeing the effects of their actions and thus feeling a sense of responsibility. And because their actions are more visible to others, they are more easily held accountable. Economic and political interactions are almost always face to face; buyer and seller have a personal connection, a connection that discourages carelessness or deceit. (...) In the traditional Ladakhi village, people have much control over their own lives. To a very great extent they make their own decisions rather than being at the mercy of faraway, inflexible bureaucracies and fluctuating markets. (2000, 50-51)

Precautionary principle/ Prevention/ Prudence

Taking all the above principles together and also bearing in mind the irreversibility of many actions, the precautionary principle has emerged as a powerful tool to enable sustainability. It basically means that *before* we launch headlong down the road of some development path or some specific technology, we should assess its likely impact. Linking back to the first principle, this will allow us to keep the future open. We have to beware of developments which reduce our options in the future and constrain our freedom to choose. Reversibility, i.e. the ability to correct mistakes, would therefore be a central aspect of a sustainable society.[72]

---

[71] There is a lot of evidence that an upper limit on the size of the community is a precondition for a functioning democracy (and society in general, for that matter) (see Lummis, 1996 and Arendt, 1970). But it is also interesting that the social sciences can only produce meaningful, in-depth results for small samples through qualitative analysis. This means that any generalised conclusions from large quantitatively analysed samples are very abstract and shaped by the scientist's perspectives. Anyone who has taken part in a large, standardised survey will know that it is impossible to fit truthfully into the restrictive grid of the standardised answers allowed (see Heinz Gutscher, Gertrude Hirsch, Karin Werner, 'Vom Sinn der Methodenvielfalt in den Sozial- und Geisteswissenschaften', in *Umweltproblem Mensch*, 1996, 43-78, here 45).

[72] Robert Jungk, 'Don't stop thinking 'bout tomorrow', in *Sie bewegt sich doch*, 1993, 130-132, here 131. – There might be more scope for reversibility than we often assume. If, for example, the hormone-disruptor theories prove true, we might have to get rid of most of our plastics. Some people

Since uncertainty plays a big role in assessing the future, we should err on the side of caution and move from end-of-pipe tampering with effects to prevention, as Holmberg *et al.* write:

> Nature is a very complex system and we probably do not know every critical function. The solution to this problem (...) is to move the focus backwards in the causal chain – from studying effects in nature to studying what constraints have to be put on the societal metabolism to avoid any systematic increases of society's impact on nature. The essential constraint is that the societal metabolism should be embeddable in the biogeochemical cycles. This involves a change from environmental pathology to societal prevention. Since problems do not originate in the environment but in society, this seems to be a reasonable change.[73]

Or, expressed with a view to biodiversity by Edmund Wilson:

> The ethical imperative should therefore be, first of all, prudence. We should judge every scrap of biodiversity as priceless while we learn to use it and come to understand what it means to humanity. We should not knowingly allow any species or race to go extinct. (...) An enduring environmental ethic will aim to preserve not only the health and freedom of our species, but access to the world in which the human spirit is born. (Wilson, 1992, 351)

This should instil a sense of awe in us which could prevent much hubris. Commoner has expressed this in his third "law of ecology" as follows: "'Nature knows best'. The reason for this is that ecosystems (...) are extremely intricate and are unlikely to be improved by random tinkering." (Carter, 1999, 21-22) Maybe we should adopt what the Six Nations of the Iroquois Confederation know as "Seventh Generation Philosophy". It instructs them to "consider the impact of their decisions on the seventh generation into the future. This way, they are to proceed cautiously, thinking of what effect their decisions will have on the welfare of their descendants."[74]

If we now turn our attention towards the foundation on which life rests and the other four spheres which define human activity, we should always bear the

---

argue that this is impossible since it would lead to a collapse of civilisation. This might be true, but only for *some* civilisations, namely the Euro-American ones. Most people on Earth live without most of our plastics, so it wouldn't make much difference to them.

[73] John Holmberg, Karl-Henrik Robèrt and Karl-Erik Eriksson, 'Socio-Ecological Principles for a Sustainable Society', in *Getting Down to Earth*, 1996, 17-48, here 18-19.

[74] Haudenosaunee, 'What is the Seventh Generation?' [http://sixnations.buffnet.net/Culture/?article =seventh_generation]. See also the 'Address to the General Assembly of the United Nations', delivered October 25, 1985 by Chief Leon Shenandoah, Tadodaho, Supreme Sachem of the Iroquois 6 Nations [http://www.earthportals.com/Portal_Messenger/shenandoah.html].

principles elaborated above in mind and use them as cross-checks with which we can clarify what sustainability means.

1. Ecology

> The organism which destroys its environment destroys itself.
> (Bateson, 2000, 457)

We know that the biosphere is a "thermodynamically closed and non-materially-growing system".[75] This means that there are definable, scientifically verifiable limits to our endeavours on Earth.[76] Let's be very clear about this: the ecological conditions of life are not man-made but established by nature, and despite human tinkering, not alterable: "We have failed to recognize that, just as in the lives of cells, the conditions of ecological systems are not established by human laws but by Nature's rules, rules which are non-negotiable and fundamentally rooted in the laws of physics." (Paul Hawken, quoted in Porritt, 2000, 103) This, based on the second law of thermodynamics, unfortunately means that humans cannot increase quality; they are bound by what nature provides:

> Contrary to what many people might believe, the rate and capacity of the Earth to create material quality depends not on human-driven activities, but on the sun. Virtually all our human activities remove or consume quality. As ingenious and important as industrial practices are, they also use up quality and order. Nature has the capacity to recycle waste and reconstitute it into new resources of concentrated material quality. However, its capacity is regulated by sunlight and photosynthesis, not by economic theory or politics. Today's extraction and processing of resources is overwhelming that capacity, while the waste from these processes systematically builds up in our water, air, soil, wildlife – and in ourselves. (Paul Hawken, quoted in Porritt, 2000, 96-97)

Whenever one talks of limits there is always, very quickly, the argument at hand that our ingenious scientists and engineers will, at the last moment, find a clever solution which will save us from having to change our way of life and our economy. Vester has brilliantly dissected this absurd claim. He wants to know *what* exactly they should find:

---

[75] Robert Costanza *et al.*, 'Integrated Envisioning, Analysis, and Implementation of a Sustainable and Desirable Society', in *Getting Down to Earth*, 1996, 1-13, here 2.

[76] In a similar formulation: "The total of resource consumption (throughput), by which the economic subsystem lives off the containing ecosystem, is limited – because the ecosystem that both supplies the throughput and absorbs its waste products is itself limited. The Earth ecosystem is finite, non-growing, materially closed, and while open to the flow of solar energy, that flow is also non-growing and finite." (Herman E. Daly, 'Consumption: Value Added, Physical Transformation, and Welfare', in *Getting Down to Earth*, 1996, 49-59, here 49) See also Vester, 1997, 32.

New thermodynamic laws, so that the *perpetuum mobile* becomes possible? A shortening of the half-life of radioactive isotopes, so that radioactivity suddenly runs out? A change of gravity and with it new tectonic mechanisms so that there aren't any earthquakes any more? Or even a new mathematics and new system laws which decree that 2 and 2 is only 3 and that life can exist without integration and interaction, without structure and communication, without matter and energy exchange, without evolution and diversity of species? (Vester, 1997, 487)

These biogeochemical limits could be disregarded for a very long period in human history because humanity wasn't in a position to make a significant impact on the global ecosystems. This has dramatically changed with the Industrial Revolution, which for the last three hundred years has handed humankind "ever bigger and handier spoons to eat up the planetary chocolate cake", as Mathis Wackernagel once remarked so poignantly.[77]

Yet because by now we know these limits, "it is increasingly clear what needs to be done to address the challenge, even if the difficulty of doing it must be acknowledged":[78]

What kind of system would be ecologically sustainable? The answer is simple – a system whose structure respects the limits, the carrying capacity, of natural systems. A sustainable economy is one powered by renewable energy sources. It is also a reuse/recycle economy. In its structure, it emulates nature, where one organism's waste is another's sustenance. The ecological principles of sustainability are well established, based on solid science. Just as an aircraft must satisfy the principles of aerodynamics if it is to fly, so must an economy satisfy the principles of ecology if it is to endure. The ecological conditions that need to be satisfied are rather straightforward. Over the long term, carbon emissions cannot exceed carbon dioxide (CO_2) fixation; soil erosion cannot exceed new soil formed through natural processes; the harvest of forest products cannot exceed the sustainable yield of forests; the number of plant and animal species lost cannot exceed the new species formed through evolution; water pumping cannot exceed the sustainable yield of aquifers; the fish catch cannot exceed the sustainable yield of fisheries.[79]

In a similar vein, *The Natural Step* has attempted to identify what it calls system conditions for sustainability, and they are so straightforward that you wonder why

[77] Mathis Wackernagel (1997), *Ecological Footprints of Nations Report and Slide Show* (Toronto, ICLEI [http://www.iclei.org/iclei/ecofoot.htm]).
[78] Paul Ekins, 'Towards an Economics for Environmental Sustainability', in *Getting Down to Earth*, 1996, 129-152, here 145.
[79] Lester R. Brown and Jennifer Mitchell, 'Building a New Economy', in *State of the World 1998*, 1998, 168-187, here 169-170.

some people claim that sustainability is a difficult concept.[80] It makes you wonder whether this has less to do with the actual parameters of sustainability and a lot more with the implications these parameters have for lifestyles and power (see chapter 2). However that may be, here are the four conditions:

- System Condition 1: Substances extracted from the Earth's crust must not systematically increase in nature. This means that, in a sustainable society, fossil fuels, metals and other materials are not extracted at a faster pace than their slow redeposit into the Earth's crust or their absorption by nature.
- System Condition 2: Substances produced by society must not systematically increase in nature. This means that, in a sustainable society, substances are not produced at a faster pace than they can be broken down and reintegrated by nature or re-deposited into the Earth's crust.
- System Condition 3: The physical basis for the productivity and the diversity of nature must not be systematically diminished. This means that, in a sustainable society, the productive surfaces of nature are not diminished in quality or quantity, and we must not harvest more from nature than can be recreated.
- System Condition 4: We must be fair and efficient in meeting basic human needs. This means that, in a sustainable society, basic human needs must be met with the most resource-efficient methods possible, including a just resource distribution.[81]

This has a number of implications which are straightforward from a scientific point of view, but have dramatic consequences for our society which, as Canadian ecologist William Rees says, is "a world addicted to growth, but in deep denial about the consequences".[82] There is no such thing as unlimited growth, and that applies to the economy, technology (even computers, see chapter 2.a), consump-

---

[80] In discussions about EfS it is customary to stress the complexity and difficulty of EfS, of finding solutions etc. (see for example *Umweltbildung im 20. Jahrhundert*, 2001, 4-6); see above footnote 53 on p. 31.

[81] The Natural Step (1999), *A Framework for Sustainability* (London: Forum for the Future) [for an online version see: http://www.mtn.org/iasa/tnssystemconditions.html]. For a more technical/scientific description of these conditions see Holmberg *et al.* in *Getting Down to Earth*, 1996, 25-27. Compare also the four basic principles of ecosystem sustainability in Nebel/Wright (2000): 1. "For sustainability, ecosystems use sunlight as their source of energy (...) which is nonpolluting and nondepletable" (73-74); 2. "For sustainability, ecosystems dispose of wastes and replenish nutrients by recycling all elements." (74); 3. "For sustainability, the size of consumer populations is controlled so that overgrazing or other overuse does not occur." (93); 4. "For sustainability, biodiversity is maintained." (101) (See also, for a similar attempt, Ekins in *Getting Down to Earth*, 1996, 142-43, or Commoner's four "laws of ecology", as quoted in Carter, 1999, 20-22)

[82] Quoted in Lester R. Brown and Jennifer Mitchell, 'Building a New Economy', in *State of the World 1998*, 1998, 168-187, here 169.

tion and population.[83] Indeed, as C.A. Bowers states, "there are no historical examples of ecologically sustainable cultures that relied on the myth of unending progress".[84] This implies nothing less than a substitution of the central plank of our current ideology, namely the myth of growth (see Dürr, 2000, 149-150) and an acceptance of the overall importance of a "steady state dynamic equilibrium" (Vester, 1997, 454). I shall look in more detail into why we have such difficulties in accepting this in chapter 2.b, but it clearly has a lot to do with a completely inappropriate notion of freedom, our postmodern idea that *everything* is just fine (in other words, the disintegration of ethical values), and with our media society which daily, hourly indoctrinates us with the view that more, new and different is always, by definition, better. But Schumacher and others have long seen that this is a violation of the truth: "An attitude to life which seeks fulfilment in the single-minded pursuit of wealth – in short, materialism – does not fit into this world, because it contains within itself no limiting principle, while the environment in which it is placed is strictly limited." (Schumacher, 1993, 16-17) Getting used to the idea that whatever we do bounces an echo back off the world's limits, in other words that we take part in a zero-sum game, is the crucial bit. That this does not only apply to resource usage, but also to the political and social dimension, has been noted by Bennholdt-Thomsen/Mies in *The Subsistence Perspective*:

> Within a limited world, aims like "unlimited growth" can be realised only at the expense of others. Or: there cannot be progress of one part without regression of another part, there cannot be development of some without underdeveloping others. There cannot be wealth of some without impoverishing others. (1999, 29)

Or, to put it in Commoner's fourth "law" of ecology: "'There is no such thing as a free lunch'. [This law] 'is intended to warn that every gain is won at some cost'." (quoted in Carter, 1999, 22) This means that we have to keep in mind that results from our actions might occur delayed, in unexpected contexts and far removed from our field of vision (Vester, 1997, 71, 86); yet still, we bear responsibility for them. Some might argue that this is talking in political, social and moral categories, but in fact, these are grounded in sound science, in the laws of thermodynamics. Vester has argued that a systemic, complex view of our world rules out growth. Zero growth is, as he states, a "cybernetic necessity" (1997, 56). Sustain-

---

[83] I know that this is a very delicate subject, and Susan George has sarcastically portrayed the many sinister driving forces behind a call for global population control (outright eugenics of 'unworthy life' [in other words, decimate the useless poor in the South, so that the rich in the North can carry on with their wasteful lifestyles] or shifting the blame for the environmental crisis from the North to the South) in her frightening *Lugano Report* (1999); on the last point see also Sachs, 1999, 82. Nevertheless, there is no running away from the fact that there is an upper limit to the world population which the planet can carry (see Vester, 1997, 452-453; Nebel/Wright, 2000, 93).

[84] C.A. Bowers, 'Changing the Dominant Cultural Perspective in Education', in *Ecological Education in Action*, 1999, 161-178, here 167.

able solutions, therefore, can be defined as those which conform with system conditions (ibid. 116). This also implies their necessarily local nature (ibid. 118, and see above p. 39).

Now, many would argue that we have to ignore these biogeochemical parameters because we cannot fit into them. Otherwise, "civilisation as we know it" would collapse. Yet, studies have shown that these predictions have more to do with scaremongering and a seriously spoilt notion of what we 'need' for a decent life. As ever, such ideas are exclusively informed by the way of life of the wealthiest ten per cent of the world population – not exactly representative, as we know. If we assume – see below – that every person on Earth has a right to the same resource consumption, the per capita energy available per annum would amount to roughly 1.5 kilowatts.[85] Now, compared to the 6 kilowatts used by a European and the 11 kilowatts consumed by an average U.S. citizen, that's not much. But if measured against the 800 watts of a Chinese or the 80 watts of an average citizen of one of the poorest countries, it is quite a bit. But even more surprising: the recommended average "would effectively represent the average living standard of a Swiss in 1969, taking modern technology into account" (Dürr, 2000, 158). However much your heart might lust after three mobile phones, an overdimensioned four-wheel-drive and a mass of electronic gadgets: this is hardly the often threatened move back into the Stone age, into hunger and destitution. It is simply reducing our excesses of overdevelopment back to a sensible and sustainable level (see chapter 2.a).

2. Empowerment

> Empower people: change will not come from governments.
> (Alicia Bárcena and Diomar Silveria)[86]

One of the most precious insights of the sustainability discourse is surely that we cannot arrive at real solutions if we limit our perspective to one aspect of the problem. This means that we now have to consider what implications the ecological preconditions of life have on the fundamental aspects of human endeavours. This, of course, also applies to the political sphere.

---

[85] Compare Strategie Nachhaltigkeit im ETH-Bereich, Wirtschaftsplattform, *2000 Watt-Gesellschaft – Modell Schweiz*, Dezember 1998 [http://www.novatlantis.ch/projects/2000W/brochure/ge_brochure.html], which starts from an upper limit of 2 kilowatts.

[86] Bárcena/Silveria, 'Envisioning Sustainable Alternatives within the Framework of the UNCED Process', in *Getting Down to Earth*, 1996, 455-463, here 462.

A number of normative reasons as well as historical experiences clearly point to the fact that self-determination is one important key to sustainability.[87] For a start, it is very difficult to argue, if we believe in the *Declaration of Human Rights* and democracy in its original sense,[88] that somebody else, be it a state, a (multinational) corporation, a party or a tradition, should decide and determine, say, the resource use of a given population (see box 2: Shrimps for life on p. 25). The logic is inescapable: the only way to decide such issues is by self-determination of the communities affected, provided they respect the limits which their impact on other people and the earth places on their self-determination. This is also borne out by experience:

> In order to integrate indigenous peoples of the Arctic into the decision-making process, in order to adhere to the rights of indigenous peoples (...) and in order to protect the Arctic environment and guarantee a sustainable management of renewable resources, it is a precondition that indigenous peoples have the organizational capacity to and are given the empowerment to attain control over their own affairs and destiny. The Arctic history shows us that any development which intends to be based on mutual understanding and respect starts with this. (Jens Dahl, quoted in Fenge, 1994)

It has also been shown that the best option to manage a forest or a fishery sustainably is to put local people in charge.[89] They cannot run away from the consequences of their actions and have a long-term interest in the survival of their local environment which after all sustains their livelihood. National and multinational

---

[87] Living systems, within the limiting confines of genes and environment, are open systems, therefore enabling self-determination. The future is not mapped out for us, we are making it, by deed or omission: living systems "are not mere passive vehicles, sandwiched between the demands of their genes and the challenges of their environments. Rather, organisms actively engage in constructing their world around them. Every living creature is in constant flux, always at the same time both *being* and *becoming*. (...) A living organism is an active player in its own destiny, not a lumbering robot responding to genetic imperatives." (Steven Rose, 'Escaping Evolutionary Psychology', in *Alas, Poor Darwin*, 2000, 247-265, here 256-257): "In short, the biological nature of being human enables us to create individual lives and collective societies whose futures lie at least in part in our own hands." (ibid. 263)

[88] "Participatory democracy means the right to participate in the making of decisions that affect one's life." (Lummis, 1996, 138)

[89] Janet N. Abramovitz, 'Sustaining the World's Forests', in *State of the World 1998*, 1998, 21-40, here 38; Anne Platt McGinn, 'Promoting Sustainable Fisheries', in *State of the World 1998*, 1998, 59-78, here 71-72. – It is also historically the case that local people utilising common land tend to look after it in a sustainable way, as opposed to landowning elites who turn crops or wood into cash crops to maximise sales, irrespective of any long-term damage (see Swiss Agency for Development and Cooperation (SDC) (2000), *Forest Development in the Swiss Alps: Exchanging Experience with Mountain Regions in the South* (Bern, SDC [CD-Rom])).

corporations have no such commitments and can relocate, often overnight (see Madeley, 1999, 8-9).

This has consequences for the way the political system ought to be built and put into practice.

Local democracy

If we are serious about self-determination, about democracy as power of the people, we have to acknowledge that only a local, transparent public sphere which is accessible to all, can do the job: "In general, democracy depends on localism: the local areas are where the people live. Democracy doesn't mean putting power some place other than where the people are." (Lummis, 1996, 18) This is a serious argument. It links to historical experience, namely that the only functioning democracies, where there was no illegitimate abuse of power by dominant groups, were small-scale communities of Native American Indians or other indigenous peoples, anarchist cooperatives in the Spanish Republic,[90] or religious communities such as many orders, for example the Shaker, Oneida or Ikaria (see Jucker, 1997, 69-70). It also tallies with the guiding principles above: only locally can we determine impact sufficiently robustly, since in a local environment even excesses and wrong developments are limited in impact and often reversible;[91] only locally can the specifics of the situation, history, tradition and culture be fully borne in mind and diversity respected in the search for solutions.[92] In a local context speed and limits are also easier to control than in highly centralised industrial systems.[93]

---

[90] See Gaston Leval (1975), *Collectives in the Spanish Revolution*, translated from the French by Vernon Richards (London, Freedom Press); José Peirats (1990), *Anarchists in the Spanish Revolution* (London, Freedom Press); Augustin Souchy and Ernst Gerlach (1974), *Die soziale Revolution in Spanien. Kollektivierung der Industrie und Landwirtschaft in Spanien 1936-1939. Dokumente und Selbstdarstellungen der Arbeiter und Bauern* (Berlin, Libertad).

[91] "Small-scale operations, no matter how numerous, are always less likely to be harmful to the natural environment than large-scale ones, simply because their individual force is small in relation to the recuperative forces of nature." (Schumacher, 1993, 22)

[92] Sachs warns convincingly of the danger of global environmental management and data systems, often based on satellites, which inform policies and which take the vantage point of the moon looking back down to Earth, thus obliterating any cultural, historical differences, being utterly incapable of visualising, let alone understanding, the real people on the ground, the power structure etc. (see 1999, 44).

[93] A good example of this is electricity production. A local grid of various types of renewable energy sources such as solar electricity (photovoltaics), biomass, wind and water is highly adaptable, versatile and open to new developments. Highly centralised systems fed by nuclear, gas, oil or coal power stations lock consumers into high dependency on expensive infrastructure, long-term financial commitments, volatile world markets and price fluctuations, entrenched corporate interests, state corruption (subsidies etc.) and outdated technology (no better example can be observed than nuclear power, its technology always questionable while economically never viable [see Myers, 1998, 69]).

We also know from the relevant research that commitment to a specific community is the only way to ensure that people live up to their responsibilities.[94] The anonymity of large masses of people leads to the "free rider symptom" and many other aspects of behaviour which are unsustainable, largely because people do not have to accept responsibility for their behaviour (see Canetti, 1998).

Another important aspect of the discussion around local democracy is subsidiarity. Schumacher defines it thus: "The Principle of Subsidiary Function implies that the burden of proof lies always on those who want to deprive a lower level of its function, and thereby of its freedom and responsibility in that respect." (1993, 204-205) In other words, whatever can be done on the local or regional level should be done and dealt with there. The bigger entities such as states, the EU or the UN should only assume functions if there is no other, local solution. This is very important with a view to the centuries of abuse by central powers over local peoples. Many peoples are only just gaining some self-control (the East Timorese); others (the Kurds, the Palestinians, indigenous peoples all over the world) are still far from any notion of self-determination.

Self-determination in all areas

If our aim is meaningful democracy and control of humans over their destiny, then it is crucial that self-determination is at work in all areas of life. In other words we are talking about "the extension of the system of fundamental rights into the economic sphere",[95] since a "strengthening of the right to self-determination for all groups of society requires industrial democracy",[96] for example through worker participation, cooperatives and/or collective leadership. Such an extension would satisfy Norberto Bobbio's criteria with which one can judge an increase in democracy, namely "whether the number of areas and spheres where you can exercise this right [of self-determination/democracy] has grown". Bobbio continues: "As long as in an advanced industrial society the two big centres representing top-down power, the corporation and administration, have not been part of a process of democratisation, this process cannot be regarded as completed."[97]

---

[94] See Andreas Diekmann, Axel Franzen, 'Einsicht in ökologische Zusammenhänge und Umweltverhalten', in *Umweltproblem Mensch,* 1996, 135-157; Huib Ernste, 'Kommunikative Rationalität und umweltverantwortliches Handeln', in *Umweltproblem Mensch,* 1996, 197-216; Urs Fuhrer, Sybille Wölfling, 'Von der sozialen Repräsentation zum Umweltbewusstsein und die Schwierigkeiten seiner Umsetzung ins ökologische Handeln', in *Umweltproblem Mensch,* 1996, 219-235; Hans-Joachim Mosler, Heinz Gutscher, Jürg Artho, 'Kollektive Veränderungen zu umweltverantwortlichem Handeln', in *Umweltproblem Mensch,* 1996, 237-260.

[95] Marc Spescha, 'Basisdemokratie ohne Volks-Mythen', in *Demokratie radikal,* 1992, 101-112, here 103.

[96] Hans Schäppi, Walter Schöni, 'Wirtschaftsdemokratie und Industriepolitik', in *Demokratie radikal,* 1992, 21-33, here 22.

[97] Norberto Bobbio, 'Die Zukunft der Demokratie', in *Sie bewegt sich doch,* 1993, 57-76, here 66.

In the United States, often hailed as the birthplace of modern democracy,[98] a very interesting idea has surfaced aimed at restoring sovereign democratic control over corporations, in particular multinational ones (TNCs), which of late have emerged as the biggest worldwide threat to people's self-determination and livelihood (see chapter 2.a). In revitalising a tradition dating back to the 17th century we are reminded that originally the citizens of a state allowed corporations to form in a certain way. In the meantime, however, most countries in the world have started to grant corporations rights which they do not grant their citizens.[99] The idea is now to reintroduce *Charters of Incorporation*: the citizens allow corporations to operate, provided they fulfil specific social and environmental standards. If the corporations violate the charters, these are withdrawn and corporate assets seized. Business owners would be liable for harms and injuries (see Karliner, 1997, 213-215). Such controls become more and more urgent since corporations, to defy attempts to curb their power, claim constantly that it would be best to solve social and ecological problems by self-regulation. This is about as absurd as asking a rapist or a kidnapper of children to fight these crimes. Even the OECD starts from the assumption that the business world will only adopt sustainable practices if they are forced to do so by citizens or governments (see Karliner, 1997, 38-50).

But there is an even more fundamental point to be considered. Lummis, in his impressive *Radical Democracy*, emphasises, in line with Hannah Arendt in *On Violence* (1970), that democracy cannot be institutionalised and then be taken for granted. Democracy depends on people, not power structures: "It is the basic idea of democracy that the people are sovereign: their power is prior to the power of the state. The state and its laws exist by their consent alone. This statement means that it is up to the people, and not to the state, to determine who 'the people' is." (Lummis, 1996, 139) Democracy is a state of mind and a way of life, which has to be recreated time and again in our actions: it can always happen but it can always be taken away from us, no matter how elaborate the institutional or legal safeguards might be: "Democratic power does not fall from above, it is generated by a people in a democratic state of mind, and by the actions they take in accordance with that state of mind." (ibid. 35) The trouble is – not really a problem, except in a society like ours where every effort is seen as negative, to be replaced by an electronic gadget which makes us even lazier – that there is a price tag: "But if the state of democracy means a state of public action, then there is no conceivable stage of history at which it can be had without effort." (ibid. 161)[100]

---

[98] But regarding the reality of that claim, see Zinn, 1996.

[99] Such as that the people responsible for the running of a corporation cannot be held responsible for crimes perpetrated by that company (for a long list, see Richard L. Grossman and Frank T. Adams (1993), *Taking Care of Business: Citizenship and the Charter of Incorporation* (Cambridge, MA, Charter Ink), 10-21).

[100] There are some conditions which seem crucial in order to enable democracy: "flourishing unions, readily accessible third parties, inexpensive media, and a thriving network of cooperatives

I do apologise for ramming this point home, but it is an absolutely crucial one. It is the reason why empowerment, after ecology, has to be the most important sphere. For if we agree that "all human beings are born free and equal in dignity and rights"[101] we have, within the bounds of nature, no other option than to organise human society according to self-determination. It follows that there are no reasons to privilege one person over another by race, sex, wealth, heritage or whatever.[102]

But if we *are* serious about this, it means that democracy, within the means of nature, is the overarching organising principle:

> If democracy is the end, all political institutions and arrangements, as well as economic systems and technologies, are means. Really to see things in this way would amount to a revolution in our understanding of those powerful words that so dominate our collective lives today: efficiency, practicality, and progress. For we often forget that these words have no fixed or absolute meanings: what is efficient depends on what effect we want to produce, what is practical depends on what practices we value, what is progressive depends on where we want to go. Taking democracy as the goal means stealing back these expressions from economics and technology, where they have been monopolized so long. It means rejecting such formulations as that there is a "trade-off" between efficiency and empowerment of the people. If empowerment were agreed on as the desired effect, any economic or technological arrangement that weakened the people would be inefficient by definition. (Lummis, 1996, 39)

If this is granted, it means not only the extension of democracy into every sphere, but also into every country (and back into the so-called democratic countries): "For the radical democrat, imperial democracy is no longer a possibility. Lest it corrupt its own spirit, the struggle for democracy must be not the struggle only for a democratic country but for a democratic world." (Lummis, 1996, 138)

Another crucial aspect of functioning democracy is citizen's dissent. The resistance to nuclear power, motorways and bypasses, genetically modified food, CFC and much more has shown that "*civil disobedience*" has become indispensable "as a legitimate policy tool of an enhanced, more than just legalistic democ-

---

and community organizations" (Ferguson, 1995, 88), as well as the "right to information including public access to ecologically relevant records, environmental impact assessment, technology risk assessment as well as the right to class action" (*Zukunftsfähiges Deutschland*, 1996, 378).

[101] United Nations, *Universal Declaration of Human Rights*, English version, 1998 (Geneva, Switzerland) [http://www.unhchr.ch/udhr/lang/eng.htm].

[102] This and the following should make it clear that by empowerment I do *not* mean government, NGO or any other 'participation' from the outside, abused as a 'development' tool to manufacture consent for "projects which serve mainly the interests of the few" (Majid Rahnema, 'Participation', in *The Development Dictionary*, 1992, 116-131, here 118).

racy": "In fact, to be free and to be capable of disobedience are inseparable."[103] This is clearly in line with the definition of democracy given by Lummis which harks back to the original meaning of the word.

Redefining freedom

But we do not only need to be reminded of the essential dimension of democracy, that it only lives in democratic action. We also need to develop social structures within which this action can take place. This means that we have to change and improve our current democratic systems. In the context of sustainability and the principle of planning for the future, we have to tackle the "functional deficiencies (...) of which short-term orientation is the most serious one" (*Zukunftsfähiges Deutschland*, 1996, 378). Parliament and governments, focusing more often than not on their re-election four or five years down the line rather than on long-term issues, need to be complemented by other bodies, such as an ecological council (similar to the elders' councils of indigenous peoples), equipped with the right of veto against governments. Long-term orientation can also be enhanced in that, similar to human rights, "ecological fundamental rights are enshrined in the constitution". This turns them into the "foundation of the community and the state", and therefore might ideally become part of people's identity (see *Zukunftsfähiges Deutschland*, 1996, 379-380).

This does imply that we have to seriously rethink our 'liberal' notion of freedom: "How much freedom – or, to be more precise, which kind – is still possible on the blue planet with its population fast approaching 8 billion?" (Martin/Schumann, 1997, 28). Currently, especially in Euro-American consumer societies, freedom is interpreted as a *carte blanche* to do whatever your fancy might suggest: "Only a culture that regards the individual as autonomous can propagate a view of individual reality within a context free from social accountability, much less accountability to the 'off-line' biotic community that sustains individual life." (Bowers, 2000, 32) But this is self-deception: "Given the physical and social limits we all experience, the very idea of absolute freedom is strictly speaking absurd. Without recognizable limits, a definition of freedom is empty and meaningless." (Marshall, 1993, 39) In a sustainability context and on the basis of the life-support system Earth, it is clearly the case that we need a notion of freedom which understands that it is limited both by the biogeochemical limits of the biosphere *and* the respective freedom of the other inhabitants and future generations. This insight, that only "self-limitation" can lead to real "self-liberation" (Sachs, 1999, 186) has been clearly described by Bakunin: "I am free only when all human beings surrounding me – men and women – are equally free. The freedom of others, far

---

[103] Erich Fromm, 'Der Ungehorsam als ein psychologisches und ethisches Problem', in Fromm, 1982, 9-17, here 14; see also Arendt, 1970.

from limiting or negating my liberty, is on the contrary its necessary condition and confirmation." (quoted in Marshall, 1993, 299)

Freedom from illegitimate power

> The conflict between humanity and nature is an extension of the conflict between human and human. Unless the ecology movement encompasses the problem of domination in all its aspects, it will contribute *nothing* toward eliminating the root causes of the ecological crisis of our time. (Bookchin, 1980, 43)

If we start from the idea, as elaborated above, that everybody is born free, then any power is first and foremost illegitimate power, or, as Chomsky says: "Structures of hierarchy and domination are fundamentally illegitimate" (1992a, 398). As with subsidiarity, contrary to a widespread assumption, the burden of proof lies not with the one who challenges power, but with power itself. If it cannot – always anew, because the circumstances might change – prove itself to be legitimate and necessary power, it forfeits its right to exist.[104]

Therefore empowerment and sustainability – as spelt out by Bookchin in the above motto – can only be turned into reality if we are aware of the three fundamental levels of power: human–human, human–nature and human–self (see Adorno/Horkheimer, 1973, 54). But taking these levels into account will also prevent us from adopting an anthropo- or even egocentric position which forgets the dependency of every human being both on nature and on human community. Or, as Plant formulates it:

> An environmental ethic needs to be sensitive to issues of gender, poverty, and global inequalities. The way humans see nature cannot be separated from the way humans see each other and to understand nature means gaining a better understanding of ourselves. (Plant, 1998, 170)

I want to make this unmistakably clear. When I talk of empowerment and self-determination throughout this book, I am *not* talking about the liberal notion of absolute, free-floating 'freedom of the autonomous individual' whereby everybody

---

[104] "I think it only makes sense to seek out and identify structures of authority, hierarchy, and domination in every aspect of life, and to challenge them; unless a justification for them can be given, they are illegitimate, and should be dismantled, to increase the scope of human freedom. That includes political power, ownership and management, relations among men and women, parents and children, our control over the fate of future generations (...) and much else." ('Noam Chomsky on Anarchism, Marxism & Hope for the Future. Interview by Kevin Doyle', in *Red & Black Revolution*, (1996) 2, 17-21 [http://flag.blackened.net/revolt/rbr/noamrbr2.html]; see also: 'Chomsky on Human Nature and all that. Interview with Michael Albert', in *Z Magazine*, 9 (1996) 3, 57-62).

can fulfil their every whim; this is essentially a negative freedom: freedom *from* responsibilities towards anybody and anything. Freedom cannot be properly understood in an anthropocentric and individually-centred way and is self-defeating as an end in itself. It is not "the well-being and self-determined interests of the individual" (Bowers, 1995, 6) that are of central concern; rather, the freedom of the individual stands in the service of and is limited by the integrity of the biosphere and the continuation of the human community supported by it.

3. Equity

> In the last instance humility is the basis for democracy, just as arrogance is the basis for the Führer/ leader principle. (Forbes, 1981, 49)

> Even what seem to us our present soundest and most final ideas of justice are noticeably cavalier and provincial and self-centered. What would we have to think of hogs who, having managed to secure justice among themselves, still and continuously and without the undertone of a thought to the contrary exploited every other creature and material of the planet, and who wore in their eyes, perfectly undisturbed by any second consideration, the high and holy light of science or religion. (Agee/Evans, 1960, 249)

If we take seriously the notion of intergenerational justice, the declaration of human rights with its proclamation that "all human beings are born free and equal", and the demands placed on a truly democratic political system, as elaborated above, we have to revitalise two old terms which the current ideology of neoliberalism has almost succeeded in eradicating: equity – meaning fair, just – and, as a consequence, equality. For on the basis of these ideas it is impossible to argue why one person should have a right to more resource use than another (apart from the small variations necessitated by geographical and climatic differences),[105] why 20 per cent of the human population should have a right to 80 per cent of the world's resources, why the richest 1 per cent of the world should own as much as the poorest 50 per cent together. If we insist, essentially, on horizontal

---

[105] There is the argument that we shouldn't start from total equity for everybody on Earth, since cultural, geographical and climatic conditions vary. This seems a weak argument, since it is often used to justify the overdevelopment of the North, thus forgetting that most of these 'different' cultural conditions in the North are historically based on colonialism, on exploitation of the South (so if anything, that would justify overdevelopment in the South and even more drastic reduction in the North).

power relationships (except where specifically justified),[106] it is hard not to accept the following as well:

The poorest/subsistence as our yardstick

One of the most important adjustments we will have to make in order to truly understand sustainability is what Gandhi once described as follows:

> Whenever you are in doubt, or when the self becomes too much with you, apply the following test. Recall the face of the poorest and the weakest man whom you may have seen, and ask yourself if the step you contemplate is going to be of any use to him. Will he gain anything by it? Will it restore him to a control over his own life and destiny? In other words, will it lead to swaraj [Sanskrit, meaning self-ruling] for the hungry and spiritually starving millions? (Gandhi, quoted in Shiva *et al.*, 1997, 178)

Despite all the rhetoric of democracy, with very few exceptions all the power structures in the world are designed with the wealthy and powerful in mind, and serve their interests. And the wealthy and powerful will do whatever it takes to keep it that way. This, more than most other things, is the driving motor for the destruction and greed we are witnessing in the world. To quote Gandhi again: "The Earth has enough for everyone's needs, but not for some people's greed." (quoted in Shiva, 2000, 19) Two points follow from this: firstly, that we have to fight the powerful and wrestle self-determination from them, or else there will never be any equity. Secondly, sustainable solutions cannot be designed with the wealthy and affluent Euro-American consumer in mind, but with a focus on those who can barely make ends meet. Everything else, whether we like it or not, from wide screen television via people carriers to transatlantic flights, has to be reassessed against this background of real, rather than manufactured, needs. A very good example which can teach us how to take the poor seriously and what flows from that in terms of critique of existing power structures is James Agee and Walker Evans' *Let Us Now Praise Famous Men* (1960, first published in 1941).[107] It is a very intense, sincere hymn to life itself, and rather than condescendingly laughing at its idealisations we should pick up on the level of truth embodied in it: it is precisely the Woods, Ricketts and Gudgers of this world from whom we have to start in order to rebuild society; it would mean focusing on self-determination and doing away with all sorts of dependences. Bennholdt-Thomsen/Mies, in their

---

[106] By horizontal relationships I mean equality as "justice or fair treatment", not the term's second meaning as "sameness or homogeneity" (see C. Douglas Lummis, 'Equality', in *The Development Dictionary*, 1992, 38-52, here 38).

[107] I am saying this despite some of the book's racist and sexist undertones, which are probably due to the times it was written in, and despite its sometimes stark romanticising and idealising.

radical study *The Subsistence Perspective*, which cuts forcefully through a whole host of contemporary myths, formulate the same insight. They state that the most important lesson is to adopt a

> view from below. This means that when we look at reality, when we want to gain clarity about where to go and what to do, we start with the perspective of women, particularly rural women and poor urban women in the South. Further, we start with everyday life and its politics, the strategies of women to keep life going. (1999, 3)

Only with such a perspective can we meaningfully talk about justice, as Wolfgang Sachs observes:

> With the emergence of biophysical limits to growth the classical notions of justice, which were devised in a perspective of finitude and not in a perspective of infinity, acquire new relevance: *justice is about changing the rich and not about changing the poor*. (...) Against the backdrop of drastic global inequality in resource use, it is the North (along with its outlets in the South) that needs structural adjustment. (1999, 173; my emphasis)

The challenge is therefore to develop political systems, social communities and consumption patterns which can satisfy the real and basic needs of everybody on Earth, in a fashion that can be generalised over time as well: "a life-style designed for permanence", as Schumacher called it (1993, 9).[108] And thinking within the perspective from below is not even *that* difficult, once we start to get the priorities right again:

> It is liberating, I think, to remind ourselves that most of the technologies that a human being really needs to live an orderly, comfortable, and healthy life are ancient. Would anyone really want to seriously argue that robots are more important to human beings than cloth woven from spun thread, or computers more important than the house with roof, walls, and windows? (Lummis, 1996, 105)

---

[108] One good starting point would be to introduce a *maximum*, rather than minimum wage: "Directors would be allowed to earn no more than a certain multiple – eight or ten perhaps – of the wages of the lowest paid member of their workforce, including subcontractors. If they wanted more money, they would have to give everyone more." (George Monbiot, 'A billion dollars? It's just a bonus', in *The Guardian*, 18.5.2000, 22) This also makes utter sense in a sustainability framework since high wages always lead to overconsumption (see Sachs, 1999, 208).

Sufficiency revolution/ Fostering new-old ethical values

> Sustainability in the last instance springs from a fresh inquiry into the meaning of good life. (Sachs, 1999, 186)

> To solve the problems of an unjust and unfair world order we need to live *simply* so that others may *simply* live. (Satish Kumar)[109]

> Communism failed because it produced too little at too high a cost. But capitalism has also failed because it produces too much, shares too little, also at too high a cost to our children and grandchildren. (...) Capitalism has failed because it destroys morality altogether. (Orr, 1994, 12)

> A marriage of the ancient idea of *commonwealth* with our presently emerging (or re-emerging) understanding of *environment* could give birth to a promising new notion of what "wealth" really is. (C. Douglas Lummis)[110]

Justice can only prevail if all of us get a comparable piece of the planetary chocolate cake. Because this cake, as we have seen on p. 42, is of a given size, it follows that the pieces cannot exceed a given size. This does mean that a lifestyle which is based on amassing material possessions and sees the meaning of life in consumption and the possession of goods cannot, in reality, be sustainable. So, however efficient we are, whatever fantastic technologies of minimal resource use we might develop in the future, a closed system will at some point cry "That's enough!" if we don't adjust accordingly:

> In fact, what really matters is the overall physical scale of the economy with respect to nature, not just the efficient allocation of resources. Herman Daly has offered a telling comparison: even if the cargo on a boat is distributed efficiently, the boat will inevitably sink under too much weight – even though it might sink optimally! Therefore, efficiency without sufficiency is counter-productive – the latter has to define the boundaries of the former. (...) In other words, an "efficiency revolution" remains without direction if it is not accompanied by a "sufficiency revolution". Nothing is ultimately as irrational as rushing with maximum efficiency in the wrong direction. (Sachs, 1999, 88)

The sufficiency revolution advocated here by Sachs is, in essence, nothing other than the old question of the good life. In other words: how can we achieve the best possible life – not the highest possible living standard, but the best quality of life – for the greatest number of people *within* the means available? On that front, I'm

---

[109] Satish Kumar, 'Simplicity for Christmas and Always', in *Resurgence*, No. 203, November/ December 2000, 3.
[110] Lummis, 'Equality', in *The Development Dictionary*, 1992, 38-52, here 49.

afraid, the jury reached a verdict a long time ago. If this verdict seems so at odds with our Euro-American view of life, it is not because the verdict is wrong, but because we are a historical aberration. Since the Industrial Revolution we have fooled ourselves into thinking that unlimited expansion is possible within a limited system. But let's hear the verdict:

> Having much obstructs living well. (...) Teachers of wisdom in the East and West (...) almost unanimously recommended adherence to the principle of simplicity in the conduct of life. That cannot just be a matter of chance. Summarizing the experience of generations, they drew the conclusion that the way towards a successful life seldom involves accumulation of possessions. (...) The opposite to a simple life is not a luxurious but a fragmented existence. An excess of things obstructs everyday existence, distracts attention, dissipates energies, and weakens capacity to find a clear-cut direction. Emptiness and dross are the enemies of happiness. (*Greening the North*, 1998, 126)

This clearly means that we have to foster new ethical values – which, for the most part, are very "old ideals of a livelihood based on love, conviviality and simplicity"[111] which we have forgotten during our historical accident. In the words of C. A. Bowers: "Long-term cultural/ecological survival will depend, in part, on our collective ability to accumulate, communicate, and renew ecologically sustainable forms of knowledge and values." (1995, 135) Such values include quality of life rather than consumption, immaterial instead of material principles, co-operation instead of competition, self-limitation instead of greed, joy rather than jealousy, community instead of egoism, pleasure instead of insatiability, a long-term rather than a short-term perspective, partnership rather than victory, solidarity instead of confrontation, wisdom instead of profit,[112] and most importantly maybe, "humility" in the sense of "tolerance and deep respect" (Forbes, 1981, 47), especially "reverence for all forms of life" (ibid. 129; see also 126ff. und 134ff.).[113]

We also need a new discussion about basic world views, since, as we shall see in chapter 2.a, our current ones – based on 'liberal' notions of 'freedom', 'democracy', 'progress', 'growth', 'development' and 'technology' – are not sustainable. This reorientation "requires us to wonder how productive society's economic output is in terms of welfare, use value, beauty and meaning. What is all this effort worth? And what do we want?" (Sachs, 1999, 181). Why is it, we might wonder, that "research into the psychology of happiness can find neither within nor be-

---

[111] Majid Rahnema, 'Participation', in *The Development Dictionary*, 1992, 116-131, here 127.

[112] See *Zukunftsfähiges Deutschland*, 1996, 208; *Factor Four*, 1997, 292-293; Hösle, 1994, 78ff.; Ruh, 1995, 67-79.

[113] This is also warranted by the following insight: "Far from leaving micro-organisms behind on an evolutionary ladder, we [human beings] are both surrounded by them and composed of them." (Lynn Margulis and Dorion Sagan, 'Marvellous Microbes', in *Resurgence*, No. 206, May/June 2001, 10-12, here 11)

tween societies any evidence that levels of satisfaction significantly increase with levels of wealth" (Sachs, 1999, 210)? On the contrary:

> Research confirms the age-old truism that money does not buy happiness. Describing the US over the past four decades, psychologist David Myers says: "We've got twice as many cars per person, we eat out two-and-a-half times as often, we enjoy all the technology that fills our lives. Yet we're slightly less likely to say we're very happy, we're more often diagnosed with depression ... the divorce rate has doubled, the teen suicide rate has tripled, the juvenile violence rate has quadrupled."[114]

Sachs makes another pertinent point in this context: it is an illusion that material and immaterial satisfaction are unconnected and can grow indefinitely side by side: "So poverty of time degrades the utility of a wealth of goods. In other words, material and non-material satisfaction cannot be maximized simultaneously; there is a limit to material satisfaction beyond which overall satisfaction is bound to decrease." (1999, 211-212) Taking everything into account, we therefore come up with the answers humankind has given time and again: simplicity,[115] respect for nature, the others and oneself, humility, love – and self-activity rather than consumption.[116] Not least Gandhi has taught us that quality of life can only be achieved with non-material values:

> Far from wishing that humankind be released from poverty, he taught that voluntary poverty on the basis of a conscious detachment from material concerns is the most blissful of all possible human situations. His entire struggle focused (...) basically on quality of life. (Woodcock, 1983, 31)

There is even quite a bit of evidence that the subsistence economy, so utterly despised in so-called developed countries as backward and unworthy, produces a "strong correlation between a modest life and happiness" (Vester, 1997, 457; see also Norberg-Hodge, 2000, 9-87). On the basis of what we have seen so far, this is not surprising, because, without wanting to idealise it, subsistence economy seems the only approach which guarantees extensive self-determination of the people in-

---

[114] Wayne Ellwood, 'Let's stop ransacking the Earth, and start searching for sustainability', in *New Internationalist*, No. 329, November 2000, 9-12, here 12.

[115] It is noteworthy that very often the best and most effective learning tools are simple, real-nature objects to be found everywhere rather than expensive equipment (see *Fool's Gold*, 2000, 58).

[116] "True satisfaction always requires an element of self-activity. In a society where people are increasingly becoming mere consumers, without the satisfaction of producing (or doing anything meaningful), whole supermarkets full of goods and all the money in the world cannot buy true satisfaction. If people can see meaning in the work they do, the things they produce, if work is not only alienated wage labour, the supposed limitlessness of our needs will be drastically reduced." (Bennholdt-Thomsen/Mies, 1999, 56)

volved, largely free of external control (see below, p. 66; Bennholdt-Thomsen/ Mies, 1999).

There is a problem connected with the values advocated above and this might well be the biggest challenge for an education for sustainability, as developed in chapter 3. On the one hand, the last thirty years of environmental education have shown that lecturing to pupils and students does indeed increase environmental awareness, but unfortunately this awareness does not automatically translate into sustainable action (see below, p. 261). On the other hand, it has equally been shown that change does take place if the fundamental values held by people are in tune with sustainability.[117] Only if you know something, love it, have an interest in it and develop responsibility towards it, will you care for it (for example the biosphere). So education, but also the community, workplace, and government have to foster *and* practise such values, otherwise all preaching will fall on deaf ears.

Box 5: Sustainability: the better moral discourse?

In strictly philosophical terms, it is not possible to provide an ultimate justification why we should act sustainably; in the end it is a decision based on one's conception of human nature. On the other hand, as we have seen in chapter 1.b, the world public seems to have accepted that we should live sustainably. In addition, one can apply a version of the Kantian categorical imperative, which makes it very difficult to argue against sustainability: Do not destroy the life-support system of other people, if you do not want other people to destroy your life-support system.

I do agree with Hirsch that we should not start from the assumption that people in favour of sustainable development are moral and those opposed to it are immoral.[118] We rather have different, competing value systems, with different priorities. Within the opponents' value system their actions undoubtedly make moral sense. But I do not believe that these different value systems are equally valid. I hope to have shown above that any value system that is not framed within a sustainability perspective can be shown to be defective and instrumental.

Yet what is clear in any case is that there is no running away from a moral framing of the issue of sustainability. Whether we like it or not in our seemingly so moral-free world, talking of inter- and intragenerational equity is a moral discussion. Indeed, our entire relationship with nature is morally framed: there are value judgements involved since we are looking at nature under the perspective of nature *for* humans, at least mostly so (see Hirsch, 1999).

Equity as sharing

> Sharing and exchange, the basis of our humanity and our ecological survival, have been redefined as a crime. This makes us all poor. (Shiva, 2000, 19)

If we take the principle of equity and apply it, as suggested above, not only to human-human relationships but also to our relationship with nature, then an im-

---

[117] Heiko Breit and Lutz H. Eckensberger, 'Moral, Alltag und Umwelt', in *Umweltbildung und Umweltbewußtsein*, 1998, 69-89, here 70, 86.

[118] Personal conversation, Zurich, 23.10.2000.

portant old insight becomes very relevant again. It is at the same time one which we seem to have forgotten through 'development', namely that you can never get something without giving something back. The externalisation of costs on all levels has tricked us into thinking that there would be "a free lunch". Shiva corrects this perspective:

> In giving food to other beings and species we maintain conditions for our own food security. In feeding the earthworms we feed ourselves. In feeding cows, we feed the soil, and in providing food for the soil, we provide food for humans. This world-view of abundance is based on sharing and on a deep awareness of humans as members of the earth family. This awareness that in impoverishing other beings, we impoverish ourselves and in nourishing other beings, we nourish ourselves is the basis of sustainability. (2000, 19)

In order to be able to adopt this perspective we need to develop respect and awe for the wonders of nature. We also need to admit our ignorance regarding many of the intricacies of how life works.[119] This should make us very cautious when it comes to tampering with life processes and natural resources. Rather than adopting the heroic position that we know everything and can control and dominate every aspect of life, we should step back, hold on and study the processes of nature which, almost without exception, far surpass anything humans have ever created in efficiency, clever design and so on (see Vester, 1997, 219-228).[120] Or where are the man-made industries which have an energy efficiency of nearly 100 per cent, produce no waste and are entirely biodegradable at the end of the life-cycle? A healthy dose of humility instead of Promethean hubris would serve us well (see precautionary principle above, p. 40).[121]

---

[119] "It is possible that there are as many as 10,000 species of bacterium in a single gram of soil, yet only 3,000 have so far been identified and named by microbiologists. Conservative estimates put the number of different species on Earth at 14 million; no one knows for sure and some have claimed that there are at least 30 million. Of these, only a few per cent – 2 million at most – have been studied, identified, named. Indeed, almost all biological research has been based on a few hundred different life forms at most." (Rose, 1998, 2) And Rose goes on: "Despite our ignorance of the overwhelming majority of life forms which exist on Earth today (indeed, most biochemical and genetic generalizations are still derived from just three organisms: the rat, the fruit fly and the common gut bug *Escherichia coli*), and our inability to do more than offer informed speculations about the processes that have given rise to them over the past 4 billion years, we biologists are beginning to lay claims to universal knowledge, of what life is, how it emerged and how it works." (ibid. 4)

[120] Vester gives a concrete example: if you travel alone in a car, "over 99 per cent of the crude oil used to move the human body evaporates as friction, heat, noise, wear and tear and toxic fumes" (Vester, 1997, 127).

[121] Sbert has reminded us of the remarkable reinterpretation of humility in capitalist ideology: "Humility turned from a saintly virtue into a rare heresy. Condemnations of greed, innate to the Christian religion and to all traditional systems of wisdom and philosophy, were transformed into

Recovery of the Commons/ Life belongs to everybody/ Redefining property

If we start from the assumption that everybody has a right to their fair share of the resources of the world to cover their basic needs as a first priority, we cannot but redefine the Euro-American notion of property (as private property). The life-support system biosphere, life itself, can then not become property because it is an inalienable entitlement of every human being. The idea of stewardship developed above on p. 34 is crucial again: something which is only lent to us so that we care for it, cannot by definition be owned. Anybody else and future generations have as much claim on it as we have. This is particularly true for patents on life:

> Patents on life are rejected by many different sectors of society on the basis of ethical convictions. Essentially the argument is always the same: we have not created life; therefore we cannot own life. In July 1999 representatives of indigenous groups demanded an absolute ban of patents on life: "A human being cannot possess his own mother. Human beings are part of Mother Earth, nature; we have not created anything and therefore haven't acquired the right to describe something as our property which doesn't even belong to us. We are forced to accept Western property rights which contradict our wisdom and values". (...) In addition, often only two or three genes of an organism are modified for a patent on life while the organism may contain 100,000 or more genes. Nevertheless the scientist claims the entire plant or animal as his invention. This is about the same as when you replace the ashtray in a car and then claim that the entire car is your invention. (*Der patentierte Hunger*, 2000, 9)

But even if we were to allow some forms of private property of natural resources, the notion of property would have to be an entirely different one: the owner would 'own' only the natural interest, the renewable yield of the resource, not the natural capital itself. He would therefore not be allowed to destroy the resource base itself (for example, the sea or the Amazon rain forest) (see Hösle, 1994, 124).

The problem is a fundamental one, at least for Euro-Americans. We have a damaging ideological heritage which Macpherson (1962) has clearly identified already in the 1960s. Liberal democracy starts from a very reductionist view of human beings, defining them essentially through property, rather than through ethical values or as social beings. For the last three hundred years this has had catastrophic consequences for the public spirit as well as for the way we treat the environment. This becomes obvious when we view our notion of property from an indigenous perspective. Property, monetary wealth, patents, property rights and the like almost call for the idea that natural resources and biodiversity are not im-

---

leniency bordering on approval toward such a sin, which is now perceived as the veritable psychological engine of material progress. So, greed and arrogance in individuals turn into prosperity and justice for nations and all mankind." (José María Sbert, 'Progress', in *The Development Dictionary*, 1992, 192-205, here 196)

portant and can be substituted by man-made surplus. We shouldn't be surprised that on this basis respect for the life-support system Earth and future generations has vanished. Common property, on the other hand, for which all community members bear responsibility – if it is not safeguarded, their survival and that of their children is threatened – does not automatically lead to, but can encourage a just and sustainable usage. It is no coincidence that historically the destruction of the commons was the condition for both the Industrial Revolution and bourgeois democracy, as Shiva *et al.* note:

> The destruction of commons was essential for the industrial revolution, to provide a supply of natural resources for raw material to industry. A life-support system can be shared, it cannot be owned as private property or exploited for private profit. The commons, therefore, had to be privatised, and people's sustenance base in these commons had to be appropriated, to feed the engine of industrial progress and capital accumulation. (Shiva *et al.*, 1997, 8)

It is therefore quite logical that in a new round of colonial imperialism – with tools such as the WTO, IMF, MAI and the World Bank – the so-called developed countries and TNCs are trying today to appropriate and privatise what rightly belongs equally to everybody and to future generations. They are attempting to turn the entire biodiversity, life itself, the genome, forests, crops etc. into property, with patents and intellectual property rights. Once this is done, others, i.e. the majority of people, will only be allowed to 'use' these things if they pay for it. This privatisation of the basis of life is, according to Shiva *et al.*, "the revolution of the rich against the poor" (1997, 13), and is only possible on the philosophical basis of capitalism, as 'legitimised' by Locke, namely "the freedom to steal" (ibid. 10).

Against this exclusion of the majority and the expropriation of common property the only help is democratic control by the community:

> Empowering the community with rights would enable the recovery of commons again. Commons are resources shaped, managed and utilised through community control. In the commons, no one can be excluded. The commons cannot be monopolised by the economically powerful citizen or corporation, or by the politically powerful state. (ibid. 9)

To enable this, Shiva *et al.* suggest the introduction of *Community Intellectual Rights* (CIR). These would guarantee that inventions would serve the community and could not be claimed for exclusive use by commercial agents who expropriate indigenous knowledge through "biopiracy" and "bioprospecting".[122] The authors

---

[122] Shiva *et al.*, 1997, 14-18 and part II, *The Recovery of the Commons* (73-178), especially the "Model Biodiversity Related Community Intellectual Rights Act" (163-174).

demonstrate that the intellectual property rights demanded by the United States and TNCs would essentially legitimate theft:

> In fact, in the free trade and trade liberalisation regime, which is supposed to end protectionism, IPR [= Intellectual Property Rights] are the main instrument of this new form of protectionism. The new protectionism for TNCs through IPRs is becoming the major means of dismantling both local and national economies as well as national sovereignty; through piracy of both material as well as intellectual and cultural resources. (...) Sharing and exchange get converted to "piracy" when individuals, organisations or corporations who freely receive biodiversity and knowledge from indigenous communities convert the freely received gifts into private property through IPR claims. This blocks the continuity of free exchange thus leading to an "enclosure of the commons". (Shiva *et al.*, 1997, 28, 31)

The barely imaginable arrogance, but also the economic motives for such moves become clear when we look at the patenting of natural medicine and organic pesticides – used for thousands of years in India – by American, Japanese and European corporations (see ibid. 35-46). Even more obvious is the case with seeds. In all rural communities everywhere on Earth it has been custom since time immemorial that farmers keep back part of the yield as seeds for next season. In addition, farmers exchange these seeds in order to improve them and breed new varieties. If it now were possible, so goes the clever plan of the seed TNCs, to patent all seeds and to forbid the farmers to keep back seeds and exchange them (or make this technically impossible), then the corporations would have an absolute monopoly over the food chain because there is no food without seeds.[123] This is by no means Orwellian science fiction, but in Europe and the US already a reality.[124] And the greed of TNCs for the US$7.5 billion market will ensure that the rest of the world will follow suit (see Shiva *et al.*, 1997, 62-70).[125] It is almost impossible not to see this as part of a neo-colonial fight for world dominance: "Further, since the majority of farmers and most biodiversity is concentrated in the south, while the seed industry is concentrated in the north, conflict of interest

---

[123] Delta and Pine Land, the world's biggest cotton seed company, recently bought by Monsanto, has, in conjunction with the US government, developed the so-called "Terminator gene". It renders the use of harvested crops as seeds impossible because it makes them sterile (see John Vidal, 'Mr Terminator ploughs in', in *The Guardian. Society*, 15.4.1998, 4-5).

[124] This is in particular the case with genetically modified seeds, where farmers have to sign contracts with Monsanto and the like that they will not use seeds for next season's sowing and will not exchange them with other farmers. In addition, Monsanto has the right to control the fields of the contract farmer for an additional three years (*Der patentierte Hunger*, 2000, 16).

[125] Other methods are obviously equally welcome: Monsanto and Novartis have threatened the Irish who didn't want to allow genetically modified sugar beet that they would stop selling them normal seeds unless they allow GM varieties (see 'Monsanto and Novartis blackmail Ireland', in *Corporate Europe Observer*, (April 1998) 1, [http://www.xs4all.nl/~ceo/observer1/blackmail.html]).

between farmers and the seed industry acquires a north/south dimension at the global level." (ibid. 70)

These conflicts can only be solved through a consequent rejuvenation of the democratic tradition:

> The philosophy of democratic pluralism recognises the anti-democratic nature of the centralised nation state on which state protectionism of the past was founded. But it also sees the emergence of corporate protectionism as the real threat to democratic rights and economic livelihoods. In this perspective, countering this **recolonisation** requires the reinvention of national sovereignty by democratic processes. (...) Self-rule of communities is the basis for indigenous self-determination, for sustainable agriculture, for democratic pluralism. (ibid. 81-82; emphasis in the original)

As far as "the enclosure of the Commons" is concerned, there is an interesting precedent with regard to the internet. Following the idea that "property rights for software damage society" countless programmes are written and improved collectively by honorary programmers. Among them are Apache, a programme with which roughly half of all the web servers are running, and Linux, a highly praised operating system, which was written and perfected by several hundred thousand programmers around the world: "Programmes are best if they are written collectively", since thousands of voluntary programmers discover mistakes far quicker than the small teams of developers in software companies.[126] The commercialisation of the software industry, on the other hand, is encouraging egotistic, uncooperative behaviour focused on quick short-term profit, completely oblivious to the long-term interests of society as a whole.[127]

4. Economy

The primacy of ecology: economy as a subsystem

The discussions of both democracy and equity have shown that it is barely possible to define these issues at the exclusion of others and so the need that participatory and equity concerns shape the discussion of a sustainable economy has already become clear. But even more so we have to remind ourselves (since eco-

---

[126] Quotes from Richard Stallman, in Ludwig Siegele, 'Programmierer aller Länder, vereinigt euch! Im Kampf gegen Microsoft gibt Netscape das Rezept seiner Internet-Software preis', in *Die Zeit*, No. 15, 2.4.1998, 45.

[127] See the chapter 'Befreiungstechnologie' in Nürnberger, 1999, 255-258 where the author encourages everybody either to buy a computer running under Linux or deleting Windows from one's harddrive and installing Linux (255).

nomic textbooks, even for today's students, still assume that the economy exists in a void)[128] that the economy is "a subsystem of the global ecosystem".[129]

> The fundamental ecological question for ecological economics is whether remaining species populations, ecosystems and related biophysical processes (i.e. critical self-producing "natural capital" stocks), and the waste assimilation capacity of the ecosphere are adequate to sustain the anticipated load of the human economy into the next century while simultaneously maintaining the general life-support functions of the ecosphere. This critical question is at the heart of ecological carrying capacity but is virtually ignored by mainstream approaches. (Wackernagel/Rees, 1996, 50-51; see also *Taking Nature into Account*, 1995, 274-275)

The biosphere provides the limits within which economic activity can take place (see p. 42ff.). We therefore need to assess the total impact of each economic activity and whether it is sustainable (see Vester, 1997, 488). In a nutshell, this requires us to free "ourselves from the deep and still unconscious legacy of the Industrial Revolution" (Bowers, 2000, 57).

The primacy of the subsistence perspective/ Self-sufficient agriculture

> We are concerned with something else: to inspire ourselves and others (...) to the realisation that Papua New Guinea or Belau or many other subsistence societies are the norm according to which the majority of people have lived for millennia and that industrial society is very young, very marginal, and cannot be generalised. It is a destructive deviation from this historically normal mode. We want to overcome the tunnel vision, the ignorance, the narrowness, the blockage of thought of our metropolitan arrogance. (Bennholdt-Thomsen/Mies, 1999, 213)

This has far reaching consequences for a sustainable economy. If self-determination and satisfying primary needs within nature's limits are the most important objectives, agriculture, indeed subsistence agriculture has to move back centre-stage. It means concentrating on the "'real economy' of water, soil and natural resources" as opposed to the "artificial economy that pulls people into a system over which they have no control" (Helena Norberg-Hodge, quoted in Kingsnorth, 2000, 37). David Orr has quoted Alan Durning to specify the parameters of a sustainable economy. It is rural, self-sufficient, local:

---

[128] See Nebel/Wright, 2000, 559-560 and Rudolf H. Strahm (1992), *Wirtschaftsbuch Schweiz* (Aarau, Sauerländer), 20-23. For a critique of economics education at universities see Bowers, 1997, 79-84.
[129] Robert Costanza *et al.*, 'Integrated Envisioning, Analysis, and Implementation of a Sustainable and Desirable Society', in *Getting Down to Earth*, 1996, 1-13, here 2.

This is an economy that returns "to the ancient order of family, community, good work, and good life; to a reverence for excellence of skilled handiwork; to a true materialism that does not just care about things but cares *for* them; to communities worth spending a lifetime in." (Orr, 1994, 183)

We can even go further and say with Norberg-Hodge that only in such "robust, local-scale economies" do "we find genuinely 'free' markets; free of the corporate manipulation, hidden subsidies, waste, and immense promotional costs that characterize today's global market" (2000, 183).

There are various dimensions to this, which have to be linked up with the previously discussed principles and spheres. The first is to acknowledge, again only possible from a perspective of the poor, that contrary to Euro-American ideology, most people, at least in so-called poor countries, are actually fed by subsistence agriculture:

It is women and small farmers working with biodiversity who are the primary food providers in the Third World and, contrary to the dominant assumption, their biodiversity-based small farm systems are more productive than industrial monocultures.[130] The rich diversity and sustainable systems of food production have been destroyed in the name of increasing food production. However, with the destruction of diversity, rich sources of nutrition disappear. (Shiva, 2000, 15)

Or to put it in other words, deconstructing another myth of neoliberal ideology: "The market economy is not the primary one in terms of the maintenance of life." (Shiva, 1992, 188) Indeed, there is more and more scientific evidence that low-input, low-impact, preferably organic agriculture is considerably more productive and resilient than its high-tech counterparts.[131] But that is precisely the problem, as George Monbiot explains:

Traditional farming has been stamped out all over the world not because it is less productive than monoculture, but because it is (...) more productive. Organic cultivation has been characterised as an enemy of progress for the simple reason that it cannot be monopolised: it can be adopted by any farmer anywhere, without the

---

[130] Vester also makes clear that monocultures, whether in agriculture, industrial production or energy generation, are far less efficient than smaller, local units (1997, 85).

[131] Peter Rosset, 'Small is Bountiful', in *The Ecologist*, 29 (1999) 8, 452-456, esp. 454. – Even Integrated Pest Management agriculture can lead to reductions of up to 75 per cent of pesticide use and 10 per cent yield increases (FAO, quoted in Madeley, 1999, 44), let alone organic production: "Disillusioned by the so-called 'green revolution' technology of TNCs, some farmers in developing countries have switched back to non-chemical or low-external-input farming. While a switch to these systems may cause an initial drop in yields, far more food can eventually be harvested from the same plot of land. Some farmers in India who have switched to permaculture ('permanent agriculture' that uses no outside inputs) are enjoying yields four times higher than they did under chemical agriculture." (Madeley, 1999, 171)

help of multinational companies. Though it is more productive to grow several species or several varieties of crops in one field, the biotech companies must reduce diversity in order to make money, leaving farmers with no choice but to purchase their most profitable seeds.[132]

The balance towards traditional and/or organic farming techniques shifts even more decisively if all energy, water and material inputs and all other costs (which are usually externalised)[133] are taken into account (see Shiva, 1991, 196)[134] and if we evaluate all the factors that determine a sustainable agriculture: ecology, empowerment, equity, together with the principles of low impact, long-term orientation and precautionary principle.[135] This means in most cases that sustainable food is local food (*Bringing the Food Economy Home*, 2000) and it shouldn't come as a surprise that the principle of "small is beautiful" applies as well: "A study in the US reveals that small farms growing a wide range of plants can produce 10 times as much money per acre as big farms growing single crops."[136] Yet again, we have to fight ignorance or outright denial:

> Worldwide examples of successful alternative agriculture exist and are growing, even while they continue to be ignored by the dominant world view of agriculture. And it is these initiatives that carry the seeds of a sustainable agriculture. Blindness to these alternatives is not a proof of their non-existence. It is merely a reflection of the blindness. (Shiva, 1991, 251)

But the other aspect clearly is self-determination and control. We are talking about "the capacity of communities to produce their life without being dependent on outside forces or agencies." (Bennholdt-Thomsen/Mies, 1999, 3-4)

> Empowerment can only be found in ourselves and in our cooperation with nature within us and around us. This power does not come from dead money. It lies in mutuality and not in competition, in doing things ourselves and not in only passively consuming. It lies in generosity and the joy of working together and not in

---

[132] George Monbiot, 'Biotech has bamboozled us all', in *The Guardian*, 24.8.2000.

[133] The total external costs of UK agriculture were £2.3 billion in 1996 (see Jules Pretty *et al.*, 'An Assessment of the Total External Costs of UK Agriculture', in *Agricultural Systems*, 65 (2000) 2, 113-136; Jules Pretty, 'The Real Costs of Modern Farming', in *Resurgence*, No. 205, March/April, 2001, 7-9).

[134] The total energy and materials usage for organic farming produce is 64 per cent less compared with similar quantity and quality of conventional farming produce (*Zukunftsfähiges Deutschland*, 1996, 317-319).

[135] See Miguel A. Altieri, Andres Yurjevic, Jean Marc Von der Weid, Juan Sanchez, 'Applying Agroecology to Improve Peasant Farming Systems in Latin America: An Impact Assessment of NGO Strategies', in *Getting Down to Earth*, 1996, 365-379, here 367-369.

[136] George Monbiot, 'Biotech has bamboozled us all', in *The Guardian*, 24.8.2000.

individualistic self-interest and jealousy. This power also lies in our recognition that all creatures on earth are our relatives. (ibid. 5)

Local subsistence farmers are – if they have free access to land: here the political question of land reform is crucial – dependent on their community and their skills as farmers. Apart from that, they are self-sufficient. Industrial farmers have delegated their knowledge to agrochemical companies, are entirely dependent on them for seeds, fertilizer, machines, pesticides and herbicides. Yet this is precisely the central point: if we acknowledge that global power, more than ever, resides in the hands of economic, rather than political power, and if we allow ourselves to understand the interests of those in power, namely staying in control, we see why there is a very real war going on here:

> Ivan Illich stated as long ago as 1982 that the war against subsistence is the real war of capital, not the struggle against the unions and their wage demands. Only after people's capacity to subsist is destroyed, are they totally and unconditionally in the power of capital. This war is a war not only to colonise subsistence work but also to colonise language, culture, food, education, thinking, images, symbols. Mono-labour, mono-language, mono-culture, mono-food, mono-thought, mono-medicine, mono-education are supposed to take the place of the manifold and diverse ways of subsistence. (Bennholdt-Thomsen/Mies, 1999, 19)

Against this background, it is hardly surprising that we Euro-Americans face an ideological barrier to acknowledging both the importance of the subsistence perspective and the primacy of agriculture in any sustainable economy. The first has to do with a relentless ideological propaganda war since 1945 which has devalued "everything that is connected with the immediate creation and maintenance of life, and also everything that is not arranged through the production and consumption of commodities", so that we have ended up with the absurd result "that the most lifeless thing of all, money, is seen as the source of life and our own life-producing subsistence work is seen as the source of death" (Bennholdt-Thomsen/Mies, 1999, 17). The latter has to do with our alienation from any sense of connectedness to and knowledge about a place.[137] Our industrially constructed lives have been utterly divorced from the knowledge that food doesn't grow on the supermarket shelf:

> As more and more food is imported from the global market into the supermarkets of the North, not only the urban consumers but also rural producers have largely lost the consciousness that their livelihood depends on their relation to the land.

---

[137] Gregory A. Smith and Dilafruz R. Williams, 'Introduction: Re-engaging Culture and Ecology', in *Ecological Education in Action*, 1999, 1-18, here 4.

> They consider money and the market as the sources of their sustenance. (Bennholdt-Thomsen/Mies, 1999, 149)

But this alienation does not only exist with regard to food and sustenance:

> Living in cities, anaesthetized by television and technology, far away from the slums of Bombay, the sweatshops of China and the copper mines of Zambia, we are alienated, physically and psychologically, from the impact of our lifestyle on the rest of the planet.[138]

This has, of course, dramatic implications on lifestyles – only for a minority of people though, but a minority which tends to generalise their world view as *the only legitimate* world view:

> This means concretely: The utopia of a socialist, non-sexist, non-colonial, ecological, just, good society cannot be modelled on the lifestyle of the ruling classes – a villa and a Cadillac for everybody, for instance: rather, it must be based on subsistence security for everybody. (Bennholdt-Thomsen/Mies, 1999, 4)

Yet, Euro-Americans also need convincing that agriculture is indeed the primary sector of any economy (it is still called that way, but not thought of as such any more): "It remains true, however, that agriculture is primary, whereas industry is secondary, which means that human life can continue without industry, whereas it cannot continue without agriculture." (Schumacher, 1993, 89)

A last, important asset of any subsistence economy is that it is not a technocrat's or futurist's dream; it is both in existence and universally possible, with one proviso, political will: "The feasibility of a modern subsistence economy does not depend on new scientific inventions. It depends primarily on the ability of a society to agree on fundamental, self-chosen, anti-bureaucratic and anti-technocratic restraints." (Illich, 1971a, 60)

The moral dimension: responsibility for life-cycle of products and consumption

We usually conceive of the economy as a moral-free zone. But in a sustainable economy we need a new moral awareness: producers as well as consumers have to face up to their responsibilities. You cannot just produce something and then forget about it once it is sold, you cannot just buy something, ignorant of how it is produced and what is going to happen once you don't use it any more. Both producers and consumers share between them the responsibility for the full life-cycle of a product. This includes not just concern for the grey energy embodied in the

---

[138] Wayne Ellwood, 'Let's stop ransacking the Earth, and start searching for sustainability', in *New Internationalist*, No. 329, November 2000, 9-12, here 12. See also Wackernagel/Rees, 1996, 132.

product through production processes and transportation, not just the material throughput necessitated to produce the item (MIPS, see Schmidt-Bleek, 1997), not just whether and how a product can be reused, recycled or decomposed after its use,[139] but also the working conditions of the people producing it, their wage levels, their safety, their level of participation, as well as their political and cultural self-determination (see Jucker, 1998). It includes awareness that we are not passively subjected to an economy, but that we drive the course of it with our decisions: "At the end of the twentieth century, the process of moral economy includes an awareness that *not only consuming but also buying is politics.*" (Bennholdt-Thomsen/Mies, 1999, 121) We have to counteract what Susan George has so aptly described as the core ethical problem of our consumer society: "The present reality is that by the time a product reaches the market, it has lost all memory of human or natural abuse." (*Lugano Report*, 1999, 177)

As opposed to that, a "moral economy" cannot be

> derived from a paradigm of private property, permanent growth and self-interest. In such a moral economy the various dimensions of life processes are not separated from one another as is the case in the compartmentalised, fragmented capitalist world market system. (...) In such an economy the concept of *waste*, for example, does not really exist. Things that cannot be consumed and things whose waste products cannot be absorbed within such a distinct eco-region cannot be produced. Such a moral economy in a particular region requires, evidently, a community that feels responsible for sustaining the self-regenerative capacities of this region. (Bennholdt-Thomsen/Mies, 1999, 153)

As we can see, the implications are that such an economy is governed by the precautionary principle. Production and consumption patterns, which cannot be accommodated within the four system conditions (see p. 44), are not acceptable.

A local, renewable-input, recyclable-output economy

Once we are clear about the foundations of a sustainable economy, i.e. that it has to rest on self-sufficient agriculture, the contours of everything else become clearer as well. A sustainable economy has to be a local economy:

> Production from local resources for local needs is the most rational way of economic life, while dependence on imports from afar and the consequent need to produce for export to unknown and distant peoples is highly uneconomic and justifiable only in exceptional cases and on a small scale. (Schumacher, 1993, 43)

---

[139] Impact assessment analysis, such as ecological footprinting, constantly reveals that reused and recycled products *do* have much less environmental impact over their lifetime (see *Sharing Nature's Interest*, 2000, 79-101, 153-161).

It also has to be a renewable-input, recyclable-output economy:

> Imagine an industrial system that has no provisions for landfills, outfalls, or smokestacks. If a company knew that nothing that came into its factory could be thrown away, and that everything it produced would eventually return, how would it design its components and products? The question is more than a theoretical construct, because the earth works under precisely these strictures. (*Natural Capitalism*, 2000, 18)

This scientific insight has already found some application, such as in the ZERI (Zero Emissions Research and Initiatives) project. Its concept is elaborated as follows:

> Zero Emissions represents a shift in our concept of industry away from linear models in which wastes are considered the norm, to integrated systems in which everything has its use. It heralds the start of the next industrial revolution in which industry mimics nature's sustainable cycles and humanity, rather than expecting the earth to produce more, learns to do more with what the earth produces. Zero Emissions envisages all industrial inputs being used in the final products or converted into value-added inputs for other industries or processes. In this way, industries will reorganise into "clusters" such that each industry's wastes/by-products are fully matched with others' input requirements, and the integrated whole produces no waste of any kind.[140]

This, in turn, would change the way we conceive of production and goods. Consumers would not buy goods, but purchase services and "manufacturers [would] cease thinking of themselves as sellers of products and become, instead, deliverers of service, provided by long-lasting, upgradeable durables." (*Natural Capitalism*, 2000, 16)

A democratically accountable economy

On the basis of the discussion on empowerment (see p. 46) we have seen that a sustainable economy, let alone a sustainable world is not possible unless all its aspects are democratically structured. This means in particular that we have to build tools for democratic accountability of the up-to-now utterly undemocratic economic sphere. Even bearing the potential for abuse in mind, we have to seriously think about international policy tools in this area. If we really want to crack down on corporations which engage in tax evasion, dumping of social and environmental standards and environmental racism, we need some form of international institutions which can enforce relevant standards.

---

[140] ZERI Foundation, 'The Zero Emissions Concept' [http://www.zeri.org/theory.htm].

Necessary (perhaps not even sufficient) steps to alter this [global insecurity, un-sustainability and injustice] include the establishment of: effective global redis-tributive mechanisms and institutions. Sustainable global development implies international institutions capable to change prevailing distributions of incomes and current distributions of access to sources of income and wealth, including en-vironmental resources and world markets.[141]

But on the other hand, civil society has to be strengthened, although only if it ad-heres to the same standards of democracy as we demanded above. Possible agents are NGOs[142] and a new International of unions which is able to counteract the "corporate International" (Martin/Schumann, 1997, 112). At the moment the latter can evade every confrontation by relocating its production to another country.[143] Such international co-operation of workers has existed, at least in part, during the strike of Liverpool dockers from September 1995 until the end of 1997: dockers all over the world refused to handle ships going to or coming from Liverpool, or they delayed the handling to incur additional costs to the shipping companies.[144] We need, in other words, a globalisation of civil society as has partly happened with the international resistance against the Multilateral Agreement on Investment (MAI), in Mexico with the Zapatistas, in Seattle, Washington and Prague, in Davos against the World Economic Forum (WEF), in India and elsewhere (despite the ritual defamation of the protesters by the world media; see chapter 2.b). This "grassroots globalization" has to show up the limitations and destructions of "corporate global-ization" which is utterly oblivious to the poor majority of the world population and its fundamental rights for self-determination, for shelter, sufficient food and clean water (see Karliner, 1997, 197-223, here 210).

5. Equipment

I have deliberately included equipment, by which I mean all technology produced by the scientific-industrial complex, as one of the five most important spheres for a sustainable society. One can easily argue that this should be subsumed under economy, but since, particularly for so-called developed countries, equipment has become not only central to everybody's daily life, but also crucial to the running

---

[141] Johannes B. Opschoor, 'Institutional Change and Development Towards Sustainability', in *Getting Down to Earth*, 1996, 327-350, here 345-346. See also Hösle, 1994, 135 and Giaco Schiesser, 'Krieg um Technologien und Kommunikationshoheit', in *Weltwoche*, No. 43, 24.10.1996, 33.

[142] But only if living up to sustainable standards themselves: for a critique of some NGOs, which are barely distinguishable from IMF or corrupt government agencies, see Toni Keppler, 'NGOs als Trostpflaster' and Barbara Unmüssig, 'Unbekanntes Wesen', both in *die solitaz*, 21./22.10.2000, 36 and 37 respectively.

[143] Hans Schäppi, Walter Schöni, 'Wirtschaftsdemokratie und Industriepolitik', in *Demokratie radi-kal*, 1992, 21-33, here 26, 28.

[144] See John Pilger, 'They never walk alone', in *The Guardian Weekend*, 23.11.1996, 14-23.

of the 'civilisation machine' – "science and technology are the driving forces behind increased economic competitiveness and societal change" (Nowotny, 2000a, 226) – I think we should focus more closely on it. Despite a history of sustained critique of the progress myth, which has at last reached such dinosaur technologies as nuclear power, there is far too little evaluation of 'equipment' such as information technology, high-tech medicine and transportation methods under a sustainability perspective. In fact, one can go as far as Bowers to say that "although both mechanical and social forms of technology permeate nearly every aspect of everyday life, technology is perhaps one of the least understood aspects of mainstream culture", with the result that few people "understand how technology influences thought and social relationships, or affects the forms of knowledge communicated from one generation to the next" (1995, 78).

But we can, as with the economy, define quite clearly the parameters within which a technology is sustainable. Today, we have a plethora of tools for environmental impact and technology risk assessment, such as Life Cycle Analysis (LCA), Material Intensity Per Service Unit (MIPS), ecological footprints and backpacks and the like. We know how crucial the evaluation of grey energy, energy usage, use of materials, reusability, recyclability and biodegradability are. All these have to be taken into account, in the light of the principles developed above. Neil Postman has put together a good list of questions we would need to ask of any new technology, in order to decide whether it is sustainable and therefore whether or not we should adopt it in the first place:

> 1. *What is the problem to which this technology is the solution?* (Postman, 1999, 42; italics in the original)
> 2. *Whose problem is it? (...) We need to be very careful in determining who will benefit from a technology and who will pay for it. They are not always the same people.* (ibid. 45)
> 3. *Which people and what institutions might be most seriously harmed by a technological solution?* (ibid. 45)
> 4. *What sort of people and institutions might acquire special economic and political power because of technological change?* (ibid. 50)
> 5. *What changes in language are being enforced by new technologies, and what is being gained and lost by such changes?* (ibid. 52)

If we take these questions seriously, what clearly emerges is the need for science and technology to become accountable, to become part of the democratic project. In social studies of science Helga Nowotny and others have argued that science has to move into "mode 2". Science should not take place at a distance from society where it is "left to make its discoveries and then make them available to society". In "mode 2" science there is "joint production of knowledge by society and science" (Nowotny, 2000a, 243). This production takes place in the public sphere, the "agora":

It is the domain (in fact, many domains) in which contextualisation occurs and in which socially robust knowledge is continually subject to testing while in the process of becoming more robust. Neither state nor market, neither exclusively private nor exclusively public, the *agora* is the space in which societal and scientific problems are framed and defined, and where what will be accepted as "solution" is being negotiated. (Nowotny, 2000a, 241)

Socially robust knowledge produced like this is superior to conventional scientific knowledge "because it has been subject to more intensive testing and anticipating the 'context of implication' as much as possible while incorporating the context into the research project" (Nowotny, 2000a, 243). This in fact means that empowerment has to enter the production of knowledge (see citizen's jury below). It demands a change in the way science is pursued, away from the elitist, hierarchical ivory tower towards a transdisciplinary co-operation of all involved, which requires openness and tolerance: "For such an equitable, 'ecological' process of communication between scientists of various disciplines or systems of thought to take place the first required step (...) is to provide space within one's own system of thought for different perspectives." (Pohl, 2000, 396)

Yet the most important aspect, due to the precautionary principle and the often irreversible nature of actions (as with the introduction of genetically modified organisms into the natural world), is that we shift the perspective from analysis of the damage after the event to prevention: the burden of proof has to be reversed. Rather than prove the damage done by particular technologies after they were introduced, any new invention, any new technology, any new gadget praised as the saviour of humankind *would have to prove its sustainability* (see Ruh, 1995, 78). We have to require "the evaluation of indirect effects and long-term consequences (...) and the interaction with other areas, before new knowledge is transformed into doable technologies and is let loose on human beings and the environment" (Vester, 1997, 488). This includes that new inventions would have to prove that they benefit humankind and that they contribute to solving the most pressing problems on Earth (hunger, poverty, environmental degradation). Ralph and Mildred Buchsbaum described this very clearly already in 1957 with regard to economic change (and most technological change is economically driven):

> The religion of economics promotes an idolatry of rapid change, unaffected by the elementary truism that a change which is not an unquestionable improvement is a doubtful blessing. The burden of proof is placed on those who take the "ecological viewpoint": unless *they* can produce evidence of marked injury to man, the change will proceed. Common sense, on the contrary, would suggest that the burden of proof should lie on the man who wants to introduce a change: *he* has to demonstrate that there *cannot* be any damaging consequences. (*Basic Ecology*, Pittsburgh 1957, quoted in: Schumacher, 1993, 109)

A good guide is to check whether first and foremost these new tools and technologies serve specific short-term interests, be they economic or political. A good example is capital-intensive high-tech research in industrialised countries which serve the interests of a tiny minority, while the real problems of the world could be solved with low-tech measures, but not without political and social changes. Such high-tech research should have to justify itself in the face of these problems; if it can't, it is simply a waste of money.[145] In other words, the assessment of the sustainability of a new invention cannot be left to the inventors or their corporate representatives. The utterly ignorant claim that GM food will "feed the world", effectively emotionally blackmailing the world into accepting a technology which has distinctly failed to pass any sort of sustainability test so far, is the most recent example. These claims were clearly used to cover up the corporate interests to monopolize the food chain and move the two-thirds of world farmers still independent from agribusiness into dependency (and, as we know from the experience of the 'Green revolution', a lot of them into destitution) (see The Cornerhouse, 1998 and Shiva, 1991). Letting the people with a direct monetary interest in a technology decide on its viability and future-proofness is like giving a fox the job of looking after the henhouse. Civil society, i.e. the people affected by a technology, is the only legitimate body to decide about the introduction of new technologies. Or as the computer guru Bill Joy, founder of Sun Microsystems, phrases it: "In any normal terms, and by any normal standards, a corporation should not be permitted to toy with fundamentals like genes, nanotechnology and the like without the full endorsement of those whose lives will be affected should anything misfire." ('Discomfort and Joy', 2000, 38)[146]

There have been very interesting experiments with citizen's juries in Great Britain and India. Composed of people across the whole social spectrum they called witnesses representing all sides of a debate. Contrary to the presumption that complicated issues should be dealt with by experts, the lay members of the jury very quickly got to grips with the issues (the one I am referring to dealt with genetically modified organisms, in Karnataka, India):

> And while most started with little idea about genetic engineering, within a short period of time they had grasped many of the issues at stake, giving the lie to the idea that "ignorant" peasant farmers could not possibly understand the wonders of GM technology. (Warwick, 2000, 52)

The most important lesson from the experiment was the insight that people at the receiving end of technology have a far more complex and realistic idea about consequences and long-term effects on all areas of life than the producers, who focus

---

[145] Hartmut Böhme, 'Wer sagt, was Leben ist? Die Provokation der Biowissenschaften und die Aufgaben der Kulturwissenschaften', in *Die Zeit*, No. 49, 30.11.2000, 41-42, here 42.
[146] See also Bill Joy, 'Why the Future Doesn't Need Us', in *Wired Magazine*, April 1999.

almost exclusively on the dimensions of novelty and financial gain (for them, that is).

The consequence is clear: if a new development cannot prove its sustainability, we have to do without it and confine it to the archives of history, under the heading of unviable ideas. We often seem to think that this is impossible; that you cannot suppress ideas. People, mostly staunchly anti-Marxist, in this context often invoke the idea of historical determinism: "there is no alternative". But this is an utterly ahistorical perspective. The history of science has clearly shown that not the necessary or best projects win the race, but the ones which fit the mood of the time and can be accommodated by the power structures (see Feyerabend, 1975; Nowotny, 2000, 2). In recent times, this has become even more pertinent. Today, those technologies which are backed by corporate or governmental interests 'make it'. It is also evident that financial backing to a large degree determines the direction of research and the technological applications. A very good example is nuclear power. Even though by now we have realised that it is an unsustainable technology which was never economically viable, it has been backed in the United States alone with direct and indirect subsidies of over US$1 trillion, yet it still "delivers less energy than wood" (Myers, 1998, 69). As recently as 1997

> in industrialized countries as a whole, governments still spend over half their energy research budgets on nuclear power, viz. $4 billion per year (by contrast with less than 10 percent on renewables). All nuclear subsidies in OECD countries amount to $10-14 billion per year. (ibid.)

This had, as we now know, nothing whatsoever to do with the quality of nuclear power or the claim that it was superior technology. It had a lot to do with the economic and military interests of the United States, with government founding, with researchers dependent on Pentagon money, with the ideology of the 1950s which was entirely uncritical of science and technology and happy to believe the lies of the Atomic Energy Commission which predicted in 1954 that by 1969 electrical energy from nuclear power "would be too cheap to meter" (quoted in Stoll, 2000, 112). Had all this money been directed into research on renewable energy, nobody today would talk about the economic viability of renewables. We have to learn that specific technologies are not a god-given fate, but the consequence of very specific political, financial and social set-ups.

This, of course, is equally true of genetic engineering. George Monbiot has described the irony very succinctly: In 1999 "the [British] government spent £52m on developing GM crops, for which demand in Britain is approximately zero, and £1.7m on research into organic farming, for which demand outstrips domestic supply by 200%."[147] No wonder a lot of people still think that organic agriculture

---

[147] George Monbiot, 'Just say no to biotech business', in *The Guardian*, 2.3.2000.

is not worthy of serious consideration. Might it be that we have here a similar constellation as with nuclear versus renewable energy and will wonder in twenty years time why on earth we threw away billions of taxpayers' money on an unsustainable technology?

In this context of supposedly historical determinism and linear growth to progress, it is important to make another observation. We are always told that there is no time, that we have to hurry up, that we cannot properly assess the impacts of a technology because we need it *now*, or else the world will collapse. This is clearly ideological talk in order to push through certain 'solutions'. As we have seen with the economy, all the technology, all the tools and skills to build a sustainable agriculture and, more generally, a sustainable economy are here. We have the knowledge. The problem lies elsewhere: with the political will, and therefore with the vested interests of those in power. We have to recognise that people have lived before, many of them with more dignity and wisdom than we do. We have to ditch the fixed idea that only the very newest technology lets us survive:

> I should like to remind you that the Taj Mahal was built without electricity, cement and steel and that all the cathedrals of Europe were built without them. It is a fixation in the mind, that unless you can have the latest you can't do anything at all, and this is the thing that has to be overcome. (Schumacher, 1993, 182)

I should modify this: we are in a hurry when it comes to solving the problems of the poor of the world. But there again, it is a matter of applying the solutions we have, rather than inventing new, unproven ones. But as far as the so-called developed countries are concerned, we need to slow down and *stop* producing ever more things. We need to start living instead.

If all this sounds too complicated and complex, don't despair. It is, admittedly, a far cry from the simplifications and mono-causal explanations with which we are daily confronted in politics and the media (see chapter 2.b), but the only reason why it seems difficult to us is that we are not used to systems thinking. But I firmly believe with Vester (1997) that we are inherently capable of complex thinking, indeed do so daily with regard to many of the decisions we have to take to live our lives. The trouble is that this human potential is not developed; on the contrary, everything is done to discourage it (see chapters 2.b and 3).

But let me end this section with Bill McKibben and his call to face up to the real world:

> What is the opposite of utopian? Let's be extremely realistic, even grim: a community, a region, a nation, a world that paid attention to limits would mute the horn of plenty, plug up the cornucopia. A community that made environmental sense *would not have all the things that we have today*. (...) *It would be poorer.* (...) I do not want to sound romantic. Careful logging is not romantic – it's hard

work. Hoeing rice in a Keralite field is not romantic. Trading garbage for food in a Curitiban slum is not romantic. The romantics are the people still gazing dreamily off into the distance at some ever shinier future – the people convinced that global warming-soil erosion-dying fisheries are some kind of small bump on the highway to global happiness. (...) Whatever evolves, it won't be idyllic; no way of life ever has been. (1997, 203-207)

Social and political imagination

What seems most important about this exploration of a sustainable society are two points: on the one hand we should acknowledge that we have a lot of the solutions, tried and tested, readily at hand and that we always should remind ourselves that we are *not* operating in an exclusively socially constructed environment. There are natural limits to human activities; there are very real consequences of human action. Whoever overstresses the social constructedness of environmental problems – an extreme example would be the position that only those environmental problems exist which people perceive as environmental problems – is in tremendous danger of falling back into the old trap of forgetting that the biosphere exists out there, that it is not created by humankind and that we utterly depend on it (see Plant, 1998, 27-28).

On the other hand, what emerged from an integrated analysis is the urgent need for social, economic and political imagination. We have developed our technical imagination to such an extent that we are not capable of understanding the implications of it any more (see below p. 179), but this development has catapulted us beyond the biosphere's limits. In contrast to this, we have been very slow in fleshing out adequate social and political models which would live up to the demands of empowerment and equity in a sustainable society. We live in very antiquated political set-ups and we need all the creative thinking we can muster and the input from existing pockets of sustainable communities around the world in order to come to terms with the challenge.

2. The Pathology of Denial

Chapter 1.c should have made two things unmistakably clear: on the basis of the principles and parameters of a sustainable society, our current Euro-American way of life, which through globalisation imposes itself on every corner of the world, is utterly unsustainable. Yet on the other hand sustainability is not a utopian dream, a fantastic scenario of lunatics; it is both possible and workable. But let's not fool ourselves: to achieve it we need a dramatic shift in all the spheres discussed above: in politics, moral and social values, the economy and science and technology. Indeed, the shift needed has to be akin to the dimensions of the Copernican revolution. Nebel/Wright make clear what happens in times of such paradigm shifts:

> Steeped in the old world view, not only did people ignore the new theory, but anyone who suggested that it had merits was vigorously attacked by the existing power structure, which was dominated by the Catholic Church and which had a vested interest in maintaining the old beliefs. (2000, 11)

That is the challenge we are facing: "The main task in the years to come will be to apply our ecological knowledge and systemic thinking to the *fundamental redesign* of our technologies and social institutions".[148] History has shown that those in power never voluntarily give up their hold on privileges. Therefore we will need to change the power structure (see empowerment above, p. 46). In order to facilitate that, we do need, as a first step, to spell out the new world view, as I have tried to do in the preceding chapter. Nebel/Wright summarize it: "In short, the new view amounts to a paradigm shift from seeing humans as the center of things, free to ride over nature in any manner possible, to seeing humans as the caretakers of nature, intricately linked to it in life processes and global systems." (2000, 11) Or, in the words of one of the global reports quoted at the very beginning: we are talking about the "broad-scale change in thinking that we need in order to cope with current environmental degradation" (*A Guide to World Resources 2000-2001*, 2000, 20). As a second step we have to show, as I intend to do in this chapter, what is wrong with the current system and why it is not viable any more. If we continue to ignore the reality and to call the destruction of the life-support system Earth 'progress', we will hit the wall even quicker.

If one asks for change, or, even worse, radical change, the trouble has already started. People asking for change are usually portrayed as leftists, as anarchists, as troublemakers, as people who want to uproot everything – as happened again and again with the mainstream media portrayal of the protesters in Seattle, Washington, Prague and Davos. The sensible people, the conservative majority, it

---

[148] Fritjof Capra, 'The challenge of our time', in *Resurgence*, No. 203, November/December 2000, 18-20, here 19.

is insinuated, resist change for good reasons since it always leads to chaos, always ends up in disaster. And, the sociobiologists tell us, change is not even possible because human nature is as it is.

This argument overlooks an important aspect of globalised Euro-American capitalism. It is the "capitalist permanent revolution"[149] which with ever-faster acceleration forces change upon us, radical, remorseless and without any regard for the victims:

> The ruling-class conservatives [of today] seek to save the branch by going after the root, and the economic and technological system they seek to conserve has eradicated more traditional techniques, customs, and institutions than has any other force in the history of the world. To give the name "conservative" to this kind of economic and technological Jacobinism is like calling a strip miner a conservationist because he conserves the institution of strip mining. (Lummis, 1996, 99)

This is not only true of the South, but also of 'developed' countries: how often did millions of workers have to re-learn a new operating system or word processor package for their computers in the last twenty years, not because it was technically necessary or because the novelties improved important aspects of work, but because the logic of capitalism demanded that the software giants should be able to sell another set of several million new operating systems, and because the producers wanted to sell a new generation of computers, alleging that the new software wouldn't run on old ones (akin to the scare stories circulated about the Year 2000 bug which helped the computer industry to achieve new sales records). The migration from Windows 3.1 to Windows 95 meant that within two years 10 million computers 'needed' replacing in Germany alone. Clifford Stoll correctly calls this "planned obsolescence" (2000, 165-169). Similar changes (and purchases) are forced upon us in all areas of consumer electronics, but also as far as food, clothing and much more are concerned.

Yet these changes are as nothing compared to the upheavals that the colonialisation of traditional cultures by global capitalism produces. I would just like to remind the reader of the almost complete re-invention of life and culture, accompanied by the abandoning of age-old customs and cultural traditions such as story-telling, which the unmediated introduction, often overnight, of Euro-American technology has brought to many indigenous cultures.[150] This shows clearly that change is not impossible at all, but rather that it depends on the context and the values society attaches to it. It does occur, quickly and unhindered, if it suits

---

[149] Volker Braun, 'Wir sind geschichtsgeschädigt', zur Verleihung des Büchner-Preises interviewt von Lothar Baier, *Die WochenZeitung*, No. 42, 19.10.2000, 21-22, here 21.

[150] See Helena Norberg-Hodge's *Ancient Futures: Learning from Ladakh* (2000) and, looking at Kurdistan, Yusuf Yeşilöz (1998 and 2000, 120ff.).

those in power and those who stand to gain from it financially or in terms of status. This we should bear in mind when we hear claims that change to a sustainable society is impossible.

But I will not only have to answer the question of what makes our present situation unsustainable (chapter 2.a), but also the question of why there is no change. This seems an important inquiry, since chapter 1 has shown an overwhelming international consensus both on the existence of the environmental crisis (1.a) and on the necessity to migrate to a sustainable future (1.b). It also showed that we do indeed know what the parameters are within which such a future can develop (1.c). Yet, why is there hardly any progress towards this future? Why, despite the almost unstoppable flow of government documents worldwide with the world "sustainability" in the title, are most people completely unfamiliar with the concept of sustainability? I discussed my research project with a wide range of people from a variety of social backgrounds. When I attempted to explain why it was entitled "Humanities and Sustainability" the most frequent reaction was "Sustainability? Never heard of it!". This has convinced me that Plant is right when he states that people are generally "unfamiliar with the idea of sustainability" (Plant, 1998, 41), a statement which is backed up by a study in Germany which found that in 1996, four years after the Earth Summit, 85 per cent of Germans had never heard of sustainable development.[151] Why on earth is that so if all the governments have their sustainability strategies and business claims it has mended its ways already? I will attempt some answers in chapter 2.b. Is the reason perhaps a combination of two things, namely the radical nature of the necessary change and the ubiquitous "pattern of denial, evidence of what Thomas Berry calls 'a deep cultural pathology'" (Orr, 1999a, 221)? Is it that we will have to reinvent our society since the industrial one is not viable (see *Taking Nature into Account*, 1995, 98)[152] (always remember, though, this radical change is only asked of the affluent consumers of the world)? Is Orr even right to put it as follows?

> And there are unspoken taboos against talking seriously about the very forces that undermine biological diversity. I am referring to our inability to question economic growth, the distribution of wealth, capital mobility, population growth, and the scale and purposes of technology. (1994, 70-71)

---

[151] See Fritz Reusswig, 'Die ökologische Bedeutung der Lebensstilforschung', in *Umweltbildung und Umweltbewußtsein*, 1998, 91-101, here 99.

[152] "In a *Baywatch* world, who will really want to slow down, to reduce expectations, to undevelop? This is a good question, and with it in mind many conscientious environmentalists have decided to concentrate on the possible: on recycling, say. Understandable as such a strategy is, it cannot ultimately claim to be 'realistic'. In the real world, the very necessary task of recycling is at best callisthenics for the marathon we must run. Realism, sadly, demands that we recognize the need for deep and fundamental change – for recycling our cars into buses and bicycles." (McKibben, 1997, 226)

One could also phrase the problem in more political terms. For most people outside the former Eastern bloc and for all the dissidents within it, it was unmistakably clear that the reality of the power structure and the economy in these countries had nothing to do with socialism; in other words that the gap between the idea which the politicians kept claiming for their actions and the reality was glaringly obvious. Today the disparity between the ideals of liberal democracy (freedom, equality), the so-called free market (free choice, best products through competition and the like) and their actual implementation is just as wide. Yet, as Orr states, our problem is that we still pretend and believe that our reality equals these ideals. We are unable to see the glaring mismatch.

a.   Our unsustainable present

> Material progress and economic growth [can] without exaggeration be called the two central obsessions of Modernity.
> (*Taking Nature into Account*, 1995, 15)

> The allegedly highly productive industrial system is, in reality, a parasite on earth, the likes of which have never before been seen in the history of humanity. It has the towering productivity of a bank robber who resorts to quick, violent attacks in an attempt to create for himself a life of prosperity at the cost of others. (Otto Ullrich)[153]

I will now try to identify the most important taboos and myths of the current global society which make it an unsustainable one. We shall see that it is primarily an ideological problem, in other words we are led to believe that all is fine and that we are on the right track with the neoliberal master ideology, even though this dogma has precious little to do with reality. This doesn't mean, though, that world views just exist in the mind. Ideology is embodied in power structures, economic organisation, institutions (such as educational ones) and language (as Bowers has shown [1995, 75]).

Consumption patterns/ 4 Earths needed

> It took Britain half the resources of the planet to achieve its prosperity; how many planets will a country like India require? (Mahatma Ghandi on the question whether India would reach Britain's standard of living after independence)[154]

The first and most important myth to dispel is the general assumption that there is a continuous and upward progression in the development of civilisation and that the United States is the pinnacle of this progression. This implies at the same time that the American way of life is the one the whole world should aspire to (see also below, p. 152). This is not just something elaborated in politician's speeches; the United States does everything in its power to force the entire world to move that way: from the inundation of the world's TV stations with Hollywood and other US-American 'cultural' output,[155] to the utilisation of the WTO, IMF, World Bank and bilateral pressure to force other countries to buy American goods (see p. 141).

---

[153] Ullrich, 'Technology', in *The Development Dictionary*, 1992, 275-287, here 280.

[154] Quoted in Robert Goodland, 'The case that the world has reached limits', in *Environmentally Sustainable Economic Development: Building on Brundtland* (1991), ed. by R. Goodland *et al.* (Paris, UNESCO), 15-27, here 15.

[155] Among the biggest exports from the US are "cultural goods", see Ignacio Ramonet, 'Wer sind die Citoyens des Cyberspace? Medienkonzentration und Pressefreiheit', in *Die WochenZeitung*, No. 49, 6.12.1996, 24 and Martin/Schumann, 1997, 37.

Yet, we have known for some time now that the resource and energy consumption such a lifestyle entails cannot be replicated for the entire world population.

> The message of Truman, Kennedy and many others to the "peoples of the world", that they could achieve the material prosperity of the West by taking over Western scientized technology, therefore, turns out to be empirically untenable. The available industrial technologies for the West are nearly all designed for plunder and the transfer of costs.[156]

The current state of the biosphere is spelt out in the *Living Planet Report 2000*: "While the state of the Earth's natural ecosystems has declined by about 33 per cent over the last 30 years, the ecological pressure of humanity on the Earth has increased by about 50 per cent over the same period, and exceeds the biosphere's regeneration rate." (2000, 1) This means that already now, with over 90 per cent of the World population *not* living an American lifestyle, we have overstepped the limits, and we are, as the economists say, eating into the natural capital base, rather than living off the interest.

The usual retort here is that the problem is not the American lifestyle but the fact that there are too many people living on the planet. Wolfgang Sachs has icily dissected this false argument (1999, 71-89). Undoubtedly numbers may be a problem at some point in the future (just remember the impact formula, see p. 39), but there are two facts to remember: firstly, overpopulation is not the root cause of our problems, on the contrary it is a *consequence* of the globalisation of the Euro-American lifestyle through the 'development model': "In fact, demographers recognize that it is *after* contact with the modern world that population levels shoot up." (Norberg-Hodge, 2000, 151). Secondly, the main reason for the world-wide strain on ecosystems is overconsumption in so-called developed countries:

> By comparing the resource consumption patterns of different countries we conclude that, in 1996, the Ecological Footprint of an average consumer in the industrialized world was four times that of an average consumer in the lower income countries. This implies that rich nations (located mainly in northern temperate zones) are primarily responsible for the ongoing loss of natural wealth in the southern temperate and tropical regions of the world. (*Living Planet Report 2000*, 2000, 1)

The richest 20 per cent of the world use 75 to 85 per cent of the world's major resources, from energy to wood, paper, ore and all the other raw materials (Wackernagel/Rees, 1996, 102). In fact, the United States alone, "with about 4.5% of the world's population (...) is responsible for about 25% of the emissions of carbon

---

[156] Otto Ullrich, 'Technology', in *The Development Dictionary*, 1992, 275-287, here 282.

dioxide" (Nebel/Wright, 2000, 150), the primary greenhouse gas leading to climate change, and the combined discharge of the US, the European Union and Japan (with 13 per cent of the world's population) add up to half the world's emissions.[157] Extrapolated with the carrying capacity of Earth in mind, we would need to advertise in the intergalactic newspaper as follows: "4 additional planet Earths needed".[158]

Rather than trying to shift the blame onto the poor people in the South, we had better face up to the truth. Sachs calls this the "home perspective" (1999, 86-89) and spells out with the necessary clarity what is needed from 'developed' countries, namely:

> to retreat from utilizing other people's nature and to reduce the amount of global environmental space it occupies. After all, most of the Northern countries leave an "ecological footprint" on the world that is considerably larger than their territories. They occupy foreign soils to provide themselves with tomatoes, rice or cattle; they carry away raw materials of all kinds; and they utilize the global commons – such as the oceans and the atmosphere – far beyond their share. Then Northern use of the globally available environmental space is out of proportion; the style of affluence in the North cannot be generalized around the globe, but is oligarchic in its very structure. (...) The principal arena for ecological adjustment is thus neither the Southern hemisphere nor the entire globe, but the North itself. (Sachs, 1999, 87)

This cuts deep into the Euro-American identity, fostered by an unrelenting advertising war ever since the 1950s, hammering in the message that new goods and more consumption are always and necessarily better. We have to come to terms here with the contradictions of a linear, single-issue thinking versus a complex, holistic approach. Huib Ernste has identified the paradox: "Rational individual action leads to irrational collective results."[159] Since we constantly forget to take into account the entire lifecycle of a product, its origin, its energy usage, knock-on and side effects, its afterlife, and, most crucially, the overall physical scale of the economy with respect to nature (see Daly/Cobb, 1989), we have the interesting phenomena of rebound effects: despite drastic cuts in the electricity consumption of household appliances, overall consumption in Germany still continues to rise because of the proliferation of new electronic devices and the incentive to use appliances more because they are so efficient. Similarly, any gains in fuel efficiency in cars over the last twenty years have been eaten up by a trend to buy heavier cars (for example, because of additional features such as air conditioning), to have

---

[157] Bernhard Pötter, 'Einheizer im Treibhaus Erde', in *die tageszeitung*, 20.11.2000, 4.

[158] See Mathis Wackernagel (1997), *Ecological Footprints of Nations Report and Slide Show* (Toronto, ICLEI [http://www.iclei.org/iclei/ecofoot.htm]).

[159] Huib Ernste, 'Kommunikative Rationalität und umweltverantwortliches Handeln', in *Umweltproblem Mensch*, 1996, 197-216, here 212.

two or more cars and to drive much longer distances; overall, an increase in consumption again.[160]

We therefore need to understand what sustainable consumption means. The starting point for any sensible concept of consumption has to be the insight that every time we buy and/or consume something – be it a tiny battery to keep our watch going or be it a TV, a car or a hamburger – we are making an impact on the social, economic and ecological environment. In the words of Anwar Fazal, former president of the International Organisation of Consumer Unions (IOCU):

> The act of buying is a vote for an economic and social model, for a particular way of producing goods. We are concerned with the quality of goods and the satisfactions we derive from them. But we cannot ignore the conditions under which products are made – the environmental impact and working conditions. We are linked to them and therefore have a responsibility for them.[161]

In other words: we cannot just take pleasure in the goods while we possess or consume them. We also bear the responsibility for the answers to the following questions: are the goods produced in an environmentally harmful or sustainable manner? Do the workers producing them get a fair wage and are they in control of their working lives? Do they have safe and healthy working conditions (see for example banana workers)? Are the substances used in production and in the product itself safe or toxic? Can the product be recycled at the end of its lifespan? If not, is it biodegradable or does it release toxic substances when landfilled or incinerated (such as PVC, which releases Dioxin, one of the most toxic substances mankind has ever invented)? Do we pay the price for all the social and environmental costs a product incurs, or are these costs shifted onto other people (e.g. Southern countries) or the public (e.g. environmental clean-up measures, usually paid for by the taxpayer)? Does the company producing the goods deal with oppressive regimes, thereby furthering human rights abuses? Is the company involved in arms production, nuclear energy, animal testing, factory farming, irresponsible marketing and/or suppression of workers' rights? Is the production company donating money to political parties?

These are some of the most important questions we should be able to answer with regard to any product we buy, if we take seriously the assumption that we shouldn't buy any products which conflict with our own moral, political, environmental and social beliefs, and if we don't want to infringe, as a consequence of our consumption, the chances of a dignified life for the poor in the South and for future generations. More concretely: if *we* are not prepared to do our jobs at low wages, in dangerous and unhealthy conditions, with long working

---

[160] Ursula Tischner, 'Öko-intelligentes Konsumieren', in *Öko-intelligentes Produzieren und Konsumieren*, 1997, 82-104, here 84-86.

[161] Quoted in Rob Harrison, 'Future Shop 3', in *EC. ethical consumer magazine*, (1996) 42, 27.

hours and in environmentally appalling circumstances, we shouldn't force *others* to do so, for the sole reason that we can buy cheap, often superfluous products, which we often quickly throw away again.

Accepting these responsibilities also means that we have to be careful with heaping all the blame on business, governments and others for the ills of our present situation. While they should get their fair share of the blame (see below), there is no running away from the fact that consumers indirectly influence the form of the economy, the way it produces. Otherwise the commitment of UK supermarkets not to use genetically modified (GM) ingredients in any of their own-brand products or the shift to organic produce cannot be understood, since neither industry nor government did anything to encourage those changes. In the words of Ralph Nader, candidate for the Greens in the 1996 and 2000 presidential elections in the US: "When the consumer finally begins to exercise the virtually untapped power of citizen action – consumers will take their logical place at the head of the economic process."[162]

One of the biggest obstacles to sustainable production and consumption is that the prices we pay for services and goods 'lie'; that is they do not include all the costs which would accumulate if ecological damage during production and costs for recycling and/or waste management (amongst other things) had to be paid for. We need to become aware of the overall impacts of products. To use some concrete examples: we might think that our car needs replacing after three years, that the two-year old computer is really too slow, that our mobile phone is outdated by two generations, that nothing but a wide-screen TV will do, that we really ought to replace that 7-year old kitchen and that the 5-year old fridge is not energy-efficient enough any more: sounds like reasonable changes all around. Yet, if we look at the bigger picture and assume – as is effectively the case in most industrialised countries – that all the world's middle-class families think along the same lines, and if we then factor in the resource and grey energy use of these goods during production and transportation, it simply becomes unsustainable. This is most clearly palpable with the waste issue, since all over the industrialised world countries are running out of landfill sites to bulldoze away the excrement of civilisation.

Today, we have a variety of tools to show us when we are clearly on the wrong track. One is Life Cycle Analysis (LCA), looking at a product's impact from 'cradle to grave'. Here is an example: the US-Microelectronics and Computertechnology Corporation (MCC) has calculated that to produce one single Personal Computer (PC) you need at least 25,000 mega joules of primary energy, you produce 60 kilos of waste, including 25 kilos of toxic waste, and you pollute

---

[162] Quoted in 'Why Buy Ethically? An Introduction to Ethical Purchasing Theory', in *The Ethical Consumer Guide to Everyday Shopping* (1993), ed. by the Ethical Consumer Research Association (Manchester, ECRA), 24.

33,000 litres of water. That is not even counting the case, keyboard, mouse, disk drive, mains adapter etc. You might think that is not too bad, but with the average life expectancy for a PC of three to four years, that actually means that a PC is consuming more than 50 per cent of its total energy requirement during its production, not its use.[163] And once you stop using it, the next problem arises. A PC is manufactured out of up to 1,000 different materials and the way they are manufactured at the moment means that most of these components cannot be recycled and are thrown into landfills, being partly toxic waste.[164] And we are not talking about low numbers here: the Silicon Valley Toxics Coalition (SVTC) has been monitoring the waste and toxic waste problem associated with computers:

> Over 50% of US households own a computer and the average life span of a computer is falling to about two years. This means computer waste is building. Last year over 12 million computers were obsolete and within four years over 315 million computers in the USA will become obsolete. Most computer scrap is currently landfilled although the majority, three-quarters of all computers ever bought in the USA, are sitting in people's homes, basements, attics or cupboards because consumers don't know what to do with them. The European Union is also facing a 6 million ton/year electronic waste problem.[165]

Another example: we throw away more than 20,000 tonnes of mixed battery waste each year in the UK alone. 20,000 tonnes of batteries equals 870 semi-trailer truck loads (@ 23t) or in other words a nose-to-tail queue of semi-trailer trucks 8 miles long. We throw these batteries away, even though we should know by now that most materials used in batteries are non-renewable, some of them like cadmium scarce, the extraction of the raw-materials produces significant amounts of carbon dioxide emissions which contribute to global warming and sulphur dioxide which causes acid rain. Additionally, zinc, manganese dioxide, alkaline, nickel and particularly cadmium are (highly) toxic substances which cannot be properly recycled or safely disposed of.[166]

Transport, packaging and storage are other factors which very often affect LCAs of a product adversely. In Germany 20 per cent (!) of the entire energy and materials consumption is used to put food onto the table of German families (that includes diesel for tractors, crops for factory farming, energy for the food industry, petrol for long distance lorries, electricity for cooling in supermarkets, energy

---

[163] Andreas Grote, 'Kupfer aus Chile, Titan aus Norwegen', in *Frankfurter Rundschau*, No. 206, 5.9.1995, 6.

[164] Silicon Valley Toxics Coalition, 'High-Tech Production' [http://www.svtc.org/hightech_prod/index.html], 'Components of a PC' [http://www.svtc.org/hightech_prod/desktop.htm].

[165] Silicon Valley Toxics Coalition, 'The Clean Computer Campaign. End of Year Report Card. Assessing hazardous materials and take-back policy of major computer corporations operating in the USA and Canada', [http://www.svtc.org/cleancc/pubs/99report.htm].

[166] Batteries main product report, in: *EC. ethical consumer magazine*, (1997) 46, 4-11.

for cooking, plus infrastructure [pipelines, motorways, factories, lorry fleets etc.]) (*Greening the North*, 1998, 121). A single average German yoghurt is transported around 8,000 kilometres before it gets to the table (*Factor Four*, 1997, 117-119). A single tomato from the Canary Islands uses 1kWh in energy to be brought to Britain. With that energy you can light an 11W low-energy light bulb (equivalent to 60W) for 91 hours or an entire month, if you have it switched on 3 hours per day.[167]

Another measuring tool is the ecological footprint as elaborated in chapter 1.a. It attempts to visualize the total resource use of a country, a specific region or an activity by translating it into the area of land needed to provide the service (see Wackernagel/Rees, 1996; *Sharing Nature's Interest*, 2000). A concrete example might help to understand what is at issue here: a commuter, doing 10 kilometres per day, uses 70 square metres by bike, 310 square metres by bus, but by using the car he takes up 1250 square metres, taking all aspects of the various modes of transport into account (use of energy [calories in the case of the cyclist], land, plus land to absorb the environmental impact of pollution) (Wackernagel/Rees, 1996, 105).

The last tool I would like to mention is the so-called "ecological back-pack". The German scientist Friedrich Schmidt-Bleek has developed this idea, trying to assess what he calls MIPS (i.e. materials intensity per service unit). MIPS takes into account that to produce a tonne of most materials, such as metal, it takes multiple tonnes of ore, hundreds of thousands of litres of water, energy and material for transport, and mountains of very often toxic overburden from mining, which destroys living areas. MIPS sums up the total material used over a lifecycle to produce a certain amount of a product (for a definition see Friedrich Schmidt-Bleek, 1997, 108). A few examples: the "ecological backpack" of a single car is roughly 15 tonnes, the catalyst alone accounts for 2-3 tonnes because of the platinum. A golden wedding ring of 10 grams produces 3.5 tonnes of excess material in the goldmine alone. One tonne of coal produces/uses 3 tonnes of overburden and water. One litre of orange juice produces up to 100 kilograms of earth and water movement. And so forth. Once the MIPS is worked out, it is interesting to compare various materials, in order to know which ones have a smaller impact. If you buy a fruit plate made from local wood its "ecological backpack" is likely to be around 4 times heavier than the weight of the plate. If you buy the same plate in copper, the backpack is around a thousand times heavier than the plate (see *Factor Four*, 1997, 242-244).

Let me add two examples from the world of transport: if you attempt to calculate the social costs of private cars – costs which have to be shared by the

---

[167] Another example: one kilo of runner beans, planted and consumed in Switzerland uses 0.1 litre of fuel; the same amount imported from Kenya uses 48 times as much (Jennifer Zimmermann, 'Labels für Lebensmittel', in *WWF-Rundbrief*, (November 2000) 4, 2-3, here 3).

whole population, not just the users who amount to around 50 per cent – you arrive at a figure of up to 14 per cent of GNP. Those costs involve, among others, accidents, loss of time through traffic jams, pollution and other environmental damage, destruction of roads and land use (*Factor Four*, 1997, 189).

Flights are getting cheaper and cheaper, worldwide air traffic is growing at a rate of 5 per cent per annum and is responsible for about 3.5 per cent of total carbon dioxide emissions. Yet, the carbon dioxide emitted during flights is at least a factor of 2.5 more damaging to the world's climate than an equivalent amount on the ground. If we start from a sustainable world average quota of 1-2 tonnes of carbon dioxide emissions per person per year (which is roughly equivalent to the 1.5 to 2 kilowatt society argued for above on p. 46), this means that every person can fly about 500 kilometres per year. Yet, the average Swiss person flies 18 times that distance: "Quite obviously we are flying our 9,000 kilometres per year at the expense of those people – predominantly in the South – who will never in their life see an aircraft from the inside."[168]

And then there is the throwaway society:

> A study for the US National Academy of Engineering found that about 93 per cent of the materials we buy and "consume" never end up in saleable products at all. Moreover, 80 per cent of products are discarded after a single use, many of the rest are not as durable as they should be. (*Factor Four*, 1997, xx)

So there we have the challenge: when we go back to the new TVs, washing machines, kitchens, mobile phones and cars which our consumer contemplated above, we have to realise that the problem does not only lie in the goods themselves (i.e. how they are produced), but also in the fact that we deem them necessary. Just walk along the pavement on those days in Switzerland when they collect bulky refuse, or walk into that kitchen which seemingly needs replacement: you will find perfectly working items. They do their job, even though they might not be the newest and might not offer the fanciest special deals. But the only thing which is really wrong with them is that they don't fit our exaggerated idea of what is necessary. This means that it is also a problem of the mind, the ideology. We have to go further than switch consumption to *new* green products. We need to reduce, re-use and recycle. Yet, if we bear in mind that around 50 per cent of all materials used and moved by mankind cannot be recycled (Schmidt-Bleek, 1997,

---

[168] Quote and data taken from Theophil Bucher-König, 'Wird Fliegen zu einer Frage des Gewissens? Die Beurteilung des Flugverkehrs auf Grundlage der Nachhaltigkeit', in *Reisen & Umwelt. Informationsmagazin von SSR Travel*, (September 1999) 7, 16-19, here 17. See also: IPCC (1999), *Aviation and the Global Atmosphere. A Special Report*, summary [http://www.grida.no/climate/ipcc/aviation/index.htm] and Josef Brosthaus *et al.* (2001), *Maßnahmen zur verursacherbezogenen Schadstoffreduzierung des zivilen Flugverkehrs* (Berlin, Umweltbundesamt [=Texte 17/01]) [http://www.umweltbundesamt.org/fpdf-k/1955.pdf].

165), the ultimate challenge is to *dematerialise* our materialistic lifestyles (see above, p. 57).[169]

Globalised capitalism: growth as a recipe for disaster

> In economics there is no concept of enough: just a chronic yearning for more, a hunger that cannot be filled. (Jonathan Rowe)[170]

> People living in the heartlands of world capitalism must pose the question of their complicity in such events [such as mining in Third World countries that lead to destruction of nature and communities]. Again and again, the rule of capital proves it to be a totalitarian system which has lost none of its totalitarian character through globalisation. (Bennholdt-Thomsen/Mies, 1999, 178)

In many areas, as we have already seen, we are thrown back onto the insight "that sustainability is at root as radical an idea as you're likely to come across".[171] This is particularly true with regard to our capitalist economy (see Steiner, 1998, 311-312). But it is also the one area where we have most difficulties in accepting it. Alex Carey, the Australian media analyst, has argued very convincingly that this is the case because of a barrage of business propaganda which started almost a century ago. The main success of this propaganda has been to immunise the capitalist economy from sustained criticism: "It is arguable that the success of business propaganda in persuading us, for so long, that we are free from propaganda is one of the most significant propaganda achievements of the twentieth century." (Carey, 1997, 21)[172] This means that we must seriously and critically rethink the way our economy works, and consider whether in fact it lives up to the promises of its proponents.

In this context, then, it is rather surprising – and at the same time frustrating that it hasn't entered mainstream consciousness – that most critics who have a close look at what sustainability would entail, and whether capitalism can achieve it, share the analysis of a socialist who wrote in 1929 "that the capitalist system was by its nature unsound: a system driven by the one overriding motive of corporate profit and therefore unstable, unpredictable, and blind to human needs" (Zinn, 1996, 377-378) – which goes to show that sometimes it is far more productive to

---

[169] For further guidelines on ethical and ecological consumption, see Jucker, 1998 [http://www.onweb.org/features/new/dematerial/dematerial.html].

[170] Jonathan Rowe, 'Eat, sleep, buy, die', in *New Internationalist*, No. 329, November 2000, 24-25, here 24.

[171] Wayne Ellwood, 'Let's stop ransacking the Earth, and start searching for sustainability', in *New Internationalist*, No. 329, November 2000, 9-12, here 11.

[172] See also Chomsky, 1989, Chomsky/Herman, 1994 and Zinn, 1996.

listen attentively to voices from the past, rather than dismiss them condescendingly as outdated. It is therefore not surprising that concepts like "natural capitalism", elaborated by Hawken/Lovins/Lovins in *Natural Capitalism* (2000), are found wanting, despite the fact that resource productivity, of course, is a worthwhile and necessary objective:

> "Natural capitalism" is a programme for change that is at best partially developed. Very little that it advocates will emerge from capitalism as it's currently structured. Both the core of its agenda – aggressive pursuit of eco-efficiencies – and the associated objectives of equity and overcoming consumerism, demand ambitious political action. When Lovins and company avoid discussion of political or cultural change they are not telling the whole story.[173]

This is blatantly clear if we just quickly return to the problem of overconsumption mentioned above. Growth, and therefore destruction of natural resources, senseless and rampant consumerism, is built into the system. Acting like this is acting according to capitalist logic. An economy *not built* on zero growth, long-term, re-use, and no-waste strategies will automatically lead to today's absurdities, such as people throwing away perfectly functioning things to buy the new generation of the same thing which is hardly different, while most of humankind can only dream of these goods. But the consumer is only partly to blame, since overconsumption is built into the system: try to find, in highly developed countries, the skilled craftsmen or -women who can fix your clothes, shoes, electrical or electronic goods, your furniture; you will be lucky if you are successful. Most consumer goods are now deliberately built so that you cannot fix them, but have to buy new ones. Just try to get almost any computer printer fixed today: just a small problem will, together with call-out costs and spare parts cost you often up to twice the price of a new printer. Or, with regard to long-term quality: why is it that manufacturers of modern washing machines, which are supposed to be so much better than the old ones, tell you that you should take out insurance against breakdown when you buy a new one? Why are they not guaranteeing the perfect functioning of the machine for 5 years and spare parts for 15 instead? They claim in the ads and brochures what a fantastic product they sell, but after the purchase they are telling you, oops, actually we don't quite know whether it will work, better take out some insurance in case it breaks down after two days. Sustainable?!?

This is so because an economy based on nothing but competition will, quite logically, be the most ruthless exploitation machine, destroying by design any morality that might have been left:

---

[173] Mary Jane Patterson, 'Natural Capitalism', in *New Internationalist*, No. 329, November 2000, 14-15, here 15.

Our societies are more and more dominated by a relentless war of competition. The economy is less and less capable of fulfilling her real task, namely to satisfy the primary needs of all people. It becomes ever clearer that the senseless race of the various national, international and multinational corporations to improve their market share is in the end won by those who manage to exploit their fellow human beings and natural resources faster, more sweepingly, more cunningly and more unscrupulously than anybody else. (Dürr, 2000, 188)

Capitalist myths deconstructed:[174]

> It is not just the image, but the promise, of consumer capitalism that lies. (David Ransom)[175]

> Capitalism, as practiced, is a financially profitable, nonsustainable aberration in human development. (*Natural Capitalism*, 2000, 5)

Myth: 'Capitalism is apolitical, anti-ideological and operates according to natural laws'

> The market is the institutionalisation of individualism and non-responsibility. (Schumacher, 1993, 29)

Reality: Very much in line with what Carey has described, there is a tendency, particularly after the end of the Cold War, to portray capitalism as an apolitical system which is not based on a particular view of human beings and their relations, but on unchangeable 'natural' laws. Therefore capitalism and the market economy are not part of a specific ideology, in fact, so it is claimed, they are ideology-free and must underpin any human society. The most extreme and publicly visible proponent of this position was Fukuyama in his article 'The End of History?' where he proclaimed that both liberal democracy and the market economy marked the "end point of mankind's ideological evolution".[176] But this conviction has driven the entire neoliberal agenda, and one of the keepers of the holy grail of this agenda, the *Neue Zürcher Zeitung*, routinely talks of the "invariability of economic laws",[177] as if economics were a hard science and not the pseudo-science a lot of it is. Rather than looking at the complex reality, mainstream economic theory is primarily concerned with self-fulfilling prophecy: shaping 'reality' in such a way that it fits the predetermined neoliberal ideology:

---

[174] For a very detailed deconstruction of the myths of capitalism (and its destructive and totalitarian consequences) see also Kurz, 1999.

[175] Ransom, 'Jeans – the big stitch-up', in *New Internationalist*, No. 302, June 1998, 7-10, here 10.

[176] Francis Fukuyama, 'The End of History?', in *The National Interest*, (1989) 16, 3-8, here 4.

[177] See, for example, Gerhard Schwarz, 'Die Marktwirtschaft als Teufelswerk', in *Neue Zürcher Zeitung*, 31.12.1998, 27.

The economic 'laws' of the classical economists were but deductive inventions which transformed the newly observed patterns of social behaviour, adopted with the emergence of economic society, into universal axioms designed to carry on a new political project. The assumption of the previous existence of economic 'laws' or 'facts', construed by economists, is untenable when confronted with what we know now about ancient societies and cultures, and even with what we can still see in some parts of the world.[178]

One of the most important examples of this imposition of ideology onto reality is that most economic textbooks still have such a limited view of the economic sphere that the dependency on the biosphere, natural input and waste output doesn't even figure in their models.[179]

Lummis has provided a most penetrating analysis of the insight that, far from being apolitical and ahistorical, the capitalist economy is a very specific and historically determined power arrangement. But he also highlights that we have even lost the words to talk adequately about this:

[The word "economy"] means a particular way of organizing power in a society, and of simultaneously concealing this power arrangement – more accurately, of concealing that it *is* a power arrangement. If this formulation seems a surprise, that is a tribute to the effectiveness of the concealing function. (...) The "economy" is a way of organizing people to work efficiently, that is, to do unnatural kinds of work under unnatural conditions for unnaturally long hours, and of extracting all or part of the extra wealth so produced and transferring it elsewhere. This process is equally true of capitalist and "socialist" countries. The economy is thus political, but pretends not to be. It is political in the most fundamental sense: it organizes power, distributes goods, and rules people. (...) Under the domination of this ideology, economics has replaced politics as the Master Science, but this political character of the economy is hidden. Through economic processes cultures are abolished or restructured, environments are destroyed or made over, work is ordered, wealth is transferred, goods are distributed, classes are formed, and people are managed. But the words for talking intelligibly about these things – words like "founding," "order," "lawgiving," "revolution," "power," "justice," "rule," "consent" – do not exist as technical terms in economic science. (...) All of these systems can be strengthened by the addition of an ideology that doing industrial labor is virtuous, or heroic, or patriotic, or a characteristic of "advanced civilization," or (for people who doubt their adulthood) mature, or (for office workers) prestigious, or (for men) macho, or (for women) liberating, or the like. The point is that to make people do unnatural kinds of work for unnaturally long hours under unnatural working conditions one must either force them or implant in their minds some ideology under which they will force themselves. The vari-

---

[178] Gustavo Esteva, 'Development', in *The Development Dictionary*, 1992, 6-25, here 19.
[179] See Nebel/Wright, 2000, 559-560 and Rudolf H. Strahm (1992), *Wirtschaftsbuch Schweiz* (Aarau, Sauerländer), 20-23.

ous "economic systems" we see in the world today are different combinations of these different sorts of force and ideology. (Lummis, 1996, 46-47)

If we want to retake control of our lives, we urgently need to bring the economy back into political discourse and vice versa. We can only liberate ourselves from alienation if we finally dispel the capitalist pretence that the economic and political spheres are two separate things and that it doesn't matter if the one is ruled in a dictatorial fashion and the other pretends to be democratic. As we have seen above (p. 72) self-determination and real democracy will elude us so long as we are not prepared to democratise the economic sphere.

Maybe the most destructive aspect of the development described by Lummis is that the economy has "moved into the very centre of public concern" and has become "the obsession of all modern societies." (Schumacher, 1993, 27)

> Economics strives to subordinate to its rule and to subsume under its logic every other form of social interaction in every society it invades. (...) 'Common sense' is now so immersed in the economic way of thinking that no facts of life contradicting it seem enough to provoke critical reflection on its character.[180]

In the context of a sustainable world view this is the most detrimental aspect. Rather than being an attempt to come to terms with the complexity of our situation, economics applies the most drastic reductionism possible (hence, its attractiveness through simplification):

> The judgement of economics, in other words, is an extremely *fragmentary* judgement; out of the large number of aspects which in real life have to be seen and judged together before a decision can be taken, economics supplies only one – whether a thing yields a money profit *to those who undertake it* or not. (Schumacher, 1993, 28; emphasis in the original)

The challenge to the predominance of economic thinking in every aspect of our lives has to go further. Because economics – limited in its scope and on shaky empirical grounds – nevertheless claims to have answers to all the problems,[181] it does not help us to understand our problems, quite the contrary:

> It would not be unfair to say that economics, as currently constituted and practised, acts as a most effective barrier against the understanding of these problems, owing to its addiction to purely quantitative analysis and its timorous refusal to look into the real nature of things. (Schumacher, 1993, 32-33)

---

[180] Gustavo Esteva, 'Development', in *The Development Dictionary*, 1992, 6-25, here 17-18.

[181] "Economists themselves, like most specialists, normally suffer from a kind of metaphysical blindness, assuming that theirs is a science of absolute and invariable truths, without any presuppositions." (Schumacher, 1993, 38)

But there is another, ethical dimension of the claim that the capitalist economy operates according to natural law, as Bennholdt-Thomsen/Mies have clearly pointed out:

> The claim that the market operates according to inherent abstract laws helps politicians, managers and bankers – and not only them – to protest their complete innocence; it allows individuals to wash their hands of responsibility for their everyday economic activities. Men and women declare themselves to have no power in relation to the market, thereby legitimating their own consumerism, their own environmental and market behaviour. This maintains the anonymity of the act of purchase – which, in our view, is one of the great obstacles on the way towards new subsistence markets and ecologically responsible behaviour. The market sphere itself is not perceived as a site of responsible behaviour. It no longer occurs to anyone in our part of the world that traders and customers might have some obligations to one another. (1999, 116)

Neoliberalism, where this irresponsibility has reached new heights, is 'liberating' itself from all accountability. The WTO and the unrelenting war for 'free' trade is not enhancing the freedom of the people, but the freedom of capital to do everything regardless of its consequences for humankind or nature: "Neoliberalism has finally 'liberated' capital from all obligations and restrictions that democratically elected governments had used so far to control capitalism." (Bennholdt-Thomsen/ Mies, 1999, 49)

Myth: 'Capitalism by necessity produces democracy'

Reality: Complementary to the neoliberal claim that markets operate according to natural laws is the allegation that the market economy by necessity produces a democratic political system. This notion has to be challenged on two levels; first on a conceptual one, second on the basis of historical evidence.

Lummis provides an interesting analogy to explain both why this myth gained so much currency in the Cold War and why it is so utterly false:

> The logic [of the view that the free market equals democracy] is simple: socialist command economy is antidemocratic, therefore the free market is democratic. This view is rather amnesiac, forgetting as it does the problem that socialism was hoped to be the solution to. An analogy is a person suffering from a deadly sickness who takes a medicine that makes him worse and then decides that if he stops taking the medicine he will be well. The original problem persists. The free market divides society into rich and poor, a division that is incompatible with democracy. Its freedom is mainly freedom for the corporation, and the capitalist corporation has itself become an antidemocratic system of rule. The question of how to democratize the main actor in the free market – the corporation – is, for the capitalists and managers, *the* subversive question. (Lummis, 1996, 17)

I think there cannot be any real doubt about the fact that the capitalist economy is utterly and inherently antidemocratic, "a system of totalitarian world market democracies", as Kurz calls it (1999, 513). The command structure, the absence of democratic accountability of the actors, the complete lack of self-determination of both the producers/workers and consumers within the production process leave no doubt as to the correctness of this analysis. That is why it is so ironic that the capitalist economy "dictates 'self-responsibility'" while simultaneously removing any "self-determination over one's own life" (Kurz, 1999, 781). Lummis is correct in likening the capitalist economy to the military; in terms of organisation and hierarchy they have more than their fair share in common (see Lummis, 1996, 133-134). When he discusses Lenin's view that capitalism is progressive, it becomes startlingly clear both why Lenin appreciated it and why it is so antidemocratic: it removes the control over economic activities from the people and places it in the hands of the capitalist corporation or the corporatist state (such as the Soviet Union), thereby turning people into passive dependents:

> Capitalism is progressive [according to Lenin] because it "separates industry from agriculture", that is, it takes farmers and makes them into industrial proletarians working in factories. It takes them from under the control of the traditions of agrarian society and places them under the control of industrial organization. It changes the nature of production by concentrating it and organizing it; it changes the nature of consumption by destroying subsistence and making people dependent on commodity consumption. (Lummis, 1996, 52-53)

In other words, it is simply illogical and absurd to assume that such an economic system – which is beyond any democratic accountability and which doesn't grant the massive majority of those who keep it running and spend most of their lifetime in it any sort of co-determination – would nurture its exact opposite, namely democratic structures, equity, solidarity and insight into the necessary respect towards other human beings and the natural life-support system. Macpherson (1962) already showed in the 1960s that liberal democracy had this paradox written into itself as a congenital defect, because it insists on the schizophrenic distinction between democracy in the political and dictatorship in the economic sphere. The basis of this is its view of human beings, defining them exclusively with regard to property, rather than as ethical or social beings.

After World War II a vast majority of people in Europe had experienced first hand the incredibly destructive force which the collaboration of political and economic dictatorship yielded in the Third Reich. This had taught them that only an extension of democracy into the economic sphere, in the form of workers' self-determination and participation, could prevent a restoration of such conditions. In Germany, there was widespread political agreement on this, even in conservative circles.[182] Only with massive repression of grass roots unions, manipulation of public opinion through business propaganda and deliberate blackmailing in the form of the Marshall Plan, was the United States, newly emerged undisputed world leader in matters economic and military, capable of suppressing this insight again.[183]

But it might be that we are slowly but surely regaining the insights so brutally inflicted on the peoples of Europe after 1945. We have begun to realise that the sacrifice of democracy on the altar of the world market is self-destructive. More or less democratically elected governments feel less and less capable of counteracting the power of democratically unaccountable corporations and financial markets:

> The Mexico crisis [in 1995] lit up with rare clarity the face of the new world order in the age of globalization. (...) In the stock exchanges and the dealing rooms of banks and insurance companies, in the investment and pension funds, a new political class has appeared on the world stage. It can no longer be shaken off by any government, any corporation, still less any ordinary taxpayer. Currency and security dealers acting on a world scale direct an ever-growing flow of footloose investment capital and can therefore decide on the weal and woe of entire nations, and do so largely free from state control. (...) What happens on the money markets follows a largely comprehensible dynamic that was actually made possible by the governments of the major industrialized countries themselves. In the name of a doctrine of salvation through totally free markets, since the early 1970s they have systematically striven to tear down all barriers which once allowed cross-border flows of money and capital to be regulated and therefore controlled. Now, like the sorcerer's apprentice, they complain that they are no longer in control of the spirits that they and their predecessors called into being. (Martin/Schumann, 1997, 45-47)[184]

---

[182] See, for example, the 'Ahlener Programm' of the centre-right Christian Democrats (CDU), agreed on the 3.2.1947: *"The capitalist economy has failed to do justice to the vital political and social interests of the German people."* (quoted in Rolf Steininger (1989), *Deutsche Geschichte 1945-1961. Darstellung und Dokumente in zwei Bänden* (Frankfurt/M., Fischer [=Fischer TB 4315]), Vol. 1, 117; emphasis in the original)

[183] The ruthless suppression of all attempts at worker co- and self-determination by the United States is historically documented, but not well known; see Carolyn Eisenberg, 'Working-Class Politics and the Cold War: American Intervention in the German Labor Movement, 1945-1949', in *Diplomatic History*, 7 (1983) 4, 283-306. On the abuse of the Marshall Plan as a means to exert political pressure in order to achieve the desired worldwide economic dominance by the US see Melvyn Leffler, 'The United States and the Strategic Dimensions of the Marshall Plan', in *Diplomatic History*, 12 (1988) 3, 277-306 and Thomas McCormick, '"Every System Needs a Center Sometimes". An Essay on Hegemony and Modern American Foreign Policy', in *Redefining the Past. Essays in Diplomatic History in Honor of William Appleman Williams* (1986), ed. by Lloyd C. Gardner (Corvallis, OR, Oregon State University Press), 195-220. For an excellent long-term perspective on the persistent (and unfortunately highly successful) efforts by business interests to suppress any attempt at workers' self-control over the last two hundred years, see Zinn, 1996.

[184] With regard to the power of transnational corporations (TNCs) we shouldn't fool ourselves: "The largest 500 companies now control 42 per cent of the world's wealth. Of the biggest 100 econo-

This means that the capitalist threat to democracy occurs on two levels. On the one hand there is the transfer of power from citizens to undemocratic, bureaucratic corporate structures. The most obvious and blatant example of this to date has been the Multilateral Agreement on Investment (MAI) which the OECD, representing the 29 richest countries of the world, started to draw up in complete secrecy from 1995 onwards, after a similar proposal, put before the WTO, was defeated by Southern countries. The public only got to know about this proposal because it was leaked to the consumer group Public Citizen, which in turn immediately publicised it on the internet.[185] Subsequently the world-wide public outcry was such that the MAI had to be withdrawn, particularly after France started to voice opposition. But because it is quite clear that the OECD countries will try to reintroduce the same issues, possibly within the WTO, it is worth giving some attention to what was put forward. The MAI is an extraordinary proposal for an international agreement since it would strategically weaken the power of local, regional and national democratic institutions in favour of a disproportionate strengthening of the power of TNCs, even though this power has already had such devastating consequences that Paul Hawken has stated: "There is no polite way to say that business is destroying the world." (quoted in Karliner, 1997, 13; see below p. 149) The three central principles of the MAI were intended to be:

---

mies, half are now corporations and half are countries. The 10 biggest companies together turn over more money than do the 100 smallest countries. Indeed, only 27 countries now have a turnover greater than the sales of Shell and Exxon combined. Shell – the world's No 2 – owns or leases some 400 million acres of land, which makes it larger than 146 countries. A few more figures: just 250 companies in Britain take almost half of everything we spend. General Motors sales revenues ($133 billion per annum) are roughly equal to the combined GNP of Tanzania, Ethiopia, Nepal, Bangladesh, Zaire, Uganda, Niger, Kenya and Pakistan. That's more than 500 million people. Twelve of the world's most important industries – including cars, aerospace, electronics, steel, oil, computers, media – are each more than 40 per cent dominated by five or fewer corporations. Their power is not just financial, though. They are now beginning to dictate the fundamentals of life. Just 10 corporations control virtually every aspect of the worldwide food chain; four control 90 per cent of the world's exports of corn, wheat, coffee, tea, pineapples, cotton, tobacco, jute and forest products. The same companies that control the commodities now handle the storage, the transport and the food processing. The growth in size and clout of some transnationals in the past two decades has been as spectacular as the fall of communism. Bizarrely, though, they are becoming more successful at centralised planning than Moscow ever was, and are beginning to make communism seem transparent. (...) So immense are they growing, and such is their skill in levering markets, so grand their resources and great their political influence, that they are now effectively units of governance. Yet they have avoided, so far, the business of having to be socially accountable, and are to all intents undemocratic and unaccountable." (John Vidal, 'The real politics of power', in *The Guardian. Society*, 30.4.1997, G2, 4-5, here 4) Karliner's *The Corporate Planet* (1997) and Madeley's *Big Business, Poor Peoples* (1999) are two excellent accounts of these phenomena.

[185] See http://www.citizen.org/trade/issues/mai/Text/articles.cfm?ID=5629.

- Non-discrimination: foreign investors must be treated as well as or better than domestically owned companies.
- No entry restrictions: governments cannot restrict foreign investment in any form or in any sector (except defence).
- No conditions: governments cannot impose so-called performance requirements, for example, to ensure local employment or control currency speculation.[186]

National or regional governments would not be allowed to implement any social or environmental clauses any more since the corporations could in that case sue the governments for limiting competition. Interestingly enough the governments wouldn't have equivalent legal leverage against the multinationals. Additionally, the OECD proved its kindness by adding a clause which wouldn't allow a country which had signed the MAI to leave it for five years. If a country then left, it would still be bound to the MAI for another fifteen years. It was a very clever idea: first, Southern countries would be more or less forced to sign the MAI by the threat that the OECD would otherwise withhold investment. Once lured in, the countries would be bound by the terms of the agreement for another 20 years.[187] The MAI, then, is an exemplary attempt by the corporate rulers of the world to prise open every last corner of the Earth for profit maximisation, without having the slightest shred of democratic entitlement to do so and without having to assume responsibility for their actions and its consequences.

Of course, the MAI is just one example of the business giants' attempt to expand their worldwide control. As mentioned before, of the one hundred most powerful economies of the world, only 49 are countries, while 51 are TNCs (Karliner, 1997, 5). I have also mentioned already that Carey (1997) has shown how corporate America has managed to determine national policy for most of the 20th century (see also Ferguson, 1995). *Europe, Inc.* (2000) traces in detail how the European TNCs have shaped the content and timetable both of the Maastricht Treaty and the Currency Union, down to phrasing whole chunks of the text of the treaties. With regard to global environmental policies there is a growing body of evidence that TNCs use every (dirty) trick in the book to torpedo a transition to a sustainable society.[188] We have just witnessed again in autumn 2000 how the oil corporations – in the run-up to the Hague conference which should have ham-

---

[186] Barry Coates, 'MAI: A leap into the dark', in *World Development Movement in Action*, December 1997, 8-9, here 9.

[187] See the extensive analysis of the MAI by Corporate Europe Observatory (CEO), 'MAIGALO-MANIA! Citizens and the Environment Sacrificed to Corporate Investment Agenda', February 1998 [http://www.xs4all.nl/~ceo/mai/] (reprinted in *Europe Inc.*, 2000, 109-122), and David Rowan, 'Meet the new world government. Corporations v states', in *The Guardian*, 13.2.1998, 15; Gerhard Klas, 'Und nun auch noch die politische Macht: Weltwirtschaft: alles neu macht das MAI', in *Die WochenZeitung*, No. 3, 15.1.1998, 13.

[188] See on the rise of corporate greenwash: Karliner, 1997, Beder, 1997, Greer/Bruno, 1997.

mered out the implementation of the Kyoto Protocol to counteract global warming[189] – skilfully orchestrated an oil shortage, driving up the oil price, which led to fuel shortages in a number of European countries. This in turn led to a public outcry against high fuel prices and almost certainly influenced the collapse of the Hague conference. The entire debate, needless to say, was completely decontextualised: no mention of the fact that we are already burning too much fuel, that current prices are far too low especially in comparison with other goods, that aviation and shipping fuels are still not taxed at all etc.: a stark reminder how a well-orchestrated PR campaign of a few multinationals can influence public opinion almost overnight.[190] But there is some consolation at least in the fact that the corporate world feels compelled to spend billions of dollars every year on greenwash.[191] The fact that they need to spend this amount of money on propaganda, manipulation and lies proves indirectly that there is at least a sizeable, genuinely democratic and enlightened world public which isn't too easily fooled (see Stauber/Rampton, 1995).

There is another thing which needs to be said on this matter: in rare moments of honesty international investors and representatives of financial markets are quite prepared to state that democracy is in fact detrimental to a favourable climate for investments and therefore business. The CEO of Shell Nigeria, for example, justified the collaboration of his company with the oppressive regime in Nigeria after the killing of Ken Saro-Wiwa: "For a commercial company trying to make investments, you need a stable environment ... Dictatorships can give you that." (Karliner, 1997, 86) Or in the words of a Wall Street observer: "Financial markets might not respond positively to increased democracy because it means increased uncertainty." (Karliner, 1997, 210)

On the other hand the process of diminishing democratic control involves a kind of Faustian pact: you hand over self-determination to the corporate world and in return you get a drug more potent than Aldous Huxley's *soma*, namely the ritualised promise of ever more consumption and economic growth.[192] Schumacher has

---

[189] For an account of how the US, Canada, Australia, Japan and New Zealand tried to reinterpret Kyoto as no-change-necessary (the tricks range from so-called carbon sinks and Clean Development Mechanisms to emission trading), a process whereby the US again holds the rest of the world hostage to meet its demands, see Simon Retallack, 'Sinking Kyoto', in *The Ecologist*, 30 (2000) 8, 58-59.

[190] For an analysis of the corporate world's attempt to torpedo Kyoto see Ross Gelbspan (1997), *The Heat is on: The Climate Crisis, the Cover-Up, the Prescription* (Boston, Perseus Books) and *The Weather Gods: How Industry Blocks Progress at Kyoto Climate Summit* (1997), ed. by Corporate Europe Observatory (CEO) (Amsterdam, CEO) [reprinted in *Europe Inc.*, 2000, 155-165].

[191] More than $1 billion per year in the US alone (Karliner, 1997, 182). See also: http://www.earthdayresources.org/publications/dont_be_fooled_2000.htm.

[192] See Charles Taylor, 'Wieviel Gemeinschaft braucht die Demokratie?', in *Transit*, (1992/93) 5, 19. Huxley's gloomy predictions seem to have fulfilled themselves by now, except that you have to replace the word "political" with "economic": "A really efficient totalitarian state would be one in

characterised our consumer society very aptly by comparing it to "a drug addict who, no matter how miserable he may feel, finds it extremely difficult to get off the hook." (Schumacher, 1993, 126) And John Berger writes:

> Publicity turns consumption into a substitute for democracy. The choice of what one eats (or wears or drives) takes the place of significant political choice. Publicity helps to mask and compensate for all that is undemocratic within society. And it also masks what is happening in the rest of the world.[193]

As regards the historical evidence for the claim that capitalism produces democracy by necessity, you just have to consider the developments in the former Eastern bloc or the so-called Asian tigers to realise the falseness of the dogma. As far as the former is concerned the introduction of the market economy has by no means led to functioning democracies, quite the contrary: the extent of corruption, the abuse of state and economic power has increased, if anything.[194] This led Martin/Schumann to the following conclusion:

> The end of the communist regime, however, brought not the end of history but an enormous speeding up of social change. (...) But what the founders of the post-war welfare states had learnt through bitter experience is again becoming ever more apparent: namely, that market economy and democracy are by no means inseparable blood-brothers who peaceably increase the well-being of all. Rather, the two central models of the old industrial nations of the West continue to be in contradiction with each other. (1997, 227)[195]

At the same time, the most successful new contenders on the world market are not democratic countries, but – at least some of them – amongst the most oppressive regimes in the world: Taiwan, Indonesia, Malaysia, Singapore, China, to name just a few.[196] To verify that fact, just have a look at the latest yearly report by Amnesty International.[197] On top of that the countries of the South have virtually no leeway to nurture processes of democratisation. They are caught between the Scylla of the

---

which the all-powerful executive of political bosses and their army of managers control a population of slaves who do not have to be coerced, because they love their servitude." (Huxley, 1977, 13-14)

[193] John Berger (1972), *Ways of Seeing* (London, Penguin), 149.

[194] See Jürgen Habermas (1990), *Vergangenheit als Zukunft* (Zurich, pendo), 111-112.

[195] See also that the US and Europe only champion democracy and human rights if this doesn't contradict economic interests, the best example being China (Martin/Schumann, 1997, 145, 147-148; and Samir Amin, 'Demokratische Herausforderung und Weltkapitalismus', in *Demokratie radikal*, 1992, 9-21, here 19).

[196] In large parts of Latin America the integration into the world market has not led to automatic democratisation, but rather to a "'modernisation of the dictatorship', the replacement of the old oligarchic and patriarchal systems with 'efficient' and 'modern' fascistic powers" (Samir Amin, 'Demokratische Herausforderung und Weltkapitalismus', in *Demokratie radikal*, 1992, 9-21, here 11).

[197] *Amnesty International Report 2001* (2001), ed. by Amnesty International (London, AI UK).

Structural Adjustment Programmes (SAP) dictated by the IMF and the Charybdis of rising debt and the pressures of a globalised market:

> Either the democratic political system accepts the subordination under the requirements of the worldwide 'adjustment' which means that it cannot enact any substantial social reforms, and democracy itself will soon be in crisis; or the people use the democratic system to enact these reforms which means that the country comes into conflict with the dominant global capitalism.[198]

Myth: 'Free market capitalism is free'

> Given existing social discrepancies, the ideals of universal liberty (...) must remain purely notional for the mass of humanity, but provide the well-to-do with a legal and ethical framework to consolidate and justify their ascendancy over their fellow-citizens. (Holmes, 1995, 140-141)

Reality: Intrinsically linked to the question of democracy is also another neoliberal claim, namely that the capitalist market is free. Yet capitalism in its embodiment as world market is totalitarian in that it excludes and destroys all other possible ways of economic exchange and organisation. Bennholdt-Thomsen/Mies have convincingly shown the marginalisation of other forms of market exchange and of the subsistence economy by the dominant capitalist paradigm, which has led to a large degree of blindness and ignorance:

> In this process, the understanding of market and subsistence underwent a fundamental transformation, so that it is now thought that the market *is* life, *is* subsistence. According to this view, subsistence disappears forever from the economy, from what is necessary to life. (1999, 112-113)

This revaluing of all values leads to the absurd end result that "the most lifeless thing of all, money, is seen as the source of life and our own life-producing subsistence work is seen as the source of death" (Bennholdt-Thomsen/Mies, 1999, 17).

This stamping out of alternatives is mirrored in the fact that the global market, despite its claims, is a rather exclusive club: "Capital is only interested in the solvent markets, represented by about 1.5 billion of the 6 billion people on earth."[199]

---

[198] Samir Amin, 'Demokratische Herausforderung und Weltkapitalismus', in *Demokratie radikal*, 1992, 9-21, here 13.

[199] Riccardo Petrella, 'Die Gefahren einer Techno-Utopie', in *Le Monde diplomatique [deutsch]*, 2 (1996) 5, 5. Also: "A total of 130 states account for just 3.6% of world exports. That means that two-thirds of all the countries in the world are of virtually no significance for the global market. They exert almost no influence on what happens there, but are highly dependent on that market for their

And, of course, free access to the world market is a myth. Protectionist measures, particularly by the rich nations,[200] subsidies of all kinds as well as unequal allocation of costs lead to anything but a level playing field for all countries.[201] On top of that there are the rules of the WTO,[202] the IMF and the World Bank, which make a mockery of freedom as equality. Not to mention the more subtle mechanisms, such as connecting the credit rates with the stability and economic power of the debtor, which usually means that smaller and poorer countries receive credits for investment on considerably worse terms than big countries and TNCs.[203] On the world market, evidently, some are freer than others, and the freest, "in other words those who call the shots", is the United States.[204] It is also clear that the freedom of the global market only means the free movement of capital, never freedom of migration;[205] and if freedom of migration, then only for a tiny ruling elite of the world

---

imports and exports. Many of the poorest countries have little more to offer than raw materials and agricultural produce, and foreign exchange earnings are often dependent on exporting one, two, or three products, which increases dependency to an even greater extent." (*Greening the North*, 1998, 202-203)

[200] The biggest economic powers (the United States, Japan, Germany) are also the most protectionist ones (see Chomsky, 1992a, 81): "Rich countries' protectionist agricultural policies cost the poorest countries £13.5 billion – equivalent to the external debt of the world's six poorest countries." (World Development Movement, 'The missing link: Debt and Trade', briefing, September 2000, 3)

[201] "Northern countries are still very versed in closing off their markets against unwelcome competition from the South, through various methods. This means that the South is more or less limited to the export of primary raw materials such as ore, fossil fuels, logs, coffee, cocoa and animal feeds. Yet world market prices for these commodities are still so low as to necessitate rapacious practices of exploitation in the South." (*Zukunftsfähiges Deutschland*, 1996, 14)

[202] See such unsound decisions by the WTO as to ban labelling of products according to their production method, say to distinguish paper from certified sustainable forestry (FSC) from paper produced from clear-cutting old growth forests or to distinguish dolphin-friendly tuna from tuna caught by other means. The argument is, again, that this labelling obstructs competition, ignoring any consumer demands for freedom of choice, openness and transparency (see Kevin Watkins, 'Dumping grounds', in *The Guardian. Society*, 4.12.1996, 4-5). See also: Martin Khor, 'Die Globalisierungspolizisten der WTO. Konferenz der Staatsfeinde', in *Le Monde diplomatique [deutsch]*, 3 (1997) 5, 22.

[203] See Martin/Schumann, 1997, 65-67. Regarding the neo-colonialism of TNCs see Edward Goldsmith, 'Neue Kolonialreiche. Das Gesetz der Multis', in *Le Monde diplomatique [deutsch]*, 2 (1996) 4, 1, 11-12. Concerning the various levels of exploitation of the South by TNCs and the North (resources, financial and profit as well as brain drain) see *Taking Nature into Account*, 1995, 128-138.

[204] "Globalization, then, is by no means just a question of 'American cultural imperialism' in the entertainment sector. (...) The United States, as the 'mass culture superpower' (Jack Lang), will not only decide which games are played but also hand out the bread." (Martin/Schumann, 1997, 37); "At least on the money markets, globalization so far means little more than the Americanization of the world." (ibid. 74)

[205] "Neoliberalism never intended to accompany its programme of liberalisation of trade and capital movements with an unrestricted liberalisation of migration of employees and is simply because of this a lie." (Samir Amin, 'Demokratische Herausforderung und Weltkapitalismus', in *Demokratie radikal*, 1992, 9-21, here 12) With regard to migration, fortress Europe and asylum seekers, a little thought experiment is instructive: if the Europeans emigrating to America, all of them 'economic

population, a freedom moreover which is based on structures that actively obstruct a transition to a sustainable world (see *Taking Nature into Account*, 1995, 161).

Myth: 'Capitalism and sustainability are compatible'

> In today's liberalized markets, by definition diversity is diminishing, and this is taking place in the name of an ideology pretending to produce diversity and freedom! (von Weizsäcker, 2000, 22)

Reality: On the basis of what we have discussed so far, it hardly comes as a surprise that few thinkers concerned with a sustainable future find that the capitalist economy, *due to* its inherent structures, is compatible with it. In fact, there is clear structural and plenty of other evidence which points to the conclusion that a capitalist economy *cannot* under any circumstances be a sustainable one. I am fully aware that saying this in today's neoliberal climate is anathema, a taboo. Yet there is no running away from the facts. Neither as far as ecology, empowerment, equity nor in fact the economy is concerned can the present system deliver.

The most important and dangerous fallacy of the current way of thinking flows from the belief that the economy is the basic system on which all else depends. Vandana Shiva has spelt out three flaws which underpin the popular notion of 'sustainable development' rather than the true meaning of sustainability: "The first is assigning primacy to capital. The second is the separation of production from conservation, making the latter dependent on capital. The third error is assuming substitutability of nature and capital." (Shiva, 1992, 189) The truth, as Wackernagel/Rees state, is different:

> Humanity needs the ecosphere, but the ecosphere does not need us. This is not to argue that economy and society are less important to humanity than ecology but rather that we need to understand the "directionality" of dependence before we can produce good policy for sustainability. Indeed, should we not be structuring the economy better to serve society rather than, as at present, restructuring society to serve the economy? (1996, 115)

This means that the hyped up conflict between economy and ecology does not exist. The question clearly is not which of these two things is more important, even if it is constantly put that way in the media. There is simply no economy without the biosphere; the former is just a *sub*system of the latter:[206]

---

refugees', had any right to conquer the continent, then all the refugees, including economic ones, coming to Europe have an equal right to stay and would even be entitled to ruthlessly slaughter all the Europeans as the settlers annihilated the native Americans in the 'New World'.

[206] See Herman E. Daly, 'Consumption: Value Added, Physical Transformation, and Welfare', in *Getting Down to Earth*, 1996, 49-59, here 57.

The scale of the growing human economic subsystem is judged, whether large or small, relative to the finite global ecosystem *on which it so totally depends, and of which it is a part.* The global ecosystem is the source of all material inputs feeding the economic subsystem, and is the sink for all its wastes. (*Taking Nature into Account,* 1995, 107; my emphasis)

Thirty years ago, Schumacher identified this misunderstanding as "one of the most fateful errors of our age" which is based on "our inability to recognise that the modern industrial system, with all its intellectual sophistication, consumes the very basis on which it has been erected". He has shown that even within the terms of an economist it is self-destructive and bad economics: "To use the language of the economist, it lives on irreplaceable capital which it cheerfully treats as income." (Schumacher, 1993, 8; see Wackernagel/Rees, 1996, 36)

If we try to recall what we have identified as the structural principles of a sustainable society in chapter 1, it very quickly becomes clear that capitalism is based on opposing principles, which also renders absurd the belief that such a system will fix our problems:

• *Capitalism rests on the domination of nature*

Capitalism not only validates pre-capitalist notions of the domination of nature by man; it turns the plunder of nature into society's law of life. (...) It requires a grotesque self-deception, or worse, an act of ideological social deception, to foster the belief that this society can undo its very law of life in response to ethical arguments or intellectual persuasion. (Bookchin, 1980, 66)

The notion that anything – humans, nature – is there to increase profits makes it impossible to acknowledge the fact that we are both dependent on nature for survival *and* that we are only stewards, not proprietors of nature. This fallacy partly stems from a very anthropocentric view of nature. The term 'environment' already suggests that there is such a thing as the centre (humans) surrounded by something else (nature) which by definition is less valuable. A horizontal view of human-nature relationships, seeing "nature as a web or a network which puts nature and humans at parity, both a part of a whole" (Plant, 1998, 29), would counteract this, but is inherently anti-capitalist. Shiva has nailed down the paradox that the very system which, due to its logic and view of nature, has created the ecological crisis in the first place, is now hailed as the saviour:

While development as economic growth and commercialization are now being recognized as being at the root of the ecological crisis in the Third World, they are paradoxically being offered as a cure for the ecological crisis in the form of 'sustainable development'. The result is the loss of the very meaning of sustainability. (...) Economic growth takes place through the over-exploiting of natural resources which creates a scarcity of natural resources in nature's economy and the people's survival economy. Further economic growth cannot help in the regeneration of the very spheres which must be destroyed for economic growth to take place. Nature shrinks as capital grows. The growth of the market cannot solve the very crisis it creates. (Shiva, 1992, 188-189)

It is very important to remember here that we are engaged in a zero-sum game (i.e. a game with a constant overall total). Resources which are exploited to facilitate "predatory capitalism", as Chomsky calls it (1992, 153), are not an addition to the ability of people to provide their sustenance, but are taken away from the overall stock. Capitalism and subsistence are mutually exclusive. I still believe that there is no better way of putting this illusion of modern *homo economicus* than Schumacher's catch-22: "Modern man does not experience himself as a part of nature but as an outside force destined to dominate and conquer it. He even talks of a battle with nature, forgetting that, if he won the battle, he would find himself on the losing side." (Schumacher, 1993, 3)

- *Capitalism rests on private property and competition*
A fundamental problem with capitalism is its underlying view of human nature, defining everything in terms of property and profit.

> The sacrifice of people's rights to create new property rights is not new. It has been part of the hidden history of the rise of capitalism and its technological structures. The laws of private property which arose during the fifteenth and sixteenth centuries simultaneously eroded people's common rights to the use of forests and pastures while creating the social conditions for capital accumulation through industrialisation. The new laws of private property were aimed at protecting individual rights to property as a commodity, while destroying collective rights to commons as a basis of sustenance. The Latin root of private property, privare, means "to deprive". The shift from human rights to private property rights is therefore a general social and political precondition for exclusivist technologies to take root in society. (Shiva, 1991, 232)

This has far-reaching consequences with regard to sustainability. Private property denies the fundamental human rights to self-determination and equity. Property is only possible at the expense of others in a zero-sum game. If there is only so much to go round, it means that if some people are appropriating certain resources in excess of their fair share these resources will be taken away from others. We have seen that sustainability can only be built on the assumption of an equitable sharing of nature's wealth amongst all people and future generations, on a cooperative attempt to provide the means of living for everybody. It follows that an economic system cannot be sustainable if is built on the maximisation of profit for some individuals at the expense of others, with complete disregard for the costs of these profits for the community and the future. The theory of the trickle-down effect – claiming that if you remove all the barriers for rich people to get even richer, eventually these riches will feed down to the poor – has so catastrophically and clearly failed in the last twenty years of neoliberal domination of the world that we should be prepared to face up to what has happened: the gap between rich and poor is constantly widening. Between 1960 and 1991, the share of global income of the richest 20 per cent rose from 70 per cent to 85 per cent, while the share of the poorest 20 per cent fell from 2.3 per cent to 1.4 per cent. The ratio between the share of the richest and the poorest therefore increased from 30:1 to 61:1 (*Human Development Report 1996,* 1996, 2). This ratio rose exponentially to reach 82:1 in 1995 (*Human Development Report 1998,* 1998, 29):

> New estimates show that the world's 225 richest people have a combined wealth of over $1 trillion, equal to the annual income of the poorest 47% of the world's people (2.5 billion). (...) The three richest people have assets that exceed the combined GDP of the 48 least developed countries. (ibid. 30)

Not a hint of trickle-down, on the contrary it seems that the present system is structured in such a way as to favour the rich and deprive the poor even further. Sachs has summed it up neatly: it is "easy to see that it [trickle down theory] derives its attraction from the promise to achieve justice without redistribution. In other words, the dedication to growth has always been fuelled by the desire to sidestep the hard questions of justice." (1999, 165)

- *Capitalism rests on growth*

> "Growth for the sake of growth," notes environmental writer Edward Abbey, "is the ideology of the cancer cell." Just as a continuously growing cancer eventually destroys its life-support systems by destroying its host, a continuously expanding global economy is slowly destroying its host – the Earth's ecosystem. (Lester R. Brown)[207]

The logic of capitalism rests on growth, on unhindered accumulation, expansion and greed. This structural feature is clearly at odds with the scientific fundamental that our biosphere, within which all economic activity takes place, is a materially non-growing system, in other words a closed system with clear limits (see Schumacher, 1993, 16-18). For that reason Schumacher rightly says that we need to seriously reconsider the essentials on which we rest our economic thinking:

> Can such a system [capitalism, private enterprise] conceivably deal with the problems we are now having to face? The answer is self-evident: greed and envy

---

[207] Brown, 'The Future of Growth', in *State of the World 1998,* 1998, 3-20, here 4.

demand continuous and limitless economic growth of a material kind, without proper regard for conservation, and this type of growth cannot possibly fit into a finite environment. We must therefore study the essential nature of the private enterprise system and the possibilities of evolving an alternative system which might fit the new situation. (1993, 222)

Vester has phrased even more drastically the self-destructive nature of the growth ideology which apart from its ideological, i.e. anti-scientific, character also shows an immensely impoverished social imagination:

Only the complete lack of imagination of a few economist apparatchiks, sworn in on their growth ethos, managed to utterly misunderstand the repeated appeals to turn away from continued growth. They didn't grasp that such change would eventually also serve their own interests. They missed the point that *development* is by no means equal to growth, that progress doesn't necessarily mean "more", "faster", "bigger", but can also stand for "different", "more beautiful", "better". (...) Let's first take the dogma to task which has prevented the timely turning away from our nearly vertically climbing growth curve: our in East and West un-contested growth ideology. For it in particular rests not on an understanding of real systems behaviour and resulting explainable facts, but on the irrational fixed idea that all growth – be it growing speed or increasing information – is from the outset desirable. This is an *a-priori* claim which nobody ever bothered to prove. Well, proof for it is in fact nowhere to be found. (Vester, 1997, 454-456)

After the end of Soviet-style state socialism, more than ever we need to transcend the false ideology of capitalism and find an economic system which is in line with, not inimical to, human needs and the survival of the biosphere.

- • *Capitalism rests on short-term profit*

Huckle has drawn attention to the fact that because of the primacy of the profit motive, capitalism is in itself incapable of accounting for the costs it imposes on society and nature as well as blind to its own long-term survival:

Capitalism requires economic growth or capital accumulation and has an inbuilt tendency to discount present and future environmental costs. It has no coordinated internal mechanisms for maintaining the conditions of production; the trend away from organic raw materials and renewable sources of energy to inorganic and non-renewable sources has hastened the arrival of ecological limits to growth.[208]

---

[208] John Huckle, 'Realizing Sustainability in Changing Times', in *Education for Sustainability*, 1996, 3-17, here 5.

This is clearly the case because capitalism rests on an abstract motive – profit maximisation at all costs – which is entirely devoid of all ethical values, responsibility and context. Capitalist ideologists tend to deny that, but Milton Friedman, the chief ideologue of the neoliberal revolution which has swept the globe in the past quarter century, has spelt it out with a clarity that leaves nothing open to discussion: "Few trends could so thoroughly undermine the very foundations of our free society as the acceptance by corporate officials of a social responsibility other than to make as much money for their stockholders as possible".[209]

An economic system based on just the 'bottom line' and overriding all other interests, cannot be concerned with human welfare, with environmental health, with equitable living conditions for all. It is structurally incapable of caring for its own survival; its prime motive constantly counteracts any long-term strategies. Bateson calls this a "paradigm for extinction by way of loss of flexibility." (2000, 509) A very good example of this are the financial markets. Despite the contrary claim by corporate exponents engaged in greenwash, these are reorientating themselves to ever-shorter reaction times. Every corporation which considers itself important now publishes quarterly results and most investment in the global financial markets is done and can be changed within seconds. This means that investors' decisions, based only on expectations of the highest returns – not on issues such as workers' conditions in production lines or environmental degradation –, can decide the fate of a country within hours, without any chance of a share in the decision-making process by the people of the country concerned: this has happened with Mexico, with the so-called South Asian Tiger economies and many African countries.

This last point clarifies again that the economic system in fact is anti-sustainability on all levels: it is anti-democratic,[210] anti-equity, anti-ecology.[211] The wealth of the winners is based on exploitation of all the spheres crucial to sustainability.

That this is so can be seen from three aspects. Firstly, even though the economy is clearly the most powerful broker in today's society, it is so narrow in its focus on short-term profit and people with high incomes (consumers and shareholders) that it neither can nor wants to encourage sustainability, "according to its current self-identity" (Dürr, 2000, 10). The second point is that capitalism focuses on standards of living instead of quality of life. With a view to the latter it

---

[209] Milton Friedman (1962), *Capitalism and Freedom* (Chicago; London, University of Chicago Press), 133. See in a newer variant: "The pressure of global competition is such that [top managers] think it unreasonable to expect a social commitment from individual businesses. Someone else will have to look after the unemployed." (Martin/Schumann, 1997, 4)

[210] "Economic development is antidemocratic in that it is the expansion of a sphere of life from which democracy is to be excluded in principle." (Lummis, 1996, 48)

[211] See Johannes B. Opschoor, 'Institutional Change and Development Towards Sustainability', in *Getting Down to Earth*, 1996, 327-350, here 328.

fails: "Economics and the standard of living can just as well be looked after by a capitalist system, moderated by a bit of planning and redistributive taxation. But culture and, generally, the quality of life, can now only be debased by such a system." (Schumacher, 1993, 220) "Human welfare and meaning" cannot be looked after by "the values of the marketplace".[212] The third aspect becomes clear if one tries to conceptualise a sustainable economy. Capitalism, with its focus on accumulation and growth, demands by necessity that whatever goods one has acquired must be thrown away and replaced with new ones, or else the system of constant expansion cannot work. Yet:

> A durable-goods economy is precisely the contrary of an economy based on planned obsolescence. A durable-goods economy means a constraint on the bill of goods. Goods would have to be such that they provided the maximum opportunity to "do" something with them: items made for self-assembly, self-help, re-use, and repair. (Illich, 1971, 63)

There is another, personal dimension to the accusation that the capitalist economy doesn't honour long-term responsibilities. This attitude has come to dominate the private sphere as well. Nowadays a lot of decisions are made in order to run away from problems, so that one does not have to face up to the consequences of what one has done. The ease with which marriages are broken up and children are dumped onto the other partner or society, the ease with which we decide to abort potentially disabled children, has very often a lot less to do with real hardship or unsolvable problems, and more with an attitude to life which cherishes an absolute notion of freedom, without any accompanying responsibilities. It is an attitude to life which wants *everything* here and now and doesn't accept the fact that confronting problems, trying to work through them and finding solutions, is harder but ultimately far more rewarding than constantly running away from them.

Myth: 'The Market is good and state intervention is bad'

Reality: Wrapped up as we are in neoliberal dogma it seems a strange proposition, but if you look at the undemocratic structure of business and the absence of any self-determination in that sphere (except for the ruling elite) the conclusion is obvious: "The more tasks the democratic state delegates to the unfree economy, the unfreer society becomes as a whole."[213] Yet this transfer of power continues and is still the central plank in the so-called SAPs, prescribed by the IMF for developing countries, if they want funds. It leads to an increase in power without

---

[212] Thomas Berry, quoted in Gregory A. Smith and Dilafruz R. Williams, 'Introduction: Re-engaging Culture and Ecology', in *Ecological Education in Action*, 1999, 1-18, here 2.

[213] Hanspeter Guggenbühl, 'Der Wert der Freiheit', in *Die WochenZeitung*, No. 40, 5.10.2000, 7.

social responsibility and makes the world more and more ungovernable. State power is unable to keep up with the range of power and wealth displayed by many TNCs. David Korten argues that the global trading system is now being worked out on the level of corporations and not governments: "A massive transfer of power is taking place from governments to corporations without so much as a by your leave to the people."[214] It is axiomatic that as corporations increase their power, according to Korten, so the power of others – individuals, governments – decreases. This is an important point to make since there is often the misunderstanding that arguing for state power is good in itself. On the basis of what I had to say about empowerment and self-determination it should be abundantly clear that state intervention is only tolerable if it is legitimised through real democratic procedures. Otherwise one illegitimate power is simply replacing the other.

It is nevertheless true that this restructuring is taking place on a global scale. The magic word is "deregulation". Yet rather than being the promised process to ensure efficiency and greater wealth, it is again just a cunning procedure to increase the wealth of the wealthy and let others foot the bill, either the poor or the ordinary taxpayer (see Martin/Schumann, 1997, 45). On the one hand national wealth and infrastructure which had been built up over decades and centuries is privatised for a fraction of the real value. At the same time the debts of these enterprises are very often cancelled, so that the new private owners start with a clean slate. In other words: they receive quasi-monopolies which allow them to increase profit margins, at the expense of the consumers, the former owners. The latter, of course, have to pay now for services which beforehand belonged to them. One of the most extreme cases of such squandering of state property, built up by many generations, is Great Britain. Large-scale privatisation, started under Thatcher but continued under Blair, very often served to finance overspending and tax cuts in order to win elections. In particular with regard to services where real competition is virtually impossible or possible only with a senseless duplication of infrastructure (such as gas, water, electricity, telecommunication or public transport) the public never benefits. The only winners are corporate shareholders who can cream off the profits.

On the other hand and rather ironically, it is of course the case that corporations and financial markets don't even dream of believing their own dogmas. The state, so they argue, has to step back whenever it stands in the way of excessive profit maximisation, whereas, of course, it has to be prepared to generously help those same corporations with subsidies, tax breaks and export credit guarantees (see Chomsky, 1992a, 108, 332). In this way, Transnational Corporations become more and more like parasites which on the one hand constantly blackmail nations into offering them tax incentives and free infrastructure with the threat of relocating elsewhere, while on the other hand cashing in on more and more subsidies and

---

[214] Quoted in John Vidal, 'The real politics of power', in *The Guardian. Society*, 30.4.1997, G2, 4-5, here 4.

making sure that, with a little help from worldwide tax coordination and evasion, they only show up profits where they don't have to pay any tax on them or where they even get rebates from the state. A recent study

> examining the profits and federal incomes of 250 of the nation's [US] largest and most profitable corporations over the 1996/98 period, reveals that 41 companies paid less than zero in federal income taxes in at least one year. (...) In those tax-free years, the 41 companies reported a total of US$25.8 billion in pre-tax US profits. Rather than paying $9 billion in federal income taxes at the standard 35 per cent rate, they enjoyed so many tax breaks that they received $3.2 billion in rebate cheques from the US Treasury![215]

Not to mention that, as a matter of course, corporations demand that all public services are accessible to themselves, their employees and children, at no charge (see Martin/Schumann, 1997, 7, 206).[216]

Yet should the market produce self-induced catastrophes its proponents naturally expect that the state and the taxpayers have to pick up the bill (see Martin/Schumann, 1997, 94-95 and Chomsky, 1992a, 332). Striking examples of this are the interventions to save the Mexican peso in January 1995 as well as similar actions by the IMF to 'save' the collapsing economies of the Southeast Asian Tigers in autumn 1997.[217] The consequence of all this restructuring is that "the state thus becomes an agency of bottom-to-top redistribution" (Martin/Schumann, 1997, 206). The German tax law for 1996 managed to equalise the cuts in public services with tax gifts to corporations and the self-employed (see ibid.).

---

[215] Study conducted by the Institute on Taxation and Economic Policy, Washington, quoted in 'No taxation but so much representation', in *The Ecologist*, 30 (2000/2001) 9, 9.

[216] Regarding tax evasion of big corporations see Martin/Schumann, 1997, 197 (BMW 198, Siemens 198 and 201, Daimler-Benz 201: in 1996 Daimler-Benz chief Jürgen Schrempp told parliamentary experts that "his company will no longer pay in Germany any taxes at all on profits. 'You won't be getting any more from us'.") and Roger Cowe, Lisa Buckingham, 'Murdoch's millions: Rupert's moving tax target', in *The Guardian*, 5.2.1998, 15 (tax evasion of 188 million Australian dollars in 1997 alone); regarding subsidies see Martin/Schumann, 1997, 201-202 (AMD, Opel, VW 202, Dow-subsidiary BSL 204, Daimler 205): "Even if one leaves out of account the traditionally subsidized sectors – agriculture, mining, housing and railways – it is cautiously estimated that subsidies to industry are costing more than 100 billion marks a year in Germany alone." (205)

[217] "The Mexico deal was both a disaster-aversion exercise (perhaps the boldest in economic history) and a brazen plundering of the tax coffers of contributor nations for the benefit of a wealthy minority." (Martin/Schumann, 1997, 70, see also 40-46) This is equally true for Southeast Asia where the intervention of the IMF cost taxpayers around the world in excess of US$130 billion (Alex Brunner, 'IMF warns Asian shockwaves worse than first feared', in *The Guardian*, 6.4.1998, 1). And all this not to help the newly unemployed masses in Thailand and so on, but to pay back the loans to the rich banks and governments of the rich countries.

Myth: *'Capitalism furthers plurality and diversity'*

Reality: This is so obviously a lie that I feel ashamed that I have to mention it. Spectacular examples are Coca Cola and McDonald's imperialism. There are no health or ecological reasons whatsoever why the wide variety of locally produced juices, drinks and foods all over the world should be replaced by surrogates scientifically proven to be harmful, nutrient-poor and sugar-rich. Yet the global economic order with only profits in mind wants it that way. This is why we are faced with a global monoculture in terms of food and drink which is harmful to people and the planet[218] *and* destroys local diversity:[219] "The expansive thrust of today's business is no more reconcilable with the principle of natural cycles, where unlimited growth is an alien concept, than economic globalism is compatible with cultural and ecological diversity." (*Greening the North*, 1998, 94)

This mechanism of destroying diversity, directly implemented by market logic, can possibly be seen most clearly with the worldwide deregulation of the media market. There are basically two ways to view the function of the media in today's societies. The first, democratic position ascribes the media the political task of providing information, of enabling citizens to form their own opinions and of controlling the powers that be, whether state or private, and of stopping them from abusing their power. This means that the media have not only rights, but also responsibilities. It is their duty, then, to represent the full spectrum of positions in a society and also to make available any information which empowers people to take control of their lives. This clearly means that such media need to be entirely independent from state or corporate control. Their chief goal is telling the truth, however difficult that may be. The other position assumes that media are no different from any other commodity, such as toothpaste or cars.[220] If you start from this latter assumption the logic of the market forces the media, dependent as they are on advertising (i.e. the corporate world), to produce more and more of the same unified pulp. Only those media products which demand nothing in terms of previous knowledge, are easy to digest and don't challenge any preconceived notions, only they can satisfy the *mainstream,* the highest possible numbers and can therefore guarantee maximisation of sales or viewing figures, i.e. profits.[221] By definition this

---

[218] See the famous McLibel case in Britain which succeeded in proving all the above claims against McDonald's. For a full documentation see http://www.mcspotlight.org/.

[219] Just two examples: "Some 97 per cent of the vegetable varieties recorded in 1903 have been lost. (...) With apples, the situation is slightly better. Of the 7,098 varieties in use in the 19th century, 'only' 6,121 or 86 per cent were lost. (...) Marketing needs (not really consumer preferences) call for a reduction of varieties." (*Factor Four*, 1997, 290)

[220] Regarding the difference between mass media as commodities and as liberating information tools see Williams, 1994, 5.

[221] "Research shows that competitive advertising-financed systems tend to compete for the centre ground of taste. Because they are selling audiences on the basis of cost per thousand they need to maximize audiences rather than plan programmes on the basis of the intensity of viewer satisfaction

means that any opinions and facts, which contravene the attempt to retain power by the current elite, are automatically marginalised.[222] Diversity and independence are sacrificed for uniformity.[223] This market logic also means that in terms of an increase in democratic control of the media we have nothing to hope from the planned expansion with the advent of digital TV. 500 or more channels do not mean more choice and better information. The intention behind it is an even more accurate carving up of the market into segments so that the advertising industry and its sponsors can reach with ever more precision the relevant consumer groups with focused temptations. The introduction of private television in Germany in 1985 has shown without a doubt that such a quantitative 'expansion' of provision does not lead to more programme diversity, but overall to a decrease in qualitative plurality: there was and is simply more of the same.[224] Similar things can be stated for the internet. On the whole the aim is not to empower citizens with adequate and in-depth knowledge about reality,[225] but to distract the 80 per cent of society who will sooner or later be amongst the losers from the current trends in globalisation –"tittytainment" is the order of the day: "Perhaps a mixture of deadening entertainment and adequate nourishment will keep the world's frustrated population in relatively good spirits." (Zbigniew Brzezinski, quoted in Martin/Schumann, 1997, 4)

---

across a range of different audiences. In comparison with public service systems there is a lower level of programme diversity." (Williams, 1994, 16)

[222] "To put it bluntly, the market is plutocracy, not democracy. Markets are run by establishments that safeguard their own freedoms, but do not confer them on others unless forced. Market-driven mass media like to speak in the name of the public but shun, marginalize, or criminalize public views not saleable to large groups of paying customers. (...) The principal challenge to democratic theory and practice today is the rise to dominance of a single, market-driven, advertiser-sponsored, and ideologically coherent media system claiming to represent diverse publics and invoking constitutional protection to pre-empt challenge to its controls." (George Gerbner, 'The USA and the Free Marketplace of Ideas', quoted in Williams, 1994, 18)

[223] "The lords of the global village have their own political agenda. Together they exert a homogenising power over ideas, culture and commerce that affects populations larger than any in history. Neither Caesar, nor Hitler, Franklin Roosevelt, nor any Pope has commanded as much power to shape the information on which so many people depend to make their decisions about, from whom to vote for, to what to eat." (Ben Bagdikian in a talk at the Berkeley School of Journalism, June 1989, quoted in Williams, 1994, 73)

[224] This was even clearly acknowledged by the Federal Constitutional Court in Germany which ruled in 1986 that private broadcasting corporations "are not capable of fulfilling the task of comprehensive information provision" since the economic necessities, i.e. market logic, force them "to broadcast programmes which are attractive to the masses and which at the same time maximise viewing and listening figures and minimise costs" (quoted in Michael Kunczik, 'Massenmedien und Gesellschaft', in *Privat-kommerzieller Rundfunk in Deutschland* (1992), ed. by Bundeszentrale für politische Bildung (Bonn, Bundeszentrale), 25f.).

[225] Which mass media, as research has shown, clearly don't provide: see Winfried Schulz, 'Medienwirklichkeit und Medienwirkung', in *Aus Politik und Zeitgeschichte*, 1.10.1993, 25.

Myth: 'Capitalism is fair'

Reality: Since we live in the post-1989 belief that "capitalism has won" we have largely forgotten that although it might have been the better system compared to Soviet-style state socialism (although that is not saying *much*), in itself it is the same old beast as ever, in other words the one which triggered workers' resistance and the rise of the socialist and other workers' movements in the first place. In a period where historical knowledge is considered superfluous since technology is catapulting us to ever more spectacular heights of development virtually every month – or so it is claimed – it is quite sobering and healthy to look back onto the struggles of our ancestors so as not to lose sight of what we are actually faced with. Capitalism, let's be clear about this, has never been a benevolent system. Whatever niceties it might have offered in the past were due to tough resistance struggles by its victims, not due to inherent generosity. Capitalism has always been the system to enrich the rich at the expense of everybody else. A very vivid account of this fact is given by Howard Zinn in his *A People's History of the United States*. At one point he describes the formation of oligopolies, created under the ideological heading of increasing competition and furthering the free market:

> And so it went, in industry after industry – shrewd, efficient businessmen build-ing empires, choking out competition, maintaining high prices, keeping wages low, using government subsidies. These industries were the first beneficiaries of the "welfare state". (1996, 251)

Fairness? Nothing could be further from the truth. The system was always built on exploitation, first at home, then during the first wave of colonialism abroad. This exploitation beyond our vision allowed the living standard to increase in 'devel-oped' countries, turning the former victims of the system into beneficiaries of the exploitation as well. Nowhere can this be seen more drastically than with the status symbol commodity *par excellence*, gold:

> Unlike any other metal, gold exemplifies a huge gulf between those who dig it from the ground and those who wear and use it. Gold has long been associated with wealth; pagan societies, for example, built goddesses made of solid gold. Around 85 per cent of gold is beaten into jewellery, and the wearing of gold is a symbol of individual wealth. Gold is also the most lucrative sector of the mining industry. Yet the conditions under which gold is often produced, and its effects on communities in gold mining areas, are a world away from the glamorous glit-ter. The oppression of gold miners is grim history, with miners being paid poor wages for working in unsafe conditions and often living in disease-prone metal shacks. (Madeley, 1999, 95-96)

The rise of the fair-trade movement has started to bring this point home. Where is the fairness, one is tempted to ask, if the worker stitching a pair of trainers for Nike (costing £59.99 in the UK) gets £2.40 whereas Nike is pocketing twice that amount in profits, spends £3.60 on advertising and Nike's CEO earns £1.1 million per annum?[226] Where is the fairness in a system which externalises all the true costs onto the poorer members of (global) society and/or nature? Where is the fairness and legitimation when one type of society declares itself the only legitimate one and destroys with utmost vengeance all others by forcing onto them

> the massive uprooting of humanity from traditional community life and work, the rendering extinct of ancient skills, values, and ways of thinking and feeling to make society into an instrument of efficient factory production – a process of which Marx said, "World history offers no spectacle more frightful". (Lummis, 1996, 55-56)?

We have already noted above that the neoliberal trickle-down effect serves the rich and widens rather than narrows the gap between rich and poor (see p. 109f.). Not only is the richest fifth of the world population the only segment that increased, in absolute terms, its income between 1960 and 1991, so that 85 per cent of the people had 15 per cent of the total world income to share between them; in the US, for example, the richest one per cent of the population increased its income share between 1975 and 1990 from 20 per cent to 36 per cent of the total (see *Human Development Report 1996*, 1996, 2). The same trend can be seen for American corporations. The mid-1990s saw a period of mild recession, with an increase in unemployment of 1 per cent, just 5 per cent more turnover, an increase in part-time employment of 13 per cent and a wage drop in real terms for employees.[227] Yet at the same time profits of American corporations rose by up to 22 per cent, the incomes of top managers rocketed by double-figure percentages and the value of the top 1,000 corporations rose 35 per cent.[228] No burden of responsi-

---

[226] For clothes it is even worse: on average, workers' wages are only 0.5% of the price we pay as customers (Labour Behind the Label (2001), *Exposed* (Norwich, NEAD), 4). See also: http://www. cleanclothes.org/campaign/shoe.htm and http://www.labourbehindthelabel.org/.

[227] "Between 1973 – when they reached their peak – and 1990, real hourly wages (leaving aside benefits) in the private business economy *fell* by 12 per cent, declining at an average annual rate of 0.7 per cent, and they failed to rise *at all* during the decade of the 1990s, up to 1997. Real hourly wages in the manufacturing sector had pretty much the same trajectory, declining at an average annual rate of 0.8 per cent, or a total of 14 per cent, between their 1977 peak and 1990, and also failing to rise at all during the 1990s. In the year 1997, real wages in the private business economy and in manufacturing were, respectively, at the same levels that they had been in 1965 and 1966!" (Robert Brenner, 'The Economics of Global Turbulence. A Special Report on the World Economy, 1950-98', in *new left review*, (May/June 1998) 229, 1-265, here 3)

[228] Noam Chomsky, 'From Containment to Rollback. Will Civilization die?', in *Z Magazine*, 9 (1996) 6, 22-31, here 25. From 1977 to 1994 the net family income of the poorest 20 per cent in the US fell 16 per cent, while the income of the richest 20 per cent rose 25 per cent and that of the richest 1 per

bility, no amount of overtime or extra work can ever provide an adequate justification for the absurd disparities in incomes which the capitalist system has by now created. In 1995 the ten best paid top managers in the US earned between 613 and 2,451 times the average yearly income of an American factory worker.[229] This means in other words that such a factory worker and his family could have lived off a single such top salary for almost two and a half millennia. Once one conceptualises these disparities in this way, it becomes immediately obvious how unjust and beyond any possible legitimation the property distribution in our societies has become. There is but one conclusion: "One may definitely conclude, writes Lester Thurow, economist at the Massachusetts Institute of Technology (MIT), that America's 'capitalists declared class war on their workers – and they have won it'." (Martin/Schumann, 1997, 119)

Thus the liberalisation and deregulation of the world economy do not actually contribute to a fairer distribution of the earth's wealth:

> So it is not at all the case that poor countries are robbing the rich of their prosperity. Actually the reverse is true. Economic globalization brings an ever greater share of the (growing) prosperity produced on a world scale to the privileged layers in North and South – wealthy owners of property and capital, highly qualified professionals – at the expense of the rest of the population. (Martin/Schumann, 1997, 153)[230]

Unfortunately this trend is unlikely to be reversed with the implementation of the Uruguay round of the GATT talks, termed "a corporate bill of rights" (Karliner, 1997, 10), which ended with the founding of the WTO (World Trade Organisation) and which is set to increase the gap between rich and poor nations even further:

> The main beneficiaries of the WTO thus continue to be the industrialized states, what are known as threshold countries, and multinational corporations. A study by the World Bank and the OECD anticipates an annual boost of US$195 billion when the GATT agreements are put into effect with over half (105 billion) going to industrial countries. These have also obtained additional advantages for them-

---

cent rose by as much as 72 per cent (Holly Sklar, 'Boom Times for Billionaires, Bust for Workers and Children', in *Z Magazine*, 10 (1997) 11, 32-37, here 34). And in 1995 British top managers awarded themselves salary increases of 19 per cent on average, while the average worker had to be happy to get an increase to level out inflation (Sarah Whitebloom, Lisa Buckingham, 'The gold diggers of Britain plc', in *The Guardian*, 1.6.1996, 40).

[229] Holly Sklar, 'Upsized CEOs. The truly greedy get greedier', in *Z Magazine*, 9 (1996) 6, 32-35, here 34.

[230] "In 1978, 54 per cent of disposable income in West Germany was allocated to wages and salaries. The rest went half to income from interest or profits, and half to pensions or social benefits. Sixteen years later, in 1994, the share of after-tax wages and salaries had fallen to just 45 per cent. Now a full third of national income goes to the paid non-labour of those who benefit from interest and corporate profits." (Martin/Schumann, 1997, 153; see also 239)

selves by means of limits on quantities, special terms, and delays in the removal of import restrictions and export subsidies. (*Greening the North*, 1998, 205)

If profit for the shareholders is the only aim, one has already isolated the fundamental problem of our current economy, as Will Hutton has shown for Britain. Squinting on the highest possible profits in the shortest possible time has led to an almost complete neglect of long-term perspectives, according to the motto "après nous le déluge". The long-term consequences of such a short-sighted policy are continuously filtered out (see Hutton, 1996). Top managers are brought in to push up the profit margin into double-digits through reorganisation, which usually means tens of thousands of sacked employees.[231] These "temp CEOs" stay for "an average term of only five years, collecting multimillion-dollar incentive packages on the way in, and multimillion-dollar golden handshakes on the way out" (Klein, 2001, 255). The state can then clear up the social misery these CEOs created which usually means that the taxpayers pick up the bill: externalisation of costs, it is called.[232] Schönborn, archbishop of Vienna, found the appropriate words for this phenomenon: it has been replicated in thousands of mergers all over the world in the past few years, but he was speaking of the merger between Sandoz and Ciba-Geigy to form the by now notorious Novartis:

"If two of the world's largest chemical corporations merge and create 15,000 job losses," he said, "even though both are in the best of shape, that is not a necessity decreed by the almighty God of the "free market" but is due to a few people's greed for dividends." (Martin/Schumann, 1997, 128)

In Britain, the privatisation of the so-called utilities (electricity, gas, water, telecommunications) has also generally led to a very fast get-richer bonanza for the top managers via disproportional salary increases and settlements, while at the same time services for the consumers either stayed the same or dropped in quality while prices soared.[233]

---

[231] "The clear message, at IBM as everywhere else, was that 'shareholder value' was the only yardstick of corporate success. (...) This logic explains why the staff at firms which make a regular profit must also be prepared for the worst." (Martin/Schumann, 1997, 121)

[232] And these funds become ever scarcer: "In our increasingly globalized economy, fewer and fewer fruits of growth are channelled through government to meet people's basic needs. Instead, while corporate profits soar, ever weaker governments have less and less resources to devote to social issues such as health, education and environment, relegating them to the brutal whims of the 'free market'." (Karliner, 1997, 44)

[233] Regarding water utilities: "Since privatisation in 1990, capital investment in new plant and distribution by the 10 water companies in England and Wales has gone down by £282 million (10%), while dividends have risen by £217.5 million (55%), profits by £510 million (36%), total boardroom salaries have shot up from £540,000 to £2.4 million (444%) and the total annual cost of providing water to you, the customers, has risen by £294 million (12%)." (Julie Stauffer (1996), *Safe to Drink?*

In Germany such irresponsible behaviour, which today – at least in business circles – seems to be the only acceptable kind, is a violation of the constitution. The Basic Law unequivocally requires that capital be used to the benefit of society as a whole: "Article 14, 2: Property imposes duties. Its use should also serve the public weal."[234] It only remains to be seen what happens first: a reorientation of the economy to take these duties into account or a change of the constitution which removes the above article completely. Until then the conclusion seems inevitable that there is a causal link between free trade and cut-backs in social security. Or in other words: "Justice is not an issue for the market to resolve; it is a question of power." (Martin/Schumann, 1997, 237)

The Quality of Your Water (Machynlleth, C.A.T. [=C.A.T. Publications New Futures; Vol. 8]), 71) Note also that Cedric Brown, former CEO of British Gas, received a salary increase of 75 per cent in 1995 (Martyn Halsall, 'Scalded fat cat exits', in *The Guardian*, 1.5.1996, 19) and a pension package worth £4.26 million (Simon Beavis, Chris Barrie, 'Gas "fat cat" gets more cream in £4m pension package', in *The Guardian*, 7.2.1996, 3). Also: Simon Beavis, 'Grid chiefs' £1.5m perk revives windfall row', in *The Guardian*, 13.1.1996, 36; Rebecca Smithers, 'Outrage over water chief's £1m retirement package', in *The Guardian*, 20.1.1996, 12; Joe Roeber, 'The real scandal is that bosses don't pay their way', in *The Observer*, 25.2.1996, Business Section, 7.

Another telling example is Railtrack, the privatised company responsible – until its recent bankruptcy – for the British rail network, renting it out to private providers. In 1996 Railtrack made £329 million in profits which were passed on to shareholders. The transport minister in charge, Labour's John Prescott, commented: "There is something seriously wrong with a system where the Government subsidised privatised rail companies to the tune of £1.5 billion, they paid fees to Railtrack, who then paid the money in profit to shareholders." (Paul Brow, 'Prescott points buses to the fast lane', in *The Guardian*, 6.6.1997, 10) At the same time Railtrack has been warned because it didn't invest as much into infrastructure as agreed when privatised (cf. Keith Harper, 'Investment warning for Railtrack', in *The Guardian*, 6.6.1997, 2), which led to the collapse of the rail system in Winter 2000/2001. All this might serve as a warning to all those who advocate (rail) privatisation in Germany, Switzerland and other countries.

[234] German original: Artikel 14, 2: "Eigentum verpflichtet. Sein Gebrauch soll zugleich dem Wohle der Allgemeinheit dienen." (*Grundgesetz für die Bundesrepublik Deutschland* vom 23. Mai 1949 (BGBl. S.1), zuletzt geändert durch Gesetz vom 19.12.2000 (BGBl. I S. 1755) [http://www. bundesregierung.de/top/dokument/Dokumentationen/Grundgesetz/ix4222_.htm/;   PDF-file: http://www. bundesregierung.de/downloads/GG.pdf]).

Myth: 'The Market guarantees fair and just prices'

> Thoreau knew what some have yet to discover: the difference between price and cost. (...) The prices we pay for food do not reflect the life we exchange for it or that which we will subsequently forfeit. This is so, in large measure, because life – biotic resources and the health of rural communities essential to a healthy agriculture and culture – is not included in our present accounting system, which instead tends to regard these "factors of production" as if they are as replaceable as worn-out machines. (...) The difference between price and cost is also a matter of honesty and fairness between those who benefit and those who, sooner or later, are required to pay. One effect of not paying full costs is that we fool ourselves into thinking that we are much richer than we really are. (Orr, 1994, 172-173)

Reality: Here we can witness another mix-up of cause and effect. If the basic economic conditions allow fair prices, then the market will produce them, but not the other way around:

> As a matter of fact, economic theory itself entails important reservations about the superiority of free trade. In his seminal book, *Trade Policy and Economic Welfare* (1974), W. Max Corden stressed that "Theory does not 'say' – as is often asserted by the ill-informed or badly taught – that 'free trade is best'. It says that, *given certain assumptions*, it is 'best'." Prominent among these assumptions is the smooth functioning of the price mechanism. Economists (...) make the point that trade is distorted, since prices do not reflect the full cost of production. Much less are prices reflecting the hypothetical costs of resource depletion and of environmental degradation. (...) Clearly, prices under the conditions of international competition are telling the exploiters, both locally and internationally, that they could make sufficient profits only when they were destroying the resource base. (*Factor Four*, 1997, 282-283; emphasis in the original)

If people took seriously the principle that the producer has to take responsibility for the full lifecycle of a product and that prices should tell the truth about all costs involved from design, raw-material extraction, production, transport, marketing and usage to disposal, the world economy would reorganize itself overnight into regional markets (see *Factor Four*, 1997, 289).

Today's reality is somewhat different. First, there are thousands of subsidies that distort the prices. The amounts of these subsidies are truly astonishing and explain to a large extent why the current system is tilted so drastically *against* sustainable economic behaviour. It is therefore worth pondering them in more detail. In the most authoritative study on worldwide subsidies to date, Myers summarises his findings as follows:

Total subsidies are estimated at around $1,900 billion per year, and perverse subsidies $1,450 billion.[235] Plainly, then, perverse subsidies have the capacity to (a) exert a highly distortive impact on the global economy of $28 trillion, and (b) inflict grandscale injuries on our environments. On both counts, they foster unsustainable development. Ironically the total of almost $1.5 trillion is two and a half times larger than the Rio Earth Summit's budget for sustainable development – a sum that governments dismissed as unthinkable.

Note that:

- The OECD countries account for two thirds of all subsidies and an even larger share of perverse subsidies.
- The United States accounts for 21 percent of perverse subsidies.
- The single sector of road transportation accounts for 48 percent of all subsidies and 44 percent of perverse subsidies. (1998, xvi)

It is worthwhile putting these figures in context, since otherwise it is difficult to gauge the region of magnitude we are talking about here. Myers again:

> *Perverse subsidies of $1.5 trillion are larger than all but the five largest national economies in the world. They are twice as large as global military spending per year, larger than the top twelve corporations' annual sales, and larger than the global fossil fuels industry or the global insurance industry.* (1998, 135)
>
> *Were just half of the perverse subsidies to be phased out, just half of the funds released would enable most governments to abolish their budget deficits at a stroke, to reorder their fiscal priorities in fundamental fashion, and to restore our environments more vigorously than through any other single measure.* (1998, 145; italics in the original)

Subsidies can have a positive effect on sustainable behaviour, for example in order to kick-start renewable energy usage. But most subsidies are set in place and then never revised, creating dependencies on unsustainable behaviour which ought to be changed. But even worse are perverse subsidies since they are not just passively, but actively undermining any reorientation towards sustainable economic activity. Their main features are summarised by Myers as follows:

- Economically they push up the costs of government, including higher taxes and prices for all. In turn, this means they aggravate budget deficits.
- They divert government funds from better options for fiscal support.
- They distort economies in numerous other ways. For instance, they undermine market decisions about investment, and they reduce the pressure for businesses to become more efficient.

---

[235] Perverse subsidies are defined as "exerting adverse effects on both the economy and the environment in the long run" (Myers, 1998, xiii).

- They tend to benefit the few at the expense of the many, and, worse, the rich at the expense of the poor.
- They can serve to pay the polluter.
- They foster many other forms of environmental degradation, which apart from their intrinsic harm, act as a further drag on economies. (1998, 140)

Let me just cite three examples to add a bit of flesh to the bone:

- Direct support for conventional energy sources is estimated at $200 billion worldwide – more than half the value of all the crude oil produced each year.[236]
- Not only are we not taxing many environmentally destructive activities, some of these efforts are actually being subsidized. More than $600 billion a year of taxpayers' money is spent by governments to subsidize deforestation, over-fishing, the burning of fossil fuels, the use of virgin raw materials, and other environmentally destructive activities.[237]
- American railway pricing that charges more to transport a tonne of recycled copper than of virgin copper; (…); over $30 thousand million a year in direct federal subsidies to the US energy system, $100 thousand million to EU agriculture, and probably more worldwide to transport – all are the tips of a whole seaful of icebergs on which even the most prosperous economy can quickly founder. (*Factor Four*, 1997, 190)

Anyone who talks about a fair and free market on this basis and for example rejects subsidies for renewable energy with the argument that all energy forms have to compete on equal terms in the market,[238] just displays either their abject ignorance or ideological stubbornness. Level playing field? I have quoted above (p. 77) the literally trillions of US dollars which have gone and the billions which still go into subsidising nuclear energy worldwide. This never made nuclear energy economically viable, but for decades it diverted funds from more sensible (and sustainable) projects such as renewable energy production. But this is not an isolated case. A German study has found that over a third of all subsidies are directly environmentally damaging whereas only 2.3 per cent actively enhance environmental protection (see *Zukunftsfähiges Deutschland*, 1996, 176).

A second area that distorts prices are tax structures which actively encourage environmentally damaging behaviour (such as company cars in the UK). This

---

[236] Christopher Flavin and Seth Dunn, 'Responding to the Threat of Climate Change', in *State of the World 1998*, 1998, 113-130, here 117.

[237] Lester R. Brown and Jennifer Mitchell, 'Building a New Economy', in *State of the World 1998*, 1998, 168-187, here 182.

[238] As has routinely happened, for example in the *Neue Zürcher Zeitung*, in the run-up to the referendum on a tax on non-renewable energy, to be used to promote the use of renewables in Switzerland (so-called Solarinitiative, rejected by a majority on 24.9.2000).

is the focus of ecological tax reform which attempts to implement the principles of producer responsibility and "the polluter pays" (see *Taking Nature into Account*, 1995).

Yet this previous point is closely linked with a more fundamental, third, problem, namely the inbuilt tendency of the current economic model to externalise costs. The true costs of environmental, social and health damage caused by, for example, cars or industry, are externalised, meaning that it is not the producers or users who are paying the costs they created, but the state, i.e. all the taxpayers. Two examples may illustrate what is meant here:

> After the catastrophe [accident of the oil tanker "Erika"] the surprised coastal inhabitants found that it was not the company Total – which owned the dangerous cargo aboard the "Erika" – that was covering the clean-up and compensation costs running into billions, but the French taxpayer. (...) Total, in the meantime, merged to form Total-Elf-Fina, and posted the biggest profits ever in its history during the first half of this year.[239]

Another example is private transport by car. Rather than have the users of private transport pay the entire costs for the building and maintenance of roads, airports and harbours as well as the environmental costs, states are, as we have seen, subsidising these activities to the tune of billions (see *Factor Four*, 1997, 288):

> In America, for example, the social costs of driving – related both to the conversion of fuel into smog and to congestion, lost time, accidents, roadway damage, land use and other side-effects of driving itself – are largely socialised. "External" (or, as Garrett Hardin called them, "larcenous") costs approaching 1 trillion dollars a year, perhaps a seventh of the American GDP, are borne by everyone but not reflected in drivers' direct costs. (*Factor Four*, 1997, 189)[240]

Sachs has clearly pointed out both the fallacy and the attraction of that kind of 'accounting':

> The power of the car excites the driver precisely because its prerequisites (pipelines, streets, assembly lines) and its consequences (noise, air pollution, green-

---

[239] Dorothea Hahn, 'Die Reeder regieren die Meere', in *die tageszeitung*, 1.11.2000, 4.

[240] We are also completely repressing the destructive history of private car transport (Winfried Wolf, 'The car as a mass murderer', in *The Guardian. Society*, 14.8.1996, 4-5). Wolf's extensive analysis in book form completely debunks every single myth (mobility, advanced technology, speed, efficiency, freedom, security, etc.) of our car society, provides the entirely destructive history of the advance of the car for city and neighbourhood development, particularly for children, the elderly and handicapped people, without having to resort to the ecological catastrophe called car. He thereby proves that there is not a single sensible justification for the private car as medium for mass transport (Wolf, 1996). See also: *Natural Capitalism*, 2000, 22-23 on the overall absurdity of the car society.

house effect) remain far beyond the view from behind the windscreen. The glamour of the moment is based upon a gigantic transfer of its cost: time, effort and the handling of consequences are shifted onto the systems running in the background of society. So the appeal of technical civilization often depends on an optical illusion. (1999, 15)

Within the current economic structure it is a fundamental law that, wherever possible, you have to pass costs onto others, and it is usually the weakest, those with no voice or without defence, who are landed with these externalised costs (see Hösle, 1994, 107). Karliner has traced the way this 'law' leads inevitably to "environmental racism": "Corporations and the U.S. government have consistently followed the path of least resistance and sited hazardous waste facilities, factories, freeways and railroads in poor communities of color." (1997, 27)

Furthermore the reductionist logic of the capitalist system and the economic theory underpinning it still doesn't allow, for example, the ecological costs connected with plundering and exploitation of resources to be brought into the balance sheet.[241] The waste and dangers created by this procedure are enormous:

Estimates by business reformer Paul Hawken indicate that the direct waste in the US economy – from subsidies to do silly and uneconomic things, to remedial costs for damage done elsewhere, to uncounted depletion and pollution, to other follies too numerous to mention – constitutes at least half the GDP. (*Factor Four*, 1997, 190)

Conclusion: as long as we are faced with "untruthful prices" (*Factor Four*, 1997, 189), as long as prices for non-renewable raw materials falsely suggest unlimited supply, and in view of the overwhelming evidence that an unrestricted market economy is not our saviour but the recipe for destruction of both quality of life and the biosphere, it seems indisputable that there need to be *Limits to competition* for ethical, political and ecological reasons (see The Group of Lisbon, 1995).

---

[241] See, as discussed above, the concept of the ecological backpack/MIPS which calculates these costs (Schmidt-Bleek, 1997, *Factor Four*, 1997, 237-244, and *Greening the North*, 1998, 75-78) or the ecological footprint (Wackernagel/Rees, 1996). A different approach was chosen by a team of scientists from all over the world. They calculated what we would have to pay for ecosystem services (such as gas, climate and water regulation, water supply, erosion control and sediment retention, soil formation, nutrient cycling, waste treatment, pollination, biological control, refugia, food production, raw materials, genetic resources, recreation and cultural services) at market prices, services which we now take for granted. It would cost us, on a conservative estimate, US$33 trillion per year globally, which amounts to 1.8 times the total global domestic product (Robert Costanza *et al.*, 1997, 253). The authors stress that many of these services cannot be adequately priced since they are vital for survival and in the true sense of the word invaluable, priceless: "The economies of the Earth would grind to a halt without the services of ecological life-support systems, so in one sense their total value to the economy is infinite." (ibid.)

Myth: 'Capitalism produces what people want and need'

> More fundamental still, the sustainability implication which
> few, if any, companies have yet considered relates to the issue
> of "need". Markets grow by inventing new needs, new wants,
> new desires. At some point, the sustainable development vector
> will intersect with this aspect of the market economy. (*Taking
> Nature into Account*, 1995, 96)

> Back to culture. Yes, actually to culture. You can't consume
> much if you sit still and read books. (Huxley, 1977, 62)

Reality: Of course it depends partially on the definitions of wants and needs, but if
we accept the Gandhian assumption quoted above that we can only start talking
about more exquisite 'needs' once the basic needs (food, shelter, clean water,
clothing) of the whole of humanity are satisfied, there is little doubt that capitalism achieves anything *but* this. On the one hand it actively destroys people's
livelihoods, their subsistence economies which enabled them to cover their basic
needs. On the other hand capitalism, due to its inherent drive to expand and grow,
is one gigantic machine of artificial need production.

Let me first concentrate on the overtly destructive first trend, even though
the latter might well be more destructive overall. Madeley (1999) and Bennholdt-
Thomsen/Mies (1999) have traced in great detail global restructuring which takes
away the means of people to provide for themselves and places them in dependency on the global market and in particular TNCs, but I would still like to give
you two concrete examples to illustrate what that means 'on the ground', as it
were: the first shows the absurd effects which follow if a functioning local or regional economy is forced to accept the rules of the global market, of how "things
are done properly":

> The forced use of packaging [in India] will increase the environmental burden by
> millions of tonnes of plastic and aluminium. The globalization of the food system
> is destroying the diversity of local food cultures and local food economies. A
> global monoculture is being forced on people by defining everything that is fresh,
> local and handmade as a health hazard. Human hands are being defined as the
> worst contaminants, and work for human hands is being outlawed, to be replaced
> by machines and chemicals bought from global corporations. These are not reci-
> pes for feeding the world, but for stealing livelihoods from the poor to create
> markets for the powerful. People are being perceived as parasites, to be extermi-
> nated for the "health" of the global economy. In the process new health and eco-
> logical hazards are being forced on Third World people through dumping geneti-
> cally engineered foods and other hazardous products. (Shiva, 2000, 17-18)

The second example goes directly to the heart of survival. Independence with regard to agricultural provision is a core element of self-determination (see p. 66),

yet globalisation, especially of the seed market – with biopirating seeds from traditional farmers and then patenting them –, pulls the rug from underneath these peasants' feet:

> The knowledge of the poor is being converted into the property of global corporations [through patents, etc.], creating a situation where the poor will have to pay for the seeds and medicines they have evolved and have used to meet their needs for nutrition and health care. (...) Patents and intellectual property rights are supposed to prevent piracy. Instead they are becoming the instruments of pirating the common traditional knowledge from the poor of the Third World and making it the exclusive "property" of Western scientists and corporations. (...) Sharing and exchange, the basis of our humanity and our ecological survival, have been redefined as a crime. This makes us all poor. (Shiva, 2000, 18-19; see also Shiva, 1997)

This leads to the well-known phenomenon that peasants formerly able to provide for themselves are locked into dependencies on seed merchants, fertilizer and herbicide/pesticide producers, and are forced to produce for the market, while starving themselves (see Shiva, 1991) or being forced into suicide.[242] In other words, the capitalist market economy is spectacularly failing to provide for the needs of these people. The reason for this, as we have seen above, is that it is not the market's intention at all to look after the needs of these people. Its job is to look after the 'needs' of the best-paying customers. That is where the inverse dependencies (with so far no consequences for power shifts) operate, leading to such ironies that countries which are amongst the poorest in the world provide many of the raw materials without which the 'wealth' of the 'developed' countries would be impossible:

> American corporations depended [in the 1970s, but equally so today – R.J.] on the poorer countries for 100 percent of their diamonds, coffee, platinum, mercury, natural rubber, and cobalt. They got 98 percent of their manganese from abroad, 90 percent of their chrome and aluminium. And 20 to 40 percent of certain imports (platinum, mercury, cobalt, chrome, manganese) came from Africa. (Zinn, 1996, 556-557)

Now let's take a closer look at those 'needs' of rich countries. First it has to be stated that even the basic needs

---

[242] "Pesticides kill. In Sri Lanka, they are taking the lives of distressed rice-paddy farmers who commit suicide by swallowing these poisons. Their desperation is fuelled by the dual burden of poverty and debt. Pesticide poisoning was the number one cause of death in four major paddy-farming districts of the country in 1997." (Lasanda Kurukulasuriya, 'At debt's door', in *New Internationalist*, No. 331, January/February 2001, 16-18, here 16)

are satisfied in industrial societies through longer technological chains requiring higher energy and resource inputs and higher creation of waste and pollution, while excluding large numbers of people without purchasing power and access to means of sustenance. (Shiva, 1992, 190)

So in terms of resource efficiency per need satisfied the so-called 'developed' countries have a dramatically worse record than poor countries. Yet the problem is the second layer of 'needs':

Affluence and overproduction generate new and artificial needs and create the impulse for overconsumption, which requires the increased exploitation of natural resources. Traditional economies are not "advanced" in the sphere of wasteful consumption, but as far as the satisfaction of basic and vital needs is concerned, they are often what Marshall Sahlins has called "the original affluent society". (Shiva, 1992, 190)

There are two vital points to be made here. Firstly, one very often hears the argument that, after all, everybody wants Coca Cola, McDonald's, Levis jeans and four-wheel drives, if offered the choice. But this is hardly surprising. If a society does whatever it can to blind its citizens to the real costs and impacts of their actions, if it is so organised that not the consumers of these products but the producers in far-away countries bear the real costs, there is little incentive *not* to want these things:

The temporal, spatial and personal separation of utilities and costs – the separation of an act committed now from the suffering that ensues, or the non-intersection between advantages that are privately consumable and disadvantages that have to be borne collectively – is an exceedingly seductive characteristic of modern scientific technologies.[243]

Yet "none of these brilliant accomplishments of industrial technology function without the massive consumption of 'free' natural resources and without the expulsion of waste, poisons, noise and stench."[244]

Or as Susan George has put it: "The present reality is that by the time a product reaches the market, it has lost all memory of human or natural abuse." (*The Lugano Report*, 1999, 177)

But there is, of course, not only this negative strategy to blind people's view. One of the most flourishing and ever more important industries in our societies is the advertising and PR industry. In America today, there are more PR people employed than there are journalists (see Stauber/Rampton, 1995, 2-3).

---

[243] Otto Ullrich, 'Technology', in *The Development Dictionary*, 1992, 275-287, here 283.
[244] Ibid. 281.

Despite the constantly reiterated claims that industry is just the servant of the consumer and the consumer calls the shots, it is absolutely clear that it is the other way round. Not consumer pressure forces Sony to throw a certain number of new products onto the market every year; it is 'market logic' and the aim of creating new markets or penetrating already saturated markets. However, real fulfilment of deep human desires is not achieved:

> The attempt to satisfy the full spectrum of human needs through the production and consumption of goods has failed. Those dimensions of life that are important to people – whether West or East, North or South – such as ties of affection with other people and a sense of esteem in society, cannot be replaced effectively by material consumption.[245]

But

> the boundless dynamic of production in industrialism is so structured that material needs are created faster than the conditions for their gratification. There arises, therefore, the phenomenon of permanently frustrated people caught in an endless spiral of needs.[246]

This in turn leads to a feeling of "impoverishment" with relation to what is on offer:

> This poverty occurs when people cannot have things they had never needed or wanted until these things were invented. Somebody invents the refrigerator, or the automobile, and succeeds in having it established as a minimum condition for ordinary living. This is a case not of meeting an existing need but of restructuring a society so as to establish a need where there had been none before, so that now the people who cannot buy this thing, including those who had never before dreamed of owing it, are to that degree impoverished. Through this process, people whose absolute standard of living does not change at all are driven deeper and deeper into "poverty" by changes that occur in distant places and over which they have no control. It is easy to see that this kind of poverty is not reduced by industrial development but is generated by it, and generated by it endlessly. Development does not bring people "freedom from want"; rather, it operates to keep people in a state of perpetual domination by want. (Lummis, 1996, 73)[247]

---

[245] Ibid. 278.

[246] Ibid.

[247] Helena Norberg-Hodge has experienced this process first hand in Ladakh: while Ladakhis would say "We don't have any poverty here" at the time when the region was first opened up to tourism (1975), she heard the same people say in 1983: "If you could only help us Ladakhis, we are so poor" (2000, 101).

If the system were consumer-driven, it would be impossible to account for the huge advertising budgets of corporations (McDonald's alone spends more than US\$2 billion per year on worldwide advertising)[248] and the fact that most TNCs today spend more on PR and advertising than on research and development:

> Few adults [let alone children] are capable of resisting, day in and day out, the relentless, sophisticated marketing ploys that some of America's most creative minds have designed, aided by professional psychologists and anthropologists paid to advise corporations on how to manipulate consumer behavior. (*Fool's Gold*, 2000, 32)[249]

If consumers really did need the products, they would go out and buy them on their own initiative. It quickly becomes clear when you listen to the corporate side talking in private that, of course, this excuse of consumer needs is the fig leaf worn by the corporations to justify unsustainable overconsumption, expansion of growth and accumulation of profits. *These* are the driving motor, not satisfaction of needs – witness the shift from products to brands, from manufacturing to marketing where companies like Nike do not own any factories any more.[250] In a talk to the American Association of Advertising Agencies, Edwin Artzt, CEO of Proctor & Gamble, one of the biggest advertisers in the world, talked freely about their strategy to constantly inundate the consumer with about six to seven ads per month. To do this, corporations need TV stations with high viewing figures and Proctor & Gamble spends about 90 per cent of its \$3 billion budget on these ads.

---

[248] The annual global spending on advertising amounted to US\$435 billion in 1998, "up sevenfold since 1950, growing a third faster than the world economy" (*Human Development Report 1998*, 1998, 63).

[249] For an account of how far marketing departments go in order to compel people to buy things they don't need, including the manipulation of visual, audio, tactile and olfactory senses, see Frank Mazoyer, 'Der Lidschlag des Konsumenten', in *Le monde diplomatique [deutsch]*, 6 (2000) 12, 2. These manipulative tools are more and more targeted at children, so as to turn them into loyal brand consumers for the rest of their lives; see here the verdict of Chief Justice Bell in the famous McLibel trial where two activists, threatened by McDonald's, used the longest libel trial in UK history to thoroughly expose to public scrutiny every possible aspect of the worldwide business practices of the fast food TNC and came down with devastating condemnations of the company, particularly on ethical and environmental issues (for more details see the 'McLibel Support Campaign', with full transcripts of all court documents and background information on all relevant issues [http://www.mcspotlight.org/]). With regard to abuse of advertising Chief Justive Bell ruled that McDonald's "exploit children by using them, as more susceptible subjects of advertising, to pressurise their parents into going to McDonald's" (Summary of the judgement, read in Open Court on 19th June 1997 [http://www.mcspotlight.org/case/trial/verdict/ verdict7_sum.html]).

[250] "After establishing the 'soul' [i.e. brand] of their corporations, the superbrand companies have gone on to rid themselves of their cumbersome bodies, and there is nothing that seems more cumbersome, more loathsomely corporeal, than the factories that produce their products." (Klein, 2001, 195-275, here 196)

Artzt went on to talk about the threat to this hegemony from the internet, pay-TV, CD-ROMs and computer games. He then suggested that the advertising industry reapply the lessons from the past. When the radio was invented the advertisers first had "to take control of the programming" and ensure that those programme types were sponsored that made the placement of their ads most effective. His conclusion: "Our job today is to take control of the electronic networks and to force the internet to operate according to our interests."[251] I for one can detect a lot of hard-nosed economic feel for the improvement of shareholder profits, but not much regard for consumers' needs. Orr has aptly spelt out the underlying poverty of arguments such as these:

> One encounters a logic that goes like this: (a) Everyone wants what we, the rich, have, which is to say wealth; (b) who are we to say they should not have it?; (c) therefore, we cannot deny "progress" and the human desire for material improvement. Not much is said of the roughly $450 billion spent worldwide on advertising each year to manufacture wants. But that quibble aside, the unstated assumption is that we can summon neither the civic and moral wisdom to create a more equitable distribution of wealth nor the wit to redefine well-being in a less stuff-oriented and ecologically destructive manner. (Orr, 1994, 76)

It is a rather bizarre feat of humankind that it can invent the most stupendous technical wonders (such as the atomic bomb or flying to the moon), yet it cannot cope with the rather straightforward task of solving the contradiction between an excessive lifestyle, propagated worldwide as the pinnacle of human achievement, and a finite world:

> The system, in its irrationality, has been driven by profit to build steel skyscrapers for insurance companies while the cities decay, to spend billions for weapons of destruction and virtually nothing for children's playgrounds, to give huge incomes to men who make dangerous or useless things, and very little to artists, musicians, writers, actors. (Zinn, 1996, 624)

It says a lot about our mental state if we believe that the virtual economy is the real economy and the basic economy which guarantees all our survival is somehow negative and should be discarded:

> When growth increases poverty, when real production becomes a negative economy, and speculators are defined as "wealth creators", something has gone wrong with the concepts and categories of wealth and wealth creation. Pushing the real production by nature and people into a negative economy implies that production of real goods and services is declining, creating deeper poverty for the millions

---

[251] Dan Schiller, 'Wer besitzt und wer verkauft die neuen Territorien des Cyperspace?', in *Le monde diplomatique [deutsch]*, 2 (1996) 5, 4-5.

who are not part of the dotcom route to instantaneous wealth creation. (Shiva, 2000, 17)

That we are not talking of real needs can be seen almost every day, whenever you walk past a billboard or open a newspaper. Only a fraction of the products advertised there are needed. And even if they are, you might need to buy them once or twice in a lifetime, say a fridge or a bicycle. Yet of course the implication is that you need the newest model, even if the old one is still working perfectly. The up to five generations of mobile phones thrown onto the market within one year speak for themselves.[252] These are signs of a saturated society which has completely lost its plot, its place in the world and its sense of responsibility for the global commons: only then can you be hooked on such a destructive lifestyle and at the same time believe that you are amongst the most advanced examples the human species has ever produced.

> The modern economy is propelled by a frenzy of greed and indulges in an orgy of envy, and these are not accidental features but the very causes of its expansionist success. (...) If human vices such as greed or envy are systematically cultivated, the inevitable result is nothing less than a collapse of intelligence. (Schumacher, 1993, 18)

Let me give a recent example of the collapse of intelligence Schumacher is talking about. In spring 2000 it became clear that the Longbridge factory of the British car manufacturer Rover, owned by BMW, would be closed. There was a massive uproar in the media, long marches of workers protesting, a good dose of the old war spirit (it's those Germans again who are responsible for all our problems), yet despite a detailed observation of the entire debate I could find only one single article which actually spelt out the truth of the matter and put it into context. In a comment in *The Guardian* George Monbiot reminded us of various facts we tend to forget in such instances, namely that Longbridge had long been subsidised by government money (£3.5 billion since 1975) and that private car transport is entirely unsustainable, costing the UK £15 billion per year caused by congestion alone. In other words, any sensible person could only applaud the closure of a car manufacturing plant, *any* car manufacturing plant. This statement immediately leaves you standing accused of inhumanity: what about the poor workers? Correct: numerous studies and actual experience show that investment into renewable

---

[252] Quite apart from the fact that there is a hefty environmental and human cost attached to mobile phones, as usually externalised. The rare precious metal columbite-tantalite (col-tan), vital for their production (as for computer chips and jet engines), is one of the main reasons behind the wars in central Africa, and its extraction wreaks havoc in the rainforests of the Congo and elsewhere (see Karl Vick, 'Vital ore plays crucial role in Congo's war', in *Guardian Weekly*, 5-11.4.2001, 37, and Dominic Johnson, 'Erzfeinde im Coltan-Rausch', in *die tageszeitung*, 22.12.2000, 4).

energy production and alternative forms of transport creates more, better skilled, more interesting, better paid and less recession prone jobs. Rather than being dependent on a few multinational car corporations the workers would then be a good step closer to direct control of their destiny.[253]

Car manufacturing is built on the assumption that ever more cars are needed and that people switch to new cars ever more quickly. This means that by definition it is an unsustainable activity, and therefore, as Schumacher shows, strictly speaking an uneconomic activity:

> From an economic point of view, the central concept of wisdom is permanence. We must study the economics of permanence. Nothing makes economic sense unless its continuance for a long time can be projected without running into absurdities. There can be "growth" towards a limited objective, but there cannot be unlimited, generalised growth. (...) The cultivation and expansion of needs is the antithesis of wisdom. It is also the antithesis of freedom and peace. Every increase of needs tends to increase one's dependence on outside forces over which one cannot have control, and therefore increases existential fear. Only by a reduction of needs can one promote a genuine reduction in those tensions which are the ultimate causes of strife and war. (Schumacher, 1993, 20)

If we want to regain control over our needs rather than let capitalist industry decide for us what we 'ought to need', we have to reorientate our values and lifestyles away from material consumption to quality of life (see p. 57).

This assertion that industry bears a crucial part of the responsibility for unsustainable overconsumption patterns shouldn't be misunderstood as an exoneration of the consumers. Just as it is clear that without the enthusiastic complicity of the German people Hitler could never have pushed through his absurd policies in the Third Reich, so it is abundantly clear that consumerism is a drug of convenience, which silences the realities of life and makes us believe that we can evade them. Combined with the constant reiteration in technology and product promotion that anything which is hard is to be avoided, it cheats us into thinking that everything has to be fun, every whim has to be fulfilled *here and now*, and that nothing in life has to be paid for by dedication, concentration and personal input. This attitude is underpinned very effectively, as Neil Postman has shown, by TV's encouragement of passivity. Everything on TV can be consumed at no cost in terms of activity, knowledge or personal input. It encourages passive consumption in the extreme (see Postman, 1987), so that the following attitudes are fostered: care for the elderly has to be delegated to homes since it is too burdensome; handicapped children have to be aborted via prenatal screening since it's such an inhuman chore to look after them for a whole life (nobody seems to ask handicapped people themselves whether they regard their life as an "unworthy

---

[253] See George Monbiot, 'Car workers rightly doomed', in *The Guardian*, 27.4.2000, 24.

life" (to use the Nazi term) as well);[254] household chores have to be done by machines since they are such hard work that it is too much to ask anybody to do things like this; walking or cycling three miles is so exhausting that nobody could possibly be asked to do it; learning in school is too demanding, so let computers do the learning for the kids; the list could go on and on. We never ask to know the costs of our deferral strategies, whether they actually are worthwhile and whether we might not lose more than we gain; our apparently unstoppable urge to gratify our lethargy is a mechanism of denial: we don't want to be confronted with the consequences and side effects of our actions, nor face up to the certainties of life (i.e. our mortality); we don't want to face up to the limits of our ecosphere, nor to the fact that all our intoxication with consumer goods cannot produce happiness, and is a running away from the crucial question of the meaning of life. But facing up to these realities of life can be unpleasant, burdensome, and hard work. And this is the reason why we so easily fall for the promises of a paradise of complete laziness, without realising that this state has a lot more to do with the inertia of death than with life:

> The decisive victory of the "Disney-colonization of global culture" rests, according to Barber [director of the Walt Whitman Center at Rutgers University in New Jersey], upon a phenomenon as old as civilization itself: the competition between hard and easy, slow and fast, complex and simple. The first term in each of these oppositions is bound up with amazing cultural achievements, while the second corresponds to "our apathy, weariness and lethargy. Disney, McDonald's and MTV all appeal to the easy, fast and simple." (Martin/Schumann, 1997, 15)

We shall see below that this attitude has devastating consequences for education in general.

Myth: 'The best tool for measuring wealth is GNP/GDP'

> The common criterion of success, namely the growth of GNP, is utterly misleading and, in fact, must of necessity lead to phenomena which can only be described as neo-colonialism. (Schumacher, 1993, 160)

Reality: One of the clearest indicators that nothing much has changed – despite the greenwashing claims quoted above by Shell and others, by people like the CEO of the largest Swiss bank that sustainability has permeated and transformed the business community – is the fact that the notion that growth is *not* sustainable has not taken hold at all, either in the economy or with the general public.

---

[254] Ahia Zemp, 'Vorgeburtliche Diagnostik', talk at the Bernhard Theater Zurich, 13.11.2000.

Certainly one point that most corporate environmentalists seem to hold as non-negotiable in their "growth equals sustainable development" scenario is that consumerism and consumption should continue to expand around the world. For all their talk of sustainable development, it is extremely rare to hear the leaders of transnational corporations speak of the need for anyone, anywhere, to consume less. Rather, they have embarked on a series of "green" marketing campaigns tailored to reassure the public that it can keep consuming with a clear ecological conscience. But this simply isn't the case. Globalized patterns of growth, geared to extending stratified consumption patterns, are diametrically opposed to ecological sustainability and social equity. (Karliner, 1997, 45)

While experts are unequivocally clear and anybody even vaguely familiar with the notion of sustainability knows that GNP/GDP is, if anything, a very adequate measurement of *unsustainability*, it is still and exclusively the measure used to talk about the 'success' of economies.

The GDP is simply a gross tally of money spent – goods and services purchased by households or government and business investments, regardless of whether they enhance our well-being or not. Designed as a planning tool to guide the massive production effort for World War II, the GDP was never intended to be a yardstick of economic progress; yet, gradually it has assumed totemic stature as the ultimate measure of economic success. When it rises, the media applaud and politicians rush to take credit. When it falls, there is hand-wringing and general alarm. (Cobb *et al.*, 1999, 1)

Even today, the league tables of economically successful countries are determined by GNP growth.[255] In fact, if you start to question GNP you encounter almost hysterical opposition: "As we have stated throughout this report, a yardstick like SNI (Sustainable National Income) represents something of a threat to a society such as ours, oriented as it has been towards traditional growth of GNP." (*Taking Nature into Account*, 1995, 208)

Yet what exactly are we measuring with this all-powerful indicator? "To the GDP, every transaction is positive as long as money changes hands. No wonder the GDP rises continuously, adding everything as a gain, making no distinction between costs and benefits, well-being or decline." (Cobb *et al.*, 1999, 2) This leads to dramatic distortions:

---

[255] Take your pick, any day of the year, and you will find a business story that either bemoans the slow growth of GNP (for example, 'US growth slowes further', in *BBC News Business*, 29.3.2001 [http://news.bbc.co.uk/hi/english/business/newsid_1249000/1249838.stm]) or celebrates high growth (Shanker KC, 'Economic Growth with a Human Face', in *The Rising Nepal*, 30.9.2001 [http://www.nepalnews.com.np/contents/englishdaily/trn/2001/sep/sep30/features.htm]).

Marilyn Waring has convincingly demonstrated that the bulk of the work done on this planet is *not* included in this indicator, namely the work of housewives and mothers, the work of subsistence peasants and artisans, most of the work in the informal sector, particularly in the South, and, of course, the self-generating activity of mother nature. All this production and work does not *count*. On the other hand all destructive work – like wars, environmental and other accidents, oil spills, arms production, trade and so on – is included in GDP, because it "creates" more wage labour, more demand and economic growth. The oil spill of the *Exxon Valdez* along the Canadian Pacific coast some years ago has resulted in the biggest rise so far in Canada's GNP – because it required an enormous amount of work to undo the damage caused by this catastrophe. (Bennholdt-Thomsen/Mies, 1999, 56-57)

There is a whole series of deeply ingrained quasi-religious beliefs which cannot be supported by facts or historical developments but which still hold their ground:

All the agendas of all the world's political and economic bodies call for growth (...), because of an unabashed conviction that this is always good for the world. But this is simply not the case: a steady growth of output does not necessarily lead to more jobs or a better environment, it does not combat famine or promote social security, neither does it improve education or public health. On the contrary, most of these aspects of welfare seem to suffer under unrestricted economic expansion, which has become a law unto itself. (*Taking Nature into Account*, 1995, 3-4)

As soon as you start to apply more realistic measurement tools as have been developed in recent years – such as the Human Development Index (HDI) (see *Human Development Report 1998*, 1998) or the Index of Sustainable Economic Welfare (ISEW), developed by Daly and Cobb (1989) – the results are, not surprisingly, rather different:

ISEW calculations for the United Kingdom show that quality of life has been declining since 1973 and by now has nearly fallen to the level of 1960 despite growing GNP – and all this with strong population growth. In the US the ISEW has stabilised since about 1970; calculated per capita it has been falling since 1980. In Italy as well, the gap between GNP and quality of life is growing, as the WWF has shown. Even though ISEW and GNP were roughly the same in 1960, today's ISEW is equivalent to Italy's GNP of 1970.[256]

---

[256] Translated from the newer German edition of Wackernagel/Rees (1996): Mathis Wackernagel/ William Rees (1997), *Unser ökologischer Fußabdruck. Wie der Mensch Einfluß auf die Umwelt nimmt* (Basel; Boston, Birkhäuser), 130. See also *Taking Nature into Account*, 1995, 149-151.

If we use the Genuine Progress Indicator (GPI), developed by Redefining Progress, we can see a similar picture: during all of the 1980s and 1990s GDP in the US rose, but GPI fell by between 1 and 2.1 per cent annually (Cobb *et al.*, 1999, 5-22). It becomes clear that we are fooling ourselves if we believe that as long as the GNP is used as a performance tool we will be able to redirect economic activity towards sustainability. The GNP figures will always mislead us. The ISEW, for example, shows us that in countries such as the US where the gap between rich and poor is increasing,[257] the overall quality of life is decreasing for the majority of people, even when GNP grows (*Factor Four*, 1997, 272). While we are hooked on a purely quantitative measure of only selected activities we will not be able to base sensible decisions on it which enhance quality; therefore we need to replace it with an indicator such as GPI which "differentiates between what most people perceive as positive and negative economic transactions, and between the costs of producing economic benefits and the benefits themselves." (Cobb *et al.*, 1999, 3)[258] It would obviously be important that not just national economies but also corporations and households start to practise full accounting which does not simply forget or externalise costs and consequences. Only then will we not be fooled any more by jubilant messages which report staggering growth rates: we will then know instantly that traditional growth equals destruction.

Box 6: The GDP is padded with fat – ours

Including agriculture, restaurants, and the like, $700 billion flowed through the U.S. food industry last year (...) The thriving food industry nourishes what Marion Nestle, the head of the Department of Nutrition at New York University, calls the 3,800-calorie-a-day problem. America's food producers produce enough food to supply 3,800 calories every day to every American. However, the average woman only needs 2,000 calories a day, the average man 2,500, and children even less. So what happens to all of this extra food we produce? Some is exported, but the food industry spends $10 billion each year in direct advertising and another $20 billion on coupons, games, and other gimmicks trying to convince Americans to absorb the surplus calories. Psychologists have found that when food is put in front of people, they will eat it – whether they are hungry or not. (...) And so we grow. As the GDP goes up, so does the percentage of overweight Americans. According to the Third National Health and Nutrition Examination Survey, over half (55%) of American adults are currently overweight or obese (severely overweight). (...) Indeed, over the last two decades, the number of overweight children has increased by more than 50%, and the number of obese children has nearly doubled to roughly 14% of children and 12% of adolescents. (...) An estimated 300,000 Americans die each year from the combined effects of an unhealthy diet and inactivity. These health misfortunes of Americans pump up the GDP as well. Medical spending on diseases associated with obesity amounted to $51.6 billion in 1995, including medical costs for obesity-related heart disease, cancer, stroke, and hypertension. These costs represent 5.7% of national health expenditures within the United States. Although this spending contributes to the GDP, it would be tough to argue that it makes us better off. (...) More than two-thirds of Americans are trying to lose

---

[257] "From 1973 to 1993, for example, while the GDP rose by 55%, real wages declined by 3.4%. In the 1980s alone, the poorest fifth of American families lost 0.5% of their income each year, while the top 5% of households increased their real income by 3.9% per year." (Cobb *et al.*, 1999, 3)

[258] See also the Sustainable National Income (*Taking Nature into Account*, 1995, 206-230).

or maintain their weight. (...) They spend $33 billion each year on weight-loss products and services. (...) While many Americans eat too many calories and spend money trying to lose the excess weight that results, 10% of Americans are going hungry or do not have enough food for an active and healthy life. (...) But the GDP is blind to how food is distributed. It goes up just as much when some go hungry and some overeat as when everyone has a reasonable amount.

Clifford Cobb, Gary Sue Goodman and Mathis Wackernagel (1999), *Why bigger isn't better: The Genuine Progress Indicator – 1999 update* (Oakland, Redefining Progress), pp. 25-28.

Myth: 'The capitalist world order and its institutions produce wealth for all through globalisation'

> In premodern times, the maldistribution of wealth was accomplished by simple force. In modern times, exploitation is disguised – it is accomplished by law, which has the look of neutrality and fairness. (Zinn, 1996, 235)

> Moreover, in a closed space with finite resources the underconsumption of one party is the necessary condition for the overconsumption of the other party. (Sachs, 1999, 168)

Reality: Globalisation has been the catchword for the last few years and mainstream politics has usually presented a positive attitude towards it as indispensable. In fact, people like Tony Blair have had recourse to the old Thatcher dictum TINA (There is no alternative) to stop any discussion about the relative merits of the concept.

If you look at globalisation from a sustainability perspective you are presented with a rather different picture:

> It is not enough only to point to the effects and symptoms of the existing unsustainable global economic and geopolitical system (e.g. desertification, pollution, poverty, injustice). It is necessary also to highlight the causes and the structural barriers to achieving ecologically sustainable, equitable human development, which include: an inequitable global economic order, reinforced by the Bretton Woods financial institutions, such as the International Monetary Fund (IMF), and the General Agreement on Tariffs and Trade (GATT); as well as unequal terms of trade, dominance over the U.N. by the G7 countries, and the global arms race which still accounts for some one trillion dollars annually. (*Taking Nature into Account*, 1995, 161)

All the main features of globalisation – free trade, removal of any special local rules for anything from goods to basic services such as water and education,[259]

---

[259] See the new campaign within the WTO to integrate all services from insurances, banking and telecommunications to education, health and basic services such as water into the free trade regime: Scott Sinclair (2000), *GATS: How the WTO's New 'Service' Negotiations Threaten Democracy* (Ottawa, Canadian Centre for Policy Alternatives [CCPA]), Maude Barlow, 'The Last

integration of every product and every country into the world market – seem to have a detrimental effect on both the poor people in the South and the health of the biosphere. One cannot talk about globalisation without realising that in the first instance this is a harmonisation process, but in the worst possible sense. It means that everything has to be dealt with according to one set of rules. This by definition is harmful to cultural, economic *and* bio-diversity. Yet, more pressingly, it also means that the most powerful call the shots and decide on the rules. No wonder that independent assessments of the process constantly point out that the real beneficiaries of globalisation are not the poor for whose help, in mysterious benevolent ways, the 'developed' countries have allegedly invented the process, but precisely the rich countries in the North, the TNCs and the rich elites in poor countries of the South. The rest, i.e. the majority, lose out. But this means that we need to keep our eyes on the most important players in the game: the US, the TNCs, the World Bank, the IMF and the WTO. In fact, if you start looking into the history of the Bretton Woods institutions (IMF and World Bank) and then try to chart the role of the WTO and the power of TNCs over the politics of the US (and most other industrialised countries – see *Europe Inc.*, 2000) you quickly start to realise that all these tools have one overarching aim: to secure the global economic (and political) hegemony of the United States. It started immediately after World War II. The Marshall Plan – still mythologised in most history books as the altruistic arm stretched out by charitable America to help the poor victims of a Europe destroyed by war – was, from its inception, a strategy that served clear economic and political aims: the economic aim was to create markets for the overproduction of US industry. In other words, rather than enable the European countries to build up independent industries, they should be locked into dependency on the US economy (see Chomsky, 1992a, 47-48, 332-348). The political aim, which was pushed through vigorously, was to build up a conservative bulwark against the Soviet bloc. All imaginable measures were used: from bribing France with the biggest share of Marshall Plan aid to accept the reindustrialisation and rearmament of Germany, to CIA support for anti-communist candidates and food distribution to non-communist constituencies, to linking the provision of aid to the condition of a non-leftwing government (as happened in France and Italy), to the crushing of grass-roots unions and their replacement by US-loyal cadres.[260]

Frontier', in *The Ecologist*, 31 (2001) 1, 38-42 and CEO/Transnational Institute, 'GATSwatch.org - Critical Info on GATS' [http://www.gatswatch.org/index.html].
[260] See Melvyn Leffler, 'The United States and the Strategic Dimensions of the Marshall Plan', in *Diplomatic History*, 12 (1988) 3, 277-306 and Thomas McCormick, '"Every System Needs a Center Sometimes". An Essay on Hegemony and Modern American Foreign Policy', in *Redefining the Past. Essays in Diplomatic History in Honor of William Appleman Williams* (1986), ed. by Lloyd C. Gardner (Corvallis, OR, Oregon State University Press), 195-220.

But the World Bank[261] and IMF always followed the same aims and "were not created with poverty alleviation in mind":

> They were designed at the United Nations Monetary and Financial Conference at Bretton Woods, New Hampshire, in July 1944, to fulfil quite another agenda. To cite Henry Morgenthau, then US Treasury Secretary and president of the conference, the purpose was, "the creation of a dynamic world economy", to sustain the domestic American economy's continuous expansion by ensuring it sufficient access to foreign markets and raw materials.[262]

And the younger WTO is no different:

> From the free market paradigm that underpins it, to the rules and regulations set forth in the different agreements that make up the Uruguay Round, to its system of decision-making and accountability, the WTO is a blueprint for the global hegemony of Corporate America.[263]

So rather than falling for the lofty reassurances, routinely reiterated by the world's mainstream media, we should be very careful about any pronouncements of altruism and benevolence, such as helping the poor through globalisation or feeding the hungry with genetically modified organisms (GMOs). Always first have a close look at the interests behind these claims, and whether, in fact, these (corporate, elite or state) interests might not be the main or sole beneficiaries of the proposed measures. There is little doubt that in both the above examples – GMOs[264] and globalisation – this is the case, in fact they are both closely linked.

   I have already mentioned that it seemed clear when the WTO was created that the rich countries would cream off US$105 billion of the expected US$195 billion created in new trade (*Greening the North*, 1998, 205). Yet at the same time

---

[261] A recent study on the IFC (International Finance Corporation), the World Bank's private sector lending arm, found that, rather than helping the poor and the environment, it was "focused more on economic growth" and much of its portfolio was "oriented toward the interest of [transnational] corporations", particularly the socially and environmentally damaging ones, such as "oil, gas and mining", "large agribusiness, coal-fired power plants (...) and luxury hotel chains" (chungyalpa *et al.*, 2000, 1). The IFC supports "many projects with dubious development benefits. Coca-Cola bottling plants in Azerbaijan, nationwide cable television in Brazil, construction of a cruise ship terminal in Mexico and industrial hog farms in China are just a few examples." (ibid. 8) Another is the positively damaging Chad-Cameroon Oil Project (ibid. 9).

[262] Edward Goldsmith, 'Editorial', in *Globalising Poverty*, 2000, 3-5, here 4; see also Karliner, 1997, 25.

[263] Walden Bello, 'WTO: Serving the Wealthy, not the Poor', in *Globalising Poverty*, 2000, 36-39, here 37.

[264] See for example the new generation of 'Terminator Technology' GM seeds where special chemicals can switch on or off the sterility of the seed, thus locking farmers into a "continuing dependence on buying [the TNCs', here Syngenta's] products" (see ActionAid *et al.*, 2000, here 6).

it was expected that Africa, the poorest continent, would be worse off. That should already give us pause. But a closer look reveals that this, indeed, is the pattern: be it SAPs (Structural Adjustment Programmes), FDI (Foreign Direct Investment), aid or any other tool which is supposed to benefit the poor, after a couple of years it always turns out that they just served to increase the gap between rich and poor and redirected wealth from South to North.[265] In other words the myth of catching up is just that, a myth:

> Even if we assume the impossible, a sustained growth rate for all the poor countries of 5%, they would still only catch up in 149 years' time. (...) In fact, the growth rate for these countries, excluding India and China, is only 0.5%. Clearly they will catch up never.[266]

This is further corroborated by evidence that the flow of wealth is going in exactly the wrong direction, in various areas:

- It now appears that between 1984 and 1990, Third World countries transferred US$178 billion to rich country commercial banks. (Franke/Chasin, 1994, xviii)
- Almost all studies that have been done on the effects of FDI have concluded that it has led "to an uneven income distribution in developing countries", according to Pan-Long Tsai. (Madeley, 1999, 9)
- In 1980 [the capital-importing developing] countries had $567 billion in debt. Over the next 12 years they repaid $891 billion in principal and also paid out $771 billion in interest. But since much of this debt servicing had to be financed by new loans, they were saddled at the end of 1992 with a much larger debt stock of $1,419 billion, two-and-a-half times larger than in 1980. (*Taking Nature into Account*, 1995, 132)
- For the period 1980-1992 total net outflow of dividends and profits from capital-importing countries was $122 billion. (*Taking Nature into Account*, 1995, 133)
- Since the UNCTAD estimate of the early 1980s, the outflows from the South on account of transfer pricing and technological payments must have increased far beyond $30-50 billion annually. (*Taking Nature into Account*, 1995, 135)
- The brain drain [migration of highly skilled manpower from South to North] is so prevalent and significant that far from the North providing technical assistance to the South, the South is, in net terms, providing much more technical aid to the North. (...) A 1982 UNCTAD study (...) reveals that the US, Britain, and Canada gave $46 billion in development aid to poor countries between 1961 and 1972, but received $51 billion of "human capital" due to the brain drain during the same period. (*Taking Nature into Account*, 1995, 135-136)

---

[265] See *Which way now? 30 Years of aid in Bangladesh*, New Internationalist, No. 332, March 2001.

[266] C. Douglas Lummis, 'Equality', in *The Development Dictionary*, 1992, 38-52, here 46.

- US Treasury Department officials even calculate that for every $1 the United States contributes to international development banks, US exporters win more than $2 in bank-financed contracts.[267]
- We thus now have a situation in which many poor countries pay more money to the World Bank and IMF each year than they receive in loans: the IMF extracted a net US$1 billion from Africa in 1997 and 1998 – more than they loaned to the entire continent.[268]

Today, we are forced to admit that the strategy of free trade is making people poorer, more dependent on almighty corporations and less capable of feeding themselves. The SAPs imposed by the IMF which force countries to privatise their state assets, reduce spending on education and health care, and increase investment in export industries to pay off the debts in US$ are, if anything, a recipe for increasing poverty.[269] Export orientation and the associated rise in cash crop production have had devastating effects on the food security of poor countries – for various, very obvious reasons. Planting cash crops such as flowers, mange doux and soybeans, or farming shrimps and cattle for the export market is already on a very direct level eating into the provisions for the poor of those countries. The best agricultural land is not used any more to grow staple crops for local people, but for the export market instead. So there is less food for the poor:

> The results of these free trade policies are already being felt in India. The export of agricultural products, above all grain, has risen 71 per cent in recent years: from a value of 21.98 billion rupees in 1988-89 to 37.66 billion in 1992-93. Since the Indian government, according to the new economic policies, no longer has a right to influence the prices of basic foodstuffs, food prices rose 63 per cent in the same period of time, which led to an immediate drop in daily consumption from 510g. to 466g. per person. In other countries of the South, the picture is similar. (Bennholdt-Thomsen/Mies, 1999, 43; see also the example above on p. 25)

These developments led to such absurd consequences as in Ethiopia where, at the height of the recent famine, more food was produced than would have been needed to feed its people, but it went abroad:

> In Amhara [a region of Ethiopia], grain production (1999-2000) was 20 per cent in excess of consumption needs. Yet 2.8 million people in Amhara (representing

---

[267] Bruce Rich, 'Still waiting: the failure of reform at the World Bank', in *Globalising Poverty*, 2000, 8-16, here 12.
[268] Carol Welch, 'A world in chains. The IMF's role in increasing the poor's crushing burden of debt', in *Globalising Poverty*, 2000, 32.
[269] See John Cavanagh, Carol Welch and Simon Retallack, 'The IMF formula: Generating Poverty', in *Globalising Poverty*, 2000, 23-25 and Ulli Diener, 'Asleep at the switch', in *New Internationalist*, No. 331, January/ February 2001, 26-28.

17 per cent of the region's population) became locked into famine zones and are "at risk" according to the FAO. Whereas Amhara's grain surpluses were in excess of 500,000 tonnes (1999-2000), its "relief food needs" have been tagged by the international community at close to 300,000 tonnes.[270]

For precisely these reasons people in poor countries actually advocate moving away from the free trade and 'liberalisation' regime of the 'developed' countries:

> Mohamed Suliman from Sudan argues that for his country the most positive aspect of ecological structural change in the industrialized countries would be the anticipated reduction of trade with the North. That would force the producers of exportable agricultural goods to rediscover the national market and work to satisfy the basic needs of their own people. That could increase the reliability of national food supplies and alleviate the massive damage which has been done to the soil. Where this is linked with land reform, subsistence farming could be revived – as could, at the same time, a well-tried way of life including barter and mutual help. In turn that could slow down migration to the towns and induce some people to return to their villages. (*Greening the North*, 1998, 207)

On a second level, the dollars earned through export very rarely end up directly in the pockets of the producing poor; more often than not the income is expropriated by the local elite or TNCs (see the example of Nike trainers on p. 119). On a third level, the uniform recipes of the IMF have led to a collapse of the worldwide commodity markets. If you go round the world and tell poor countries that the best thing is to grow export crops such as coffee or cacao, any economist looking at the global market can foretell the consequences: if masses of people start to produce the same commodities the prices will tumble,[271] and therefore the earnings will not enable the poor countries to pay back their debt, as originally intended. This debt, as a consequence, has reached ridiculous levels: "In 1997, external debt payments made up 92.3 per cent of the GDP of countries of low development" (*Human Development Report 1999*, 1999):

> Just as total debt in developing countries rises again (by $150 billion in 1998 to a total of almost $2.5 trillion), a plunge in commodity prices – to which many developing countries have been made highly vulnerable by the IMF's policy of

---

[270] Michel Chossudovsky, 'The real cause of famine in Ethiopia', in *Globalising Poverty*, 2000, 26-28, here 26.

[271] This in fact is what has happened, and dramatically so: "Between 1980 and 1992, nonfuel commodity prices had fallen by 52 percent on average. In other words, on average, whilst in 1980, 100 units of a Southern country commodity export could buy 100 units of manufactured product imported from the North, the same 100 units of commodity export could only pay for 48 units of the same imported manufactured product in 1992." (*Taking Nature into Account*, 1995, 129)

export-led growth – has further crippled the poorest countries' ability to repay foreign debt.[272]

The forth truth, of course, is that nobody needs those cash crops and export goods except the rich consumers in 'developed countries', bearing in mind what 'need' means in a Euro-American context. The aim is *not* to feed the poor, but to stuff the already overweight Northerners with extra protein from shrimps they don't need, to allow them to expand their already unsustainable consumption levels.

But above all the current regime of globalisation is a profit maximisation scheme for multinationals. They use the world market to extend their control to every corner of the globe, to dominate their market segments. The extent of this domination has been described by Madeley:

> TNCs dominate world markets in internationally traded agricultural commodities, with a small number of companies accounting for a large percentage of the trade. Six corporations handle about 85 per cent of world trade in grain; fifteen TNCs control between 85 and 90 per cent of world traded cotton; eight TNCs account for between 55 to 60 per cent of world coffee sales; seven account for 90 per cent of the tea consumed in Western countries; three account for 83 per cent of world trade in cocoa; three again account for 80 per cent of bananas, while five firms buy 70 per cent of tobacco leaf. Food products account for three-quarters of agricultural trade and raw materials the remaining one-quarter. (1999, 36)

And Karliner sums it up: "All told, the transnationals hold 90 percent of all technology and product patents worldwide and are involved in 70 percent of world trade." (1997, 5) But it is not just the degree of domination which leaves no room for self-determination to the affected people, it is the financial muscle and the unfair practices with which the TNCs force the world to serve their interests, while they exploit it. One of their prime strategies is transfer pricing:

> Under transfer pricing, the parent TNC sells materials to a subsidiary in another country at an artificially high price. Such materials are then used in a manufacturing process or service industry. Having to pay these high prices reduces the profits of the subsidiary company, and means that it pays less tax in the country where it operates, which is therefore cheated out of tax revenues. Transfer pricing is tax avoidance. (Madeley, 1999, 12)

Boxes 1, 3 and 4 at the beginning of the book have already exposed some of the other malpractices of TNCs (see pp. 24-26). Because of their global reach, they can thoroughly exploit local markets, once they are prised open by the IMF or the World Bank, because they can outperform any local competitor in terms of size,

---

[272] Carol Welch, 'A world in chains. The IMF's role in increasing the poor's crushing burden of debt', in *Globalising Poverty*, 2000, 32.

flexibility, financial muscle and pressure they bring to bear on the local governments.

Transnational corporations have used their money, size and power to influence the policies of governments and change the rules of the game in their favour. They have pushed the idea of privatisation and taken over much of the economic role that government once played. They have used their position to influence international negotiations and their muscle in different ways to effectively cause hardship for the poor – the invisible in corporate eyes. And they have used the power of public relations to assure us that all is well. In some cases they have been funded by aid schemes. (Madeley, 1999, 168)

Box 7: Export Processing Zones (EPZ) – free trade gone mad

Jobs in the zones typically represent no more than 5 per cent of total employment in the manufacturing industries of developing countries and are tiny compared with the estimated 300 million people who work in the "informal sectors". Yet governments of developing countries have nonetheless allocated substantial amounts of scarce funds to attract companies into the zones. A country might typically offer companies a free building, a five-year "tax-free" holiday, low-wage labour and other perks. To develop an EPZ at Bataan, in the Philippines, for example, the Filipino government offered companies 100 per cent ownership, permission to impose a minimum wage lower than in the capital, Manila, tax exemptions on imported raw materials and equipment, exemption from export tax, low rent for land, plus other inducements. The first six months of employment at Bataan are a probationary period, paid at 75 per cent of the "minimum" wage. "Some plants terminate employment after this period has elapsed and replace workers by fresh trainees", reveals an ILO survey.

John Madeley (1999), *Big Business, Poor Peoples* (London, Zed Books), p. 113.[273]

Box 8: Tourism – helping the poor?

Tourism is cultural prostitution (Haunani-Kay Trask of Hawaii) (...) Tourism is the second largest foreign currency earner for developing countries (next to oil) and one of few economic sectors that is thriving. (...) But while tourism promises much, especially for governments of countries with natural resources that might attract foreign holiday-makers, the promise is an illusion. Most of the foreign exchange that developing countries appear to earn from international tourism goes to transnational corporations, through their ownership of hotels, airlines and tour operators, and while TNCs reap the benefits [this "leakage" of earnings ranges between 30 and 80 per cent], the industry often harms the environment and the poor in the countries that play host to the tourists. (...) Hawaii's culture has suffered acutely from tourism. (...) Huge hotel development has put severe strains on water supplies. Reefs and fishing grounds have been destroyed because of hotel sewage runoff and golf course irrigation.

John Madeley (1999), *Big Business, Poor Peoples* (London, Zed Books), pp. 128, 140.

---

[273] For a more detailed account of the abysmal conditions and the cynical methods of exploitation in the EPZs see Klein, 2001, 195-229.

Box 9: Pharmaceutical TNCs

The record of the pharmaceutical TNCs in developing countries is one of putting the pursuit of profit before peoples' health, even when that profit is tiny compared to their overall profits. Greed for every cent they can get has pushed TNCs to irrational behaviour and arguments. It is what these TNCs are omitting to do, as well as what they are doing, that is of concern. A materially poor parent whose child is sick should be able to buy a locally produced, inexpensive medicine. In many developing countries, she or he does not have the choice of doing that because the TNCs have persuaded governments that locally produced drugs are not necessary. Choice has been denied the poor because of what is effectively a corporate veto.

John Madeley (1999), *Big Business, Poor Peoples* (London, Zed Books), p. 157.

It is an endless race to the bottom where long-term consequences, social and cultural context play no part whatever as long as the profits are right:

> They [TNCs] are increasingly mobile, manoeuvring technology, finance, information, goods and services throughout vast regions and across the globe, seeking to exploit the optimum combination of low wages, skilled non-union workers, burgeoning markets, abundant natural resources and lax environmental regulations. (Karliner, 1997, 12)

As soon as the 'bottom line' doesn't live up to shareholder and financial market expectations any more, TNCs relocate elsewhere, leaving unemployment, environmental degradation and social misery in their wake. The financial markets are the driving force behind this race. They are possibly the most striking opposition to sustainability: rather than focussing on long-term survival, they are interested only in the highest possible profit in the shortest possible time. These international markets, and through them the shareholders, are the ones applying pressure on corporations to yield profits of 15 or, even better, 25 per cent per annum, achievable only if you go for the lowest wages, don't pay for the raw materials and shed as many employees as possible. If we look at the structure of the global financial markets, it becomes clear that they act as a money-printing machine for shareholders, not to create real goods and services to be used by real people: "In 1995 worldwide daily transfers on the financial markets amounted to US$1,300 billion. The total worldwide trade in real goods and services amounted to US$4,800 billion for the entire year, representing just 1.6 per cent of financial transactions."[274]

The financial power in the hands of large shareholders is staggering and has a tremendous and devastating influence on the course of economic activity. The figures are almost "incomprehensible to ordinary mortals": Pension funds, insurance companies and financial houses hold combined assets of US$21,000

---

[274] Gertrud Ochsner, 'Sand ins Getriebe werfen. Die Tobin-Steuer gegen die schnellen Gewinne der Spekulanten', in *Die WochenZeitung*, No. 41, 8.10.1998, 13-14, here 13.

billion, "half of which belongs to US sources": "To put it in perspective, it amounts to more than the combined annual GNP of all the industrialised countries, or US$3,500 for every man, woman and child alive today on the planet." (*The Lugano Report*, 1999, 177) Yet this is not wealth creation for all, this is expropriation of the common wealth for the benefit of the few and can only be described using an old word which of course never passes the lips of TNC CEOs: it is "colonialism by companies, rather than countries" (Madeley, 1999, 20). And this colonialism has far-reaching consequences not just for the South, but also for the North:

> These corporations effectively play the role of Earth brokers in the global economy – buying and selling the planet's resources and goods, as well as deciding what technologies will be developed and used, where factories will be built and which forests will be cut, minerals extracted, crops harvested and rivers dammed. Their inordinate power puts them in the position of mediating the future of local, regional and global ecosystems in the interests of their own bottom lines [i.e. profits] and an antiquated version of economic growth that is wreaking havoc on the world's ecology. (Karliner, 1997, 13)

Lummis has rightly stressed the anti-democratic nature of this situation. It seems that self-determination is one of the biggest victims of globalisation, by necessity:

> Today imperial power is incarnated in three bodies: pseudo-democracy at home, vast military organizations, and the transnational corporations that are seeking to put all of humankind and nature under their managerial control. The result is that "major decisions which affect the lives of millions of people are made outside their countries, without their knowledge, much less their being consulted." (Muto Ichiyo) (Lummis, 1996, 138)[275]

If "participatory democracy means the right to participate in the making of decisions that affect one's life" (ibid.), then it is clear that this cannot happen when the board of the TNC ripping open the land on a Philippine island is sitting in New York and taking the decisions there. It is clear that globalisation counteracts sustainability, most notoriously when it comes to empowerment and ecology. It is, as Sachs has most eloquently described, a concerted strategy to completely free the economy of any responsibilities whatever; it is the attempt to put the exclusive striving for profits, which Milton Friedman demanded, finally into practice:

---

[275] "The categorical imperative of world market competition repeatedly thwarts attempts to organize societies creatively and differently. Mobilizing for competition means streamlining a country; diversity becomes an obstacle to be removed. (...) There is scarcely a country left today that seems able to control its own destiny." (Sachs, 1999, 98-99)

The emergence of the globe as an economic arena where capital, goods and services are able to move without much consideration for local and national communities has delivered the most serious blow to the idea of a polity built on reciprocal rights and duties among citizens. As corporations attempt to reach out to gain foreign markets, they feel held back by the weight of domestic responsibilities. The same holds true for the elites. As they aspire to catch up with the vanguards of the international consumer class, their sense of responsibility for the disadvantaged sections of their own society withers away because, instead of feeling superior with respect to their compatriots, they now feel themselves to be inferior with respect to their global reference group. Globalization thus undercuts social solidarity. Through transnationalization, capital is in the position to escape any links of loyalty to a particular society. It prefers not to be bothered by things like paying taxes, creating jobs, reinvesting surplus, keeping to collective rules or educating the young, because it considers them mere obstacles to global competition. (Sachs, 1999, 163)

To counteract this we need two things. Opschoor, as quoted already (see p. 73), has spelt out the first: "Sustainable global development implies international institutions capable to change prevailing distributions of incomes and current distributions of access to sources of income and wealth, including environmental resources and world markets."[276] Thürer has pointed out that TNCs, due to their power and their supranational operations, are functioning either outside the law (see box 10) or in a legal framework which these corporations have largely written themselves.[277] The biggest challenge, he writes, is to create "an effective world rule of law". This includes a binding integration of corporations into the human rights framework and an international economic order "which is adequately guided by the democratic will of the people and the primacy of justice and public interests".[278]

Box 10: Flexing the muscles – TNCs

Transnational corporations have the highest access to the most senior policy-makers; they can call presidents, prime ministers and heads of key international agencies to put their case, and their call will be put through. They also know that government ministers can often be persuaded more easily of a TNC's claims if palms are suitably oiled when the need arises. While the large corporations

---

[276] Johannes B. Opschoor, 'Institutional Change and Development Towards Sustainability', in *Getting Down to Earth*, 1996, 327-350, here 345-346; see also Hilary F. French, 'Assessing Private Capital Flows to Developing Countries', in *State of the World 1998*, 1998, 149-167, here 166.

[277] The fact that the business community managed to suppress any reference to and commitment for itself from *Agenda 21* has led Karliner to conclude: "If nothing else, the Earth Summit clarified the fact that global corporations have the power and capacity to seriously influence the focus and trajectory of international agreements on environment and development." (Karliner, 1997, 57) See also *Europe Inc.*, 2000.

[278] Daniel Thürer, 'Globalisierung – Wirtschaftsmacht und Menschenrechte. Gedankenanstösse und Grundsatzfragen', in *Neue Zürcher Zeitung*, No. 134, 10./11.6.2000, 101.

have the money to do this, developing countries are often broke. In a poverty-stricken country, especially, ministers may not be averse to a deal that gives them a degree of personal security. (...) Corruption has become a widespread global problem, with TNCs sometimes paying huge bribes to win business. (...) A bribe, typically between 10 to 20 per cent of the cost of a deal, may be paid to government ministers and officials and added, at least in part, to its cost. A TNC may win the contract, and a tiny number of people in a developing country will gain from the bribe, but the country as a whole pays more than it should. This means that less money is available for other purposes, such as health care and education. (...) "Once a decision maker has a personal interest in placing an order with a firm that is willing to pay a bribe, his judgement goes out of the window" (George Moody-Stuart). This means that priorities get distorted. Goods and services may be purchased that are not needed (armaments are a classic case) and a project which has attracted a large bribe may get priority over others, which are possibly more useful. This can lead to projects going ahead that make economic nonsense.

John Madeley (1999), *Big Business, Poor Peoples* (London, Zed Books), p. 158-159.[279]

But secondly, we need to bury the notion of globalisation, and replace it with an answer that is in tune with Gandhi's principle of concern for the poorest: the strategy needed "must not encourage further downward spirals caused by ruthless competition, but must oppose the very principle of economic globalisation and instead facilitate the building up and diversification of local economies globally."[280] Concentrating on such "cosmopolitan localism", as Sachs calls it, would mean facing up to one's duties within a sustainable world: "It implies instead that each country puts its own house in order in such a way that no economic or environmental burden is pushed onto others that would constrain them in choosing their own path." (Sachs, 1999, 107; see also Bennholdt-Thomsen/Mies, 1999, 202-203) Or in Karliner's words, we "must confront the essential paradox and challenge of the twenty-first century by developing ways of *thinking and acting both locally and globally at the same time*." (Karliner, 1997, 199; emphasis in the original)

---

[279] Through an undercover operation, the independent Indian media company Tehelka has recently documented extensive bribery in Indian politics (see the Tehelka Tapes on defence deals [http://www.tehelka.com/]).

[280] Colin Hines, 'Globalisation's cruel smokescreen', in *Globalising Poverty*, 2000, 48-49, here 49.

Myth: 'Progress through development'

> The idea of development stands today like a ruin in the intellectual landscape, its shadows obscuring our vision. (Sachs, 1999, 3)

> Development is stimulating dissatisfaction and greed; in so doing, it is destroying an economy that had served people's needs for more than a thousand years. (Norberg-Hodge, 2000, 141)

> The so-called "disadvantages" of "underdevelopment" or "poverty" should actually be viewed as advantages and opportunities. (Madhu Suri Prakash and Hedy Richardson)[281]

> What is quite clear is that a way of life that bases itself on materialism, i.e. on permanent, limitless expansionism in a finite environment, cannot last long, and that its life expectation is the shorter the more successfully it pursues its expansionist objectives. (Schumacher, 1993, 121)

Reality: Next to growth, this is quite likely the most unchallenged and powerful myth of our capitalist system: "Development occupies the centre of an incredibly powerful semantic constellation. There is nothing in modern mentality comparable to it as a force guiding thought and behaviour."[282] It is important, therefore, to deepen the analysis. Globalisation, just like the development model, starts from what we know today to be the erroneous notion that there is one linear mode of development, leading from non- or under-development necessarily upwards to the fully developed lifestyle of a US citizen. In other words, it starts from very firm notions of what development is, of a singular path to the endpoint of this development (an industrialised liberal democracy of the Euro-American type).[283]

There are two very serious problems associated with this idea, which still have scarcely penetrated the public mind (just as the associated notions of pro-

---

[281] Prakash/Richardson, 'From Human Waste to Gift of Soil', in *Ecological Education in Action*, 1999, 65-78, here 75.

[282] Gustavo Esteva, 'Development', in *The Development Dictionary*, 1992, 6-25, here 8. – Even amongst the keepers of the ideological Holy Grail, such as the *Neue Zürcher Zeitung*, doubts are accumulating (see Mamadou Diawara, 'Globalisierung, Entwicklungspolitik und lokales Wissen. Warum technokratische Lösungsansätze zum Scheitern verurteilt sind', in *Neue Zürcher Zeitung*, No. 264, 11./12.11.2000, 101-102). But still, ten years after the publication of *The Development Dictionary* (1992), which comprehensively consigned the concept of 'development' to the rubbish heap of history, the term is used as if nothing had happened, the latest example being the UK government's white paper on international development entitled *Eliminating World Poverty: Making Globalisation work for the World's Poor* (London, Department for International Development, 2000) which is saturated with traditional, i.e. counterproductive, 'development' talk.

[283] Yet even this notion of historical development is flawed. Various studies have shown that in terms of purchasing power a carpenter from the 14th century would have earned much more than his 19th-century counterpart, living in industrial society (see Kurz, 1999, 15-21).

gress, growth and GNP are still used despite the fact that they have proved them-selves to be utterly misleading and outright destructive concepts). The first is that we usually don't know that the idea of 'development' as more or less the sole goal of humankind is not a timeless concept,[284] but one which has a very clear histori-cal context of creation. It was from the beginning, just like the Bretton Woods in-stitutions, a specific ideological tool used by the US to gain and maintain world hegemony. Whereas before there were different cultures, traditions, and civilisa-tions, with one stroke there was suddenly only one model. Rather unsurprisingly the US, having invented the concept, emerged as the star pupil and "as a conse-quence, sure enough, all other cultures suddenly appeared to be deficient, even defective" (Sachs, 1999, xi) (see box 11). As with some of the other simplistic concepts we have already dealt with (growth, free trade, market), developmental-ism reduces everything to an economic problem, or even to a problem of the level of industrial production. The discourse excludes any complex assessment of a country's situation. The model cannot cope with "the idea that poverty might also result from oppression and thus demand liberation, or that a culture of sufficiency might be essential for long-term survival, or, even less, that a culture might direct its energies towards spheres other than the economic." (Sachs, 1999, 9)

Box 11: The origin and ideological context of the 'development model'
Wind and snow stormed over Pennsylvania Avenue on 20 January 1949 when, in his inauguration speech before Congress, US President Harry Truman defined the largest part of the world as "un-derdeveloped areas". There it was, suddenly a permanent feature of the landscape, a pivotal con-cept that crammed the immeasurable diversity of the globe's south into a single category: underdeveloped. For the first time, the new world view was announced: all the peoples of the earth were to move along the same track and aspire to only one goal – development. And the road to follow lay clearly before the president's eyes: "Greater production is the key to prosperity and peace." After all, was it not the USA that had already come closest to this Utopia? According to that yardstick, nations fall into place as stragglers or lead runners. And "the United States is pre-eminent among nations in the development of industrial and scientific techniques". Clothing self-interest in generosity, Truman outlined a programme of technical assistance designed to "relieve the suffering of these peoples" through "industrial activities" and "a higher standard of living". (...)

---

[284] Tim Ingold has shown that this myth also has no biological basis: "Any catalogue of alleged human universals tends to project the image that people of affluent, Western societies have of themselves. Thus where *we*, as privileged members of such societies, can do things that *they* – people of 'other cultures' – cannot, this is typically attributed to the greater development, in our-selves, of universal human capacities. But where they can do things that we cannot, this is put down to the particularity of their cultural tradition. This is to apply just the kind of double stan-dards that have long served to reinforce the modern West's sense of its own superiority over 'the rest', and its sense of history as the progressive fulfilment of its own ethnocentric vision of human potentials. Once we level the playing field of comparison, however, only one alternative remains: that all human beings must have been genotypically endowed, at the dawn of history, with the 'ca-pacity' to do everything that they ever have done in the past, and ever will do in the future." ('Evolving Skills', in *Alas, Poor Darwin*, 2000, 225-246, here 232)

A new world view had found its succinct definition: the degree of civilisation in a country could be measured by the level of its production.

Wolfgang Sachs (1999), *Planet Dialectics. Explorations in environment and development* (London, Zed Books), p. 3-4.

Hidden in Truman's muddled prose one can see the basic outlines of the newly emerging ideology of development. Of course, Truman was not seriously proposing that the functioning of capitalism could be changed by persuading capitalists to develop instead of exploit. In fact, the sentences do not say that capitalists should do something different; they say that we should stop calling what they do "exploitation" and start calling it "development". (...) But it was with the Point Four Program that "development" took its full post-World War II form, to mean a conscious project of the industrial capitalist countries aimed at the *total transformation of societies, primarily in the Third World*, allegedly directed at curing a malaise called "underdevelopment".

C. Douglas Lummis (1996), *Radical Democracy* (Ithaca; London, Cornell University Press), p. 59; my emphasis.

It is impossible to dissociate the idea from its ideological context and its origin in the Cold War: development was and is a strategy to fulfil the self-interests primarily of the US, but also of the industrialised nations in general.[285] It is *not* a humanitarian strategy to free anybody from misery, whatever the rhetoric might claim. This can be seen historically, since the development strategy has demonstrably failed. The 'development model' has led to all sorts of things, but it has not led to the abolition of poverty, misery, hunger and the unnecessary death of millions of children from easily preventable diseases. To put it clearly: "The crisis [in the Third World] has demonstrated that conventional development strategies are fundamentally limited in their ability to promote equitable and sustainable development."[286]

But, this is my second point, the concept is flawed in principle within a limited system such as the biosphere. Quite apart from the fact that it turned out that the "identification of social progress with economic growth was pure fiction" (Sachs, 1999, 6), the development model has changed from the anticipated solution into the most pressing problem. In a limited, materially non-growing system linear, unlimited growth is, as we have seen, a collective attempt at suicide. And it

---

[285] Historically, most 'development aid' was military aid: "From 1952 on, foreign aid was more and more obviously designed to build up military power in non-Communist countries. In the next ten years, of the $50 billion in aid granted by the United States to ninety countries, only $5 billion was for non-military economic development. When John F. Kennedy took office, he launched the Alliance for Progress, a program of help for Latin America, emphasizing social reform to better the lives of people. But it turned out to be mostly military aid to keep in power right-wing dictatorships and enable them to stave off revolutions." (Zinn, 1996, 430)

[286] Miguel A. Altieri, Andres Yurjevic, Jean Marc Von der Weid and Juan Sanchez, 'Applying Agroecology to Improve Peasant Farming Systems in Latin America: An Impact Assessment of NGO Strategies', in *Getting Down to Earth*, 1996, 365-379, here 366.

turns out that rather than the 'underdeveloped' poor countries being the problem, it is the "overdevelopment" of the North which is at the root of our environmental problems today:

> Both the crisis of justice and the crisis of nature necessitate looking for forms of prosperity that would not require permanent growth, for the problem of poverty lies not in poverty but in wealth. And equally, the problem of nature lies not in nature but in overdevelopment. (Sachs, 1999, 89)

Jack D. Forbes puts the spotlight on this when he notes that ironically "those peoples and human beings tend to be categorised as 'underdeveloped' and 'uninteresting' who do not subjugate others and who do not accumulate vast amounts of stolen goods" (1981, 15).

The Euro-Americans need structural adjustment to correct their overdevelopment

> Come, then, comrades, the European game has finally ended; we must find something different. (Fanon, 1967, 251)

Given our limited resources and the fact that "since 1980, the global economy has tripled in size" (*A Guide to World Resources 2000-2001*, 2000, 6), we are therefore confronted with two insights: one is that the problem, rather than being population growth or 'underdevelopment' in poor countries, lies squarely at the door of the so-called developed countries. The other is that we had better face up to what the 'development model' really is, judged within a sustainability framework: a destructive colonialist system of exploitation.

The first point can best be summarised in the revealing term "overdevelopment". As we have already cited (see p. 87), we need to adopt the "home perspective" (Sachs, 1999, 86-89), i.e. rather than shifting the blame onto 'the others' we should finally face up to the fact that *our* lifestyle is not *and cannot be* sustainable[287] and therefore cannot under any circumstances be a model for the rest of the world:

> According to UN statistics, the 1.1 billion people who live in affluence consume over three-quarters of the world's total output. The remaining 4.7 billion people – 80 percent of the population – survive on less than a quarter of world output. (...)

---

[287] Particularly in the US, this has not yet sunk in: in their popular *Environmental Science* textbook Nebel and Wright write that "we have no chance of achieving sustainability in other areas until we achieve population stability" (2000, 135), thereby continuing a long tradition of thought by environmentalists (cf. Sachs, 1999, 82). This is all the more surprising because such a statement directly contradicts the impact formula they use (Nebel/Wright, 2000, 144), where "affluence of lifestyle" is a negative factor equal to "population size". Yet in a 600-page textbook there is just one full text page devoted to "Lifestyle Changes" (ibid. 595-597).

> Consumption by the affluent 1.1 billion people alone claims more than the entire carrying capacity of the planet. (Wackernagel/Rees, 1996, 102; see also *Taking Nature into Account*, 1995, 99)

However, this generalisation of the lifestyle of the rich was and is exactly the promise and the core of the 'development model'. Yet if we start to ask the hard questions, such as "Is sustainable development supposed to meet the needs for water, food and minimal purchasing power or the needs for air-conditioning and university studies abroad?" the answer becomes clear, namely "that there will be no sustainability without restraint on wealth", because overdevelopment "produces both poverty and biospherical risk" (Sachs, 1999, 160-161). If we define a sustainable society as one whose "demands on nature do not exceed the environmental space it is entitled to use", then the crucial question is whether "the rich countries [are] capable of living without the surplus of environmental space they appropriate today?" (Sachs, 1999, 172) Because all the indications clearly point to the fact that the industrial nations are overexploiting nature, the focus should be on their excessive overconsumption and how to lower it. Yet the development discourse – which was unfortunately revitalized in new disguise, but on the same shaky foundations, with the notion of 'sustainable development' (see footnote 46 on p. 29) – has always done exactly the opposite:

> In designing strategies for the poor, developmentalists work towards lifting the bottom, rather than lowering the top. The wealthy and their way of producing and consuming remain entirely outside the spotlight, as always in the development discourse where the burden of change is solely heaped upon the poor. (Sachs, 1999, 173)

We know today that this strategy starts at the wrong end. Within the limits of a finite system, "justice is about changing the rich and not about changing the poor". If anybody, then "it is the North (along with its outlets in the South) that needs structural adjustment" (ibid.). So rather than promoting the unrealisable notions of ever more consumption and overproduction,

> sufficiency in resource consumption is now bound to become the axis around which any post-developmentalist notion of justice will revolve. In future, for industrialized countries and classes, justice will be about learning how to take less rather than how to give more. (Sachs, 1999, 174)

It is now clearer than ever that "luxury for all", the Cadillac, villa and Concorde society for everybody, is a myth and a physical impossibility, yet it is still promoted, in order to keep going the short-term profit-generating machines of overconsumption and overproduction. Morosini has shown the hypocrisy of this notion of "luxury for all": "In the globalised society 'all' nearly always means 'all members of the elite'."[288] A classic example is the Concorde: "the most expensive marketing experiment in history"[289] and "a glorious anachronism since it first flew" cost the British and French taxpayers at least £20 billion and was so uneconomic that "each Concorde passenger has been subsidised to the tune of approximately £3,300".[290] This leads Morosini to ask:

> How much modern technology and projects of general public interest – say in the field of renewable energy – could have been financed with the sums that the British, French and Soviet citizens had to cough up? In addition the "social" balance sheet of the Concorde has to be taken into account. (...) It is guided by the principle of: "All pay so that very few very rich people can fly very fast for half the real costs".[291]

But it is not just the real costs, it is a matter of the overall throughput as well: "And if 'all' should really mean 'all' simple maths allows us to see the risks and impossibilities of certain options":[292]

> Certain aspects of industrial country lifestyles are particularly environmentally damaging, including high levels of private motoring or aircraft travel, diets intensive in meat and packaged beverages, home environments requiring high inputs of fossil or nuclear energy for heating, cooling or power, or a fashion-driven approach to living that requires a high turnover of consumer goods and appliances. Such behavior is not only occupying more than its fair share of environmental space. It is also foreclosing options for future generations, while simultaneously stimulating the consumer appetites of the 80% of the world's population that cannot yet do these things, but is increasingly desirous of doing them in the future. Turning towards sustainability requires that consumers start to feel the full implications of these aspects of their lifestyles and to find ways of feeling good about sustainable alternatives to them.[293]

Part of understanding the full impact of Euro-American lifestyles is that we need to realise that they are possible only on the basis of massive exploitation, both of the South and of nature. It also means facing up to our personal responsibilities rather than shifting them onto governments, industry and the South.

---

[288] Marco Morosini, 'Ein tot geborener Traum. Die Concorde ist nicht nur technisch gescheitert', in *Die WochenZeitung*, No. 40, 5.10.2000, 24.

[289] Sarah Hall, 'Flying legend built on speed, safety and style', in *The Guardian*, 26.7.2000.

[290] Jonathan Glancey, 'Nightmare ending to 40-year dream', in *The Guardian*, 26.7.2000.

[291] Marco Morosini, 'Ein tot geborener Traum. Die Concorde ist nicht nur technisch gescheitert', in *Die WochenZeitung*, No. 40, 5.10.2000, 24.

[292] Ibid.

[293] Paul Ekins, 'Towards an Economics for Environmental Sustainability', in *Getting Down to Earth*, 1996, 129-152, here 146.

Another part of the illusion of development, clearly to be recognised if impact considerations enter the fray, is the view that the introduction of technology automatically leads to development. This idea is obviously false, but deeply ingrained in our contemporary ideology. But if we don't start to penetrate that illusion, our 'solutions' will lead to the opposite effect: "Technological processes can lead to higher withdrawals of natural resources or higher additions of pollutants than ecological limits allow. In such cases they contribute to underdevelopment through destruction of ecosystems." (Shiva, 1991, 233)

Development ideology produces poverty and neo-colonialism by necessity

> In short, elegant power does not force, it does not resort either to the cudgel or to chains; it helps. (Marianne Gronemeyer)[294]

We have to be very clear about one fundamental issue: it is the development model itself which *produces* poverty and scarcity in the Euro-American, economic sense.

> Millions go hungry around the world – nearly one fifth of humanity. At the same time, immense surpluses resulting from hyper-technologisation are being held or destroyed in other parts of the world. Yet the UN's Food and Agriculture Organisation is not embarrassed to call for more technology (genetic engineering), and further moves to an open world market, as the way to combat hunger in the Third World. The reality, on the contrary, is that hunger is a result of technologisation. The US grain surpluses used for so-called famine relief have destroyed the indigenous millet market for African peasants. The boring of deep wells in the Sahel in order to raise the productivity and therefore the profitability of cattle-breeding has dangerously lowered the water table and compounded the desertification effects of over-pasturing. The FAO's much-heralded Green Revolution, with its technologically generated maximum yields, has led in India, Thailand, Mexico and elsewhere to the concentration of land among those with the most capital, and to a veritable army of landless peasants. (Bennholdt-Thomsen/Mies, 1999, 82)

Only through the values of the continually expanding Euro-American consumer society does the notion of lack of material goods even enter the discourse. It is a particular ideology, with its associated behaviour and technology, which has assumed world hegemony and has produced the notion of deficiency, of needing ever more: "Frugality is a mark of cultures free from the frenzy of accumulation. (...) Scarcity derives from modernized poverty." (Sachs, 1999, 10-11) This implies that the Euro-American societies which are responsible for the spread of this un-

---

[294] Gronemeyer, 'Helping', in *The Development Dictionary*, 1992, 53-69, here 53.

sustainable ideology also bear the responsibility to take it back, to correct it, to provide and live sustainable alternatives.

Poverty production is not just a freak incident of the development model, but a necessity in a capitalist economy. Since we are operating in a zero-sum game the wealth of the rich must be taken away from somewhere. It is an old insight, but nevertheless true today that capitalism lives off exploitation and inequality. It is structurally inbuilt into the system:

> The world economic system does not produce inequality accidentally, but generates it systematically. It operates to transfer wealth from poor countries to rich countries. A big part of the "economic development" – that is, the wealth – of the rich countries *is* wealth imported from the poor countries. From where could wealth be imported to create the same condition for all? (...) The myth that [economic development will lead to ultimate prosperity for all] is, of course, "functional": providing the fuel for the great engine that drives development forward; providing the spectacle that enthrals, transfixes, and draws the attention of the world's people from the real inequality generated by the world economy; providing the legitimation for the vast development industry that keeps many goodhearted people in it along with the development carpetbaggers. (Lummis, 1996, 69-70)

If we start to think more clearly about what it means to be rich, what conditions have to be met, it becomes obvious that only an equitable distribution of wealth would make all 'rich'. Any other system is based on exploitation (see box 12). This, of course, means that we need to change the economic system.

Box 12: Rich and poor

We think a person rich who has enough purchasing power to control the labor of a large number of other people. This control can take the form of directly hiring workers and servants or of arranging through the "service" industry to have other people do your work for you. We think a country rich when it has enough purchasing power to have a portion of its work done in other countries by "cheap labor." (...) So the old saying "The rich get richer and the poor get poorer" is not some kind of ironic paradox but an economic law as trim and tidy as Newton's Third Law of Motion: the rich get richer *when* the poor get poorer, and vice-versa. Economic-development mythology is a fraud in that it pretends to offer to all a form of affluence which *presupposes* the relative poverty of some. Movies, television, and advertising originating in the overdeveloped countries idealize the lives of people who do less than their share of the world's work (because others do more), who consume more than their share of the world's goods (so that others must do with less), and whose lives are made pleasant and easy by an army of servants and workers (directly or indirectly employed). (71; my emphasis)

A person is poor who is one of those whose poverty generates the rich people's richness, whose labor generates their leisure, whose humiliation generates their pride, whose dependency generates their autonomy, whose namelessness generates their "good names". (73)

Richness means, exactly, economic power over other people. (76)

C. Douglas Lummis (1996), *Radical Democracy* (Ithaca; London, Cornell University Press).

Development ideology is antidemocratic

> The Utopia of affluence undercuts the Utopia of liberation.
> (Sachs, 1999, 193)

C. Douglas Lummis, in his *Radical Democracy*, and Gustavo Esteva have shown with penetrating precision the extent to which the development ideology is incompatible with the principle of empowerment and self-determination. The antidemocratic nature of economic development actually encompasses all levels, from consciousness to politics to economics.

On the level of consciousness it teaches people to view their way of life, their values and culture not for what they are worth in themselves, but only on the trajectory of 'development', in relation to what is supposed to be best. It inflicts a bad conscience on them so that they "see themselves as 'obstacles to development'". Already here the development model exerts a dictatorial power of definition: it is "robbing peoples of different cultures of the opportunity to define the forms of their social life".[295]

On the political level economic development is profoundly antidemocratic "in that it is the expansion of a sphere of life from which democracy is to be excluded in principle." (Lummis, 1996, 48) What is meant here is that the capitalist economy, which is totalitarian in its structure and has no democratic accountability let alone self-determination of the producers,[296] becomes the one sphere, the "Master Science" (Lummis, 1996, 46), which dominates and determines all the others. This goes hand in hand with an Orwellian redefinition of classical political demands which were originally meant as tools for empowerment, but turn into PR exercises instead: "freedom becomes the free market; equality becomes equality of opportunity;[297] security becomes job security; consent becomes 'consumer sovereignty'; and the pursuit of happiness becomes a lifetime of shopping." (Lummis, 1996, 48)

And lastly, on the economic level, capitalist development "*generates* inequality and it *runs on* inequality".[298] Because of the antidemocratic nature of the capitalist economy itself and the ever more total grip this economy exerts on our lives, be it as employees or as consumers, this sphere determines the overall antidemocratic nature of our societies, even if they are nominally democratic:

---

[295] Gustavo Esteva, 'Development', in *The Development Dictionary*, 1992, 6-25, here 9.

[296] This can be most drastically seen in the EPZs which are hailed as the embodiment of free trade, yet in reality are "miniature military state[s]" (Klein, 2001, 204).

[297] In capitalism "the object of the game is precisely to produce inequality. The idea is that the division of society is fair if it takes place under fair rules. Equality of opportunity can thus be seen as a device for legitimizing economic inequality." (C. Douglas Lummis, 'Equality', in *The Development Dictionary*, 1992, 38-52, here 43)

[298] C. Douglas Lummis, 'Equality', in *The Development Dictionary*, 1992, 38-52, here 46.

Economic development means mobilizing more and more people into hierarchical organizations in which their work is disciplined under the rule of maximization of efficiency. And it means mobilizing more and more people as consumers, that is, people whose livelihood is dependent on the things produced by those big organizations. Both trends are antidemocratic. So even in a society with a "democratic" constitution, elections, free speech, and guaranteed human rights, economic development places a kind of antidemocratic Black Hole at the center of each person's life. (Lummis, 1996, 75)

Through the gigantic smokescreen of advertising and PR, which constantly tell us how free we are, we forget that we are becoming more and more dependent on industrial goods, big corporations and the state, and therefore progressively unfree, so that we "accept unquestioningly [our] human condition as one of dependence on goods and services".[299] You can very easily find this out if you try to break free from the senseless consumption cycles most people in the so-called developed countries are now locked into. Go out and try to buy just locally produced organic fruit and vegetables, or buy some wooden toys made of local, sustainable hardwood, or do any of your shopping without getting packaging you neither want nor need. You will find it nigh-on impossible, not to mention growing and sharing vegetables on communal land. Or try to bring up your children in a way respectful of the earth and their own inner potentials while at the same time trying to keep them away from the commercial pressures of the Pokémons, Barbies, Teletubbies and Harry Potters, which have at best *very* limited educational, let alone aesthetic and sensory value, but absurdly high profit-return value for the corporations.

The fact that the 'development model' is an ideology means that it, just like its brother globalisation, is colonising the minds of people. It has brought upon the earth a tremendous loss of cultural diversity, let alone biodiversity. It has acquired tremendous power over the imagination of people and has resulted in a dramatic impoverishment:

The mental space in which people dream and act is largely occupied today by Western imagery. (...) Moreover, the spreading monoculture has eroded viable alternatives to the industrial, growth-oriented society and dangerously crippled humankind's capacity to meet an increasingly different future with creative responses. (...) Knowledge, however, wields power by directing people's attention; it carves out and highlights a certain reality, casting into oblivion other ways of relating to the world around us.[300]

The international mainstream media do everything in their power to keep it that way so that we are increasingly faced with a paradox: the less self-sufficient and

---

[299] Ivan Illich, 'Needs', in *The Development Dictionary*, 1992, 88-101, here 89.
[300] Wolfgang Sachs, 'Introduction', in *The Development Dictionary*, 1992, 1-5, here 4-5.

self-determined we become, the more our "desires, fuelled by glimpses of high society, spiral towards infinity" (Sachs, 1999, 11), wrecking the earth and the poor peoples' livelihoods in the process.

Destruction of the livelihood of the poor and their resistance

There is another truth to be uncovered from underneath the debris of years and years of development ideology. There are many cultures, peoples and traditions on Earth which are in fact living a sustainable lifestyle:

> And if the rule of just distribution is to give to each his or her due, we need to understand that there have been in the world communities that organized themselves so as to give the land its due, the forest its due, the fish, birds, and animals their due. These communities, defined by development economics as at the absolute extremity of poverty, actually maintained in this way a vast "surplus", the great common wealth that was the natural environment in which they lived. (Lummis, 1996, 77)

Such lifestyles are certainly different from ours. Yet they are achievable by means of a reorientation of our values to basic needs and immaterial, spiritual fulfilment (as described in chapter 1.c on p. 57). That is a message we don't like, because we have been told all along that our way of life, our technology, our ideology, our political system, our culture is the "endpoint of human development" (Fukuyama). Suddenly, rather than being the winner of the prize for "best civilisation", it looks as if we are the fools and in fact guilty of some of the most hideous crimes in history. It seems that, rather than developing a deep understanding of our relationship with nature, of how nature works, of how we can fit into our life-support system, rather than developing free, self-determined forms of communal life, we have ridden high on our ideological horses, drawing the entire world into two world wars, immense ecological, cultural and political destruction, steam-rolling everything with our big technological tools, with no understanding of what we were and are effectively doing. Psychologically it is clear why we don't want to face up to that truth: if you have lived a lie all the time, it is very difficult to step back and accept responsibility for what you have created. It would mean acknowledging the dark side of the "Century of Development":

> If we can tear our gaze away from the fantasies of futurology and look at the real world around us, what we see are unprecedented forms of mass poverty, unprecedented forms of mass killing, unprecedented methods of regimentation, unprecedented pollution, destruction, and uglification of the earth, and unprecedented concentration of wealth and power in the hands of the few. (...) When we think of modernization and development, we tend to think of the International Style of the Bauhaus, high steel-and-glass buildings, quiet-running engines, airports, computers, and so on. We must recognize this image as a self-deception if we truly

are to look at things scientifically, and in a world-systems perspective. If development is a world-scale phenomenon, then everything that it has produced, and not just those parts that are pleasing to the eye or to the moral sense, must equally be called modern and developed. "Modern architecture" must be seen as precisely what virtually every major city in the Third World actually has today: steel-and-glass high-rise buildings plus slums built by squatters. For the slums are just as new as the high-rises, or newer. (...) Modernization and development never meant the elimination of poverty; rather it means the rationalization of the relationship between the rich and the poor. In this sense development includes not only the development of poverty but the development of the technology of management and oppression necessary to keep people in their position of relative poverty, quietly generating the surplus value that keeps the rich people rich. (Lummis, 1996, 66-67)

So rather than having brought paradise to the entire world, the "development of underdeveloped countries" is "neo-colonialism" in the disguise of a saviour which has "launched the most massive systematic project of human exploitation, and the most massive assault on culture and nature which history has ever known" (Lummis, 1996, 60). Indeed, this second wave of colonialism has destroyed most of what had been sustainable before:

- To the argument that international capital is necessary for the development of third and fourth world economies, [Daly and Cobb] respond that "we have come, as have many others, to the painful conclusion that very little of First World development effort in the Third World, and even less of business investment, has been actually beneficial to the majority of the Third World's people. ... For the most part the Third World would have been better off without international investment and aid [which] destroyed the self-sufficiency of nations and rendered masses of their formerly self-reliant people unable to care for themselves." (Orr, 1994, 167)
- In many countries environmentally sound practices carried on for generations by local communities have been overturned by interference from outside in the name of development.[301]
- Misery is frequently the result of enclosed or destroyed commons. Wherever communities base their subsistence on the renewable resources of soil, water, plant and animal life, the growth economy threatens nature and justice at the same time; the environment and people's life-support are equally degraded. In that context, for many communities sustainability means nothing less than resistance against development. To protect both the rights of nature and the rights of people, the enclosure of extractive development, a federal state with village democracy and an affirmation of people's "moral economies" are

---

[301] Margaret Macintosh, 'Development Education and Environmental Education: Working Towards a Common Goal', in *Sustainability, Development and Environmental Education: potentials and pitfalls*, 1994, 2-4, here 4.

called for. Searching for sustainable livelihoods in this sense means searching for decentralized, and not accumulation-centred, forms of society. (Sachs, 1999, 86-87)[302]

Box 13: Lake Victoria – development on the back of locals

The collapse of the lake's biodiversity was caused primarily by the introduction of two exotic fish species. (...) Although the introduced fishes devastated the lake's biodiversity, they did not destroy the commercial fishery. In fact, total fish production and its economic value rose considerably. Today, the Nile perch fishery produces some 300,000 metric tons of fish, earning \$280-400 million in the export market – a market that did not exist before the perch was introduced. Unfortunately, local communities that had depended on the native fish for decades did not benefit from the success of the Nile perch fishery, primarily because Nile perch and tilapia are caught with gear that local fishermen could not afford. And, because most of the Nile perch and tilapia are shipped out of the region, the local availability of fish for consumption has declined. In fact, while tons of perch find their way to diners as far away as Israel and Europe, there is evidence of protein malnutrition among the people of the lake basin.

A Guide to World Resources 2000-2001. People and Ecosystems: The Fraying Web of Life (2000), ed by UNDP, UNEP, World Bank and WRI (Washington, WRI), p. 7 [http://www.wri.org/wr2000/index.html].

One of clearest outcomes of the "Century of Development" is that the powerful in North and South – states, capital and industry – collaborated to rob the people of the world of what autonomy they had left. In the South this took the form that "the new nation-states, heavily committed to development, found in [Euro-American] science an attractive instrument for their project of remaking their people in the image of what they believed was an advanced form of man". This produced an increasing gap between the state, assuming "the role of developer", and non-Euro-American people. Of the latter, "science-fuelled developmentalism" "actually demanded greater sacrifices, more work, and more boring work, in return for a less secure livelihood. It required the surrender of subsistence (and its related autonomy) in exchange for the dependence and insecurity of wage slavery". Yet "the modern state does not understand, much less accept, the right of people not to be developed".[303]

All the above leads to the only conclusion that we need to abolish this destructive, monocultural notion of development and find new, local, self-determined ways of living a fulfilled life (rather than a life full of goods, which might well be an empty life in terms of meaning). No wonder that more and more people in the South are starting to actively and massively resist this destruction of their livelihood, this process of making them unfree. One example is the mass environmental movement of poor peasants in India because they are directly affected

---

[302] For a very vivid example of the destruction of a sustainable, self-sufficient community/way of life through electricity and TV, see Yeşilöz, 2000.

[303] Claude Alvares, 'Science', in *The Development Dictionary*, 1992, 219-232, here 226.

by the destruction, which led to 100,000s of people burning genetically modified crops at a "cremate Monsanto" day. A second are the people of Ladakh in the Himalayas (Norberg-Hodge, 2000). A third is the south Indian state Kerala (*Kerala: The Development Experience*, 2000, Richard W. Franke and Barbara H. Chasin, 1994 and McKibben, 1997, 117-169). A fourth example would be the resistance of the women of Maragua in Kenya against a loss of control over their produce through an export-led production regime (Bennholdt-Thomsen/Mies, 1999, 214-217) and a fifth the defence of their commons by the people of Papua New Guinea (Bennholdt-Thomsen/Mies, 1999, 145-149; see also the numerous other stories in the same book).

Political System

> [Democracy] is another fiction. As ever it is an oligarchy, and democracy has never functioned differently. There are the few who live at the expense of the many. (Müller, 1994, 191)

> We have been seduced into becoming secret accomplices in our own evisceration as active citizens. Two centuries after the battle cries of Liberty, Fraternity, and Justice, we remain as obedient as ever to a corporate state that is largely deaf to the genuine needs of people. And we have forfeited our identity as "producers" who are collectively responsible for our lives. (John Friedmann, quoted in Orr, 1994, 165)

> Our society is not really based on public participation in decisionmaking in any significant sense. Rather, it is a system of elite decision and periodic public ratification. (Noam Chomsky)[304]

If we recall what we have said about the necessary parameters of empowerment in a sustainable society (see p. 46), then it must have become clear how illusory the view is that 'liberal' democracy is the "final form of government" (Fukuyama). There are two main obstacles to real democracy and they are both considerable. Lummis sums up both of them: "Democracy will continue to have little staying power until the democratic movement has succeeded in establishing a democratic civil society and, in particular, in democratizing the world of work." (Lummis, 1996, 141-142)

Let's start with the latter. As we have discussed above (see p. 72), there is a inbuilt structural fault in a political system which claims to enable people to determine their own lives (after all democracy means literally "power to the people"), yet for the majority leaves one of the most determining aspects of their lives (their involvement with the economy as employees) under totalitarian rule without

---

[304] In James Peck, 'Interview with Noam Chomsky', in Chomsky, 1992, 1-55, here 42.

any chance of co- or self-determination. Now you could argue that this is not that bad as long as you have two totally separated spheres and at least one of them, the political one, is democratic. Yet this is not so. The antidemocratic structure of the economic sphere and the more and more total dominance of this sphere over all other spheres leads to a limitation and perversion of the notion of democracy even in the political sphere. Chomsky has aptly put this commodification of the public space: "In a perfectly functioning capitalist democracy, with no illegitimate abuse of power, freedom will be in effect a kind of commodity; effectively, a person will have as much of it as he can buy."[305] If that is the case, then the proclaimed liberty for all becomes not just a mockery but also an ideological justification for the power of the rich: "Given existing social discrepancies, the ideals of universal liberty (...) must remain purely notional for the mass of humanity, but provide the well-to-do with a legal and ethical framework to consolidate and justify their ascendancy over their fellow-citizens." (Holmes, 1995, 140-141) We can see here that this is an ideological problem *par excellence* in the way Marx has defined it: the specific interests of the few have to be "portrayed as the common interests of all members of society".[306] To understand why this is a problem we need to expose another myth of 'liberal democracy'. This notion starts from the assumption that all members of a democratic society are ideal citizens – i.e. fully informed and educated about all aspects of society, its functioning, its power structure, its economic system, its impact on the environment – and self-determined human beings capable of seeing through attempts by vested interests to influence decisions. If you don't assume this, you cannot in earnest talk about a level playing field in politics and therefore real democracy.

Yet reality in Euro-American-style democracies couldn't be further from the truth. Studies in Switzerland – which is all too often idealised as the embodiment of real democracy due to its instruments of direct democracy (initiatives and referenda) – have shown that only between 16 per cent and at the most 48 per cent of eligible voters "can be judged as adequately competent to decide on the issues put before them in ballots".[307] It is highly unlikely that the situation in other countries is much different. Hannah Arendt has identified already in 1950, when travelling through Germany, that part of the problem is the erroneous view that it doesn't matter which 'opinion' you hold, regardless of whether it might tally with any facts or not:

---

[305] Noam Chomsky, 'Equality. Language Development, Human Intelligence, and Social Organization' (1976), in *The Chomsky Reader*, 1992, 183-202, here 189; see also Ferguson, 1995, 25-26.
[306] Karl Marx, Friedrich Engels, *Deutsche Ideologie*, in Marx/Engels (1958), *Werke* (Berlin, Dietz), Vol. III, 47.
[307] Hanspeter Kriesi, 'Bürgerkompetenz und Direkte Demokratie', in *Demokratie radikal*, 1992, 92-100, here 92-93.

In all fields there is a kind of gentlemen's agreement by which everyone has a right to his ignorance under the pretext that everyone has a right to his opinion – and behind this is the tacit assumption that opinions really do not matter. This is a very serious thing, not only because it often makes discussions so hopeless (one does not ordinarily carry a reference library along everywhere), but primarily because the average German honestly believes this free-for-all, this nihilistic relativity about facts, to be the essence of democracy.[308]

This problem, which is usually hidden under the term pluralism and which mistakes ignorance as a virtue, has been exacerbated by postmodernism which insisted, at least in the most dominant strands, that "everything goes", that all opinions are equally valid and that there is no such thing as truth or facts. How devastating such a world view – which declares the ozone hole or biogeochemical limits of the Earth as mere social constructions (see more on that in chapter 2.b) – is for a sustainability perspective is clear. Alan Sokal has shown in his experiment with *Social Texts* how untenable such a position is and he has clearly spelt out the consequences:

Intellectually, the problem with such doctrines is that they are false (when not simply meaningless). There *is* a real world; its properties are *not* merely social constructions; facts and evidence *do* matter. What sane person would contend otherwise? (...) Theorizing about "the social construction of reality" won't help us find an effective treatment for AIDS or devise strategies for preventing global warming. Nor can we combat false ideas in history, sociology, economics and politics if we reject the notions of truth and falsity.[309]

Yet of course it is not by chance that such theories gain predominance in academia (at least the humanities) and that in political discourse people now very willingly accept the notion that *any* opinion is just as good as any other. It serves a very direct political purpose and decades of active propaganda have made sure that we think that way:

Ironically, even while corporate propaganda overwhelms democracy, it is able to create an ever-strengthening popular belief that the free-enterprise system which sponsors it is some kind of bulwark and guarantor of a democratic society: that is, a society where official policies and values are realistically within the free choice

---

[308] Hannah Arendt, 'The Aftermath of Nazi Rule: Report from Germany', in Hannah Arendt (1994), *Essays in Understanding 1930-1954*, ed. by Jerome Kohn (New York; San Diego; London, Harcourt Brace), 248-269, here 251-252.

[309] Alain D. Sokal, 'A Physicist Experiments With Cultural Studies', in *Lingua Franca*, May/June 1996, 62-64, here 63-64 [http://physics.nyu.edu/faculty/sokal/lingua_franca_v4/lingua_franca_4.html]. See for an extensive analysis of the higher nonsense produced by leading postmodern French philosophers: Sokal/Bricmont, 1998.

of a majority of ordinary citizens. Indeed it remains, as ever, an axiom of conventional wisdom that the use of propaganda as a means of social and ideological control is distinctive of totalitarian regimes. Yet the most minimal exercise of common sense would suggest a different view: that propaganda is likely to play at least as important a part in democratic societies (where the existing distribution of power and privilege is vulnerable to quite limited changes in popular opinion) as in authoritarian societies (where it is not). It is arguable that the success of business propaganda in persuading us, for so long, that we are free from propaganda is one of the most significant propaganda achievements of the twentieth century. (Carey, 1997, 20-21)

Carey provides a very thorough examination of how business propaganda has progressively managed to shape and determine public opinion in Europe and America since the 1920s, following the success of turning around public opinion in the US, predominantly opposed to war, with a tremendous propaganda effort and building quite unashamedly on Goebbels' propaganda triumphs during the Third Reich. It is his contention that the people in Euro-American democracies have been fooled, rather than enabled to lead a self-determined life:

> The "common man", instead of emerging triumphant, has never been so confused, mystified and baffled; his most intimate conceptions of himself, of his needs, and indeed of the very nature of human nature, have been subject to skilled manipulation and construction in the interests of corporate efficiency and profit. (Carey, 1997, 12)

Only if we un-learn this "brainwashing by liberalism" (Kurz, 1999, 783), this ideological view of human nature and re-emerge capable of seeing through this veil of propaganda will we be able to build anything resembling a democratic society, let alone a sustainable one. Until then it is simply ludicrous to assume that the power structure in Euro-American-style democracies "reflects the popular will in any significant way" (Carey, 1997, 23). Even classical liberal theorists like Walter Lippman had to concede that in bourgeois democracies the importance of propaganda actually *increases*, not decreases in comparison to a dictatorship:

> The manufacture of consent ... was supposed to have died out with the appearance of democracy. But it has not died out. It has, in fact, improved enormously in technique ... Under the impact of propaganda, it is no longer possible ... to believe in the original dogma of democracy. (Lippman, quoted in Carey, 1997, 23)

The reason for this is obvious. In totalitarian regimes, there is always the threat of physical violence and oppression to limit dissent with the powers-that-be. In democracies, however, the people have theoretically the chance to disempower the business elite and take over democratic control. In order to prevent them from doing so, you need a strong dose of propaganda to make the majority believe that in fact the interests of the few equal theirs.[310]

That these are not mere theoretical claims of an outdated counterideology can be easily shown. It is particularly ironic that the American presidential elections 2000 hung in the balance for more than a month. In the German-speaking press, the one statement you read most was that this result in no way meant that the model democracy of the US wasn't working; on the contrary, it still, so we were told, has to serve as a model for all other countries in the world.[311] What are they talking about?

Thomas Ferguson has conclusively shown that the US democracy is a "money driven system" in which the main political parties pay scant regard to the will of the people.

> The fundamental market for political parties usually is not voters. As a number of recent analysts have documented, most of these possess desperately limited resources and – especially in the United States – exiguous information and interest in politics. The real market for political parties is defended by major investors, who generally have good and clear reasons for investing to control the state. (Ferguson, 1995, 22)

If the entire system is money driven, if the real power lies with wealth, then quite obviously the political elite, determined to get into power or to stay in power, will prioritise the interests of those who finance the system. Ferguson calls this the investment theory of politics:

> *The investment theory holds that voters hardly count unless they become substantial investors. When the ranks of significant investors are limited to relatively small numbers of elite actors commanding disproportionate shares of politically mobilized resources, mass voting loses most of its significance for controlling public policy. Elections become contests between several oligarchic parties, whose major public policy proposals reflect the interests of large investors, and which minor investor-voters are virtually incapable of affecting, save in a negative sense of voting (or nonvoting) "no confidence".* (Ferguson, 1995, 28; emphasis in the original)

If we take this insight together with the fact that – ever more so with the advent of the internet – information is a commodity which depends strategically on how much money, time and education you can spend on acquiring it, it becomes clear that mere economic power as wealth immediately and necessarily translates into political power. Knowledge is power:

---

[310] Chomsky has shown that US policymakers have always well understood this (1992a, 12, 331).

[311] *Die tageszeitung*, for example, writes that the US is "ruining their good reputation worldwide" (10.11.2000, 1).

"Whereas [in commonly held assumptions about democracy] citizen-voters are expected to have well developed opinions about a wide range of issues, a focus on information costs leads to the expectation that only some voters – those who must gather the information in the course of their daily lives or who have a particularly direct stake in the issue – will develop a detailed understanding of any issues. Most voters will only learn enough to form a very generalized notion of the position of a particular candidate or party on some issues, and many voters will be ignorant about most issues." (Popkin) As a consequence it is not necessary to assume or argue that the voting population is stupid or malevolent to explain why it often will not stir at even gross affronts to its own interests and values. Mere political awareness is costly; and, like most of what are now recognized as "collective goods", absent individual possibilities of realization, it will not be supplied or often even demanded unless some sort of subsidy (at least in the form of advertising) is supplied by someone. (Ferguson, 1995, 25-26)

This has tremendous implications for the political system. In the absence of cheap access to relevant information and "strong channels that directly facilitate mass deliberation and expression", "a party system that is competitive (...) cannot prevent a tiny minority of the population – major investors – from dominating the political system. The costs that the voters must bear to control policy will be literally beyond their means." (Ferguson, 1995, 29; see also 87)

That this is not just the case in the US can again be seen in Switzerland, the other country that prides itself as the democratic paradise on Earth. There are possibly more institutionalised democratic tools in Switzerland than in most countries. Swiss citizens have the possibility of amending the constitution, introducing new laws and challenging most decisions of parliament through popular vote. And there is a constant stream of initiatives which are put before the people's vote. Yet studies have clearly shown that on the one hand the electorate, often unable to cope with the complexity of the issues at stake, are influenced by demonstrably false propaganda, and that on the other hand there is a very clear correlation between the amount of money one side can muster for the advertising campaign before a vote and the result of that vote. In other words, if you have enough money to spend on propaganda and advertising you can decide almost certainly the outcome of a vote.[312]

This is relatively easy to explain. If a large proportion of the electorate is genuinely ignorant about the issues at stake, it is rather easy to influence these people with horror scenarios, emotional blackmail and the most ludicrous, but utterly unfounded promises. In Switzerland there is a whole string of examples for this strategy of blackmailing people with lies. I will cite just two:

---

[312] Marc Spescha, 'Basisdemokratie ohne Volks-Mythen', in *Demokratie radikal*, 1992, 101-112, here 102.

- In 1984 two initiatives were put before the people which attempted to ease the transition of the Swiss energy provision towards renewables and to stop the non-sustainable use of nuclear power. Quite beyond any scientific basis the Swiss were threatened with slogans like "If you vote yes, it will be dark in your living room tomorrow", "You don't want to watch TV any more?" or "You don't want to take the train tomorrow?", even though it was proven beyond doubt that no such things could have happened. A similar scenario was rolled out in 2000 when an initiative to subsidise renewable energy for a limited period of time was rejected on the basis of false or misleading claims about the size of the subsidy, carefully avoiding any mention of the massive subsidies thrown at non-renewable energy (incl. nuclear energy) every year (see Myers, 1998).

- In 1998 an initiative to stop the abuses of genetical engineering and biotechnology was rejected by the population after an advertising and disinformation campaign by the powerful lobbies of chemical TNCs and the medical establishment succeeded in emotionally blackmailing the "politically mature" electorate in the desired direction. The most expensive advertising campaign to date used slogans like "Can you make peace with your conscience when you deny this cancer patient her only hope?" This against the background that not one single gene therapy has borne palpable results on the ground as yet.[313]

I am even allowing for the fact that there might not have been good arguments for or against these initiatives. But the crucial point here is that complex issues are reduced to "simplistic Yes/No alternatives, which don't allow differentiated points of view",[314] and these Yes/No votes are then determined, through political influence, to suit the vested interests with the most resources available, with little or no regard to facts. And the reason why this can be so is directly connected to the "dictatorship of the majority",[315] as Hans Henny Jahnn has called Euro-American democracy. Hösle has clearly put the case: "In general terms it is crucially important to understand that a democratic decision is not correct because it is taken by a

---

[313] According to a dpa (German Press Agency) news item gene therapy has been tested for ten years. So far 691 people have been treated, 4 have died as a direct consequence and none has been healed (widely reprinted, see for example 'Tod bei Gentherapie', in *Salzburger Nachrichten*, 8.5.2000). At the World Health Congress in Hanover experts cautioned against too much hope for gene therapy. Science, it was said, was only just starting to understand the issues, so realistically there was no hope for effective therapy for a long time to come ('Experten dämpfen Hoffnung auf Gentherapie', in *Rheinische Post*, 7.8.2000).

[314] Ruedi Epple-Gass, 'Volkssouveränität statt Systemlegitimierung', in *Demokratie radikal*, 1992, 113-124, here 116.

[315] In Switzerland, because of voter apathy, all the 'majorities' even in highly contested popular votes over the last decades were, in fact, minorities of the total population entitled to vote (Hans Tschäni (1990), *Das neue Profil der Schweiz. Konstanz und Wandel einer alten Demokratie* [Zurich, Werd], 142-143).

majority of the population or of parliament, but it becomes correct or wrong because of factual arguments." (Hösle, 1994, 132)

The other main obstacle to real democracy is the fallacy that democracy can be institutionalised and once you have 'free elections' and the like you have a democracy. Lummis described this problem when he called for a "democratic civil society" (1996, 142). If we agree, as argued above on p. 46ff., that democracy is the aim, not the means, that it is a state of mind and action which essentially is lived self-determination, it is clear that institutionalisation cannot be the end. It might provide some means, some tools to help and enable democracy, but 'democratic' institutions do not guarantee democracy. Lummis has explained this by using one of those icons which almost seem to embody democracy in common understanding:

> Free elections are an important democratic method – under some circumstances. In other circumstances elections may be a way for demagogues or rich landowners to take power. In the United States today, where election campaigns have been taken over by the marketing industry, they have little to do with empowerment of the people. The Nicaraguan election of 1990 was a parody of the free election: Vote for A and we will make war on you, vote for B and we won't. When someone sticks a gun in your ribs and says, "Your money or your life!" – that's a "free choice" too. (Lummis, 1996, 18)

He therefore concludes:

> It is more than a tautology to say that in order to hold the power the people must become a body by which power can in principle be held. Power cannot be held by people who live unresisting under the lie of state propaganda. It cannot be held by people who are convinced that dog-eat-dog competition is a doom from which human beings cannot escape, that the best we can hope for is a courteous state of nature. It is an illusion to think that an institutional change that drops power into the lap of the people stuck in such a state of mind will bring about democracy. The result may be as effective as pouring water into a sieve – unless, as can and does happen, the institutional change triggers a change in state of mind. But even to say that is misleading: democratic power does not fall from above, it is generated by a people in a democratic state of mind, and by the actions they take in accordance with that state of mind. It is the possibility of this change of state that is the power of the powerless. (Lummis, 1996, 35)

There is no democracy unless people actually take their fate into their own hands, rather than delegate it to representatives or institutions. Ferguson, in his study about our money-driven political systems quoted above, arrives at exactly the same conclusion. If you want to break the power of the powers-that-be then you have to become a power yourself:

Unless significant portions of [the electorate] are prepared to try to become major investors in their own right, through substantial expenditure of time and (limited) income, there is nothing any group of voters can do to offset this collective investors' dominance. (Ferguson, 1995, 28)

This is very important to state because it means that we can only solve both the problem of democracy and the problem of sustainability if we face up to *the* crucial question, the question of power. No amount of discussion and well-meaning pleading will ever change anything unless we address this question. Susan George, in her afterword to the *Lugano Report*, has made this unequivocally clear:

More times than I care to count I have attended events ending with a rousing declaration about what "should" or "must" occur. So many well-meaning efforts so totally neglect the crucial dimension of power that I try to avoid them now unless I think I can introduce an element of realism that might otherwise be absent. (...) Assuming that any change, because it would contribute to justice, equity and peace, need only be explained to be adopted is the saddest and most irritating kind of naivety. Many good, otherwise intelligent people seem to believe that once powerful individuals and institutions have actually *understood* the gravity of the crisis (any crisis) and the urgent need for its remedy, they will smack their brows, admit they have been wrong all along and, in a flash of revelation, instantly redirect their behaviour by 180 degrees. While ignorance and stupidity must be given their due, most things come out the way they do because the powerful want them to come out that way. (*The Lugano Report*, 1999, 181)

Bearing this in mind it is clear that the problem is not what to do, but mustering the political will to do it and that depends on the power structure:

Everybody knows perfectly well what "should" or "must" be done if fairer income distribution, an end to hunger, and so on are really the goals. The problem is not to persuade those who stand in the way of these outcomes that their policies are mistaken but to get power. The problem is not to repeat mindlessly what "should" or "must" occur but to begin by asking two simple questions:
- Who is responsible for the present crisis?
- How can we make them stop? (*The Lugano Report*, 1999, 182-183)

This also means, to pre-empt a finding from chapter 3, that any EfS that fails to address this question of power might as well abolish itself, since it will degenerate into an exercise in perpetuating the current state of affairs.

Yet even beyond these crucial limitations of the current political system, which mainly stem from the fact that the economy is the "Master Science", there are some obstacles within the political sphere which prevent a reorientation of our societies towards sustainability. They also make clear that we have to start reconceptualising democracy, to develop our "social imagination" (*Greening the North*, 1998, 189) in order to find the necessary sustainable solutions.

One of the most important reasons why our political system is less and less able to deal with the real problems of our world, let alone offer solutions,[316] lies in the fact that the terms of office for most of our ministers and MPs more or less forces all their focus onto short-term solutions and re-election, while the economic, social and ecological problems we face would necessitate a long-term perspective, ranging from fifty to a hundred or even more than 25,000 years: the half-life of plutonium is 24,360 years. At a time when the currently working nuclear power stations will be radioactive ruins, when the plundering of non-renewable resources will hit the limits, when today's children's children will have grown up and the consequences of the short-term measures taken today will become evident, most of the politicians responsible for these decisions will not be alive any more, let alone will the future generations be able to call them to account for what they have done.[317] But this means that our form of democracy runs into trouble:

> This argument [for democracy], as is well known, states that those people who are affected by a decision should take this decision themselves; only thus could it be avoided that their legitimate interests would be suppressed in the long run. This argument is not to be doubted, it just follows from it that the current form of democracy is illegitimate, because we take decisions on a daily basis which have consequences far beyond the territorial borders of our countries or the temporal boundaries of our generation. (Hösle, 1994, 127)

The (environmental) sins committed in the last decades just to buy the electorate for a (re-) election are quite possibly irreversible, and the visualisation of politics for the mainstream media, where good looks and well rehearsed single line slogans have become more important than concepts and solid solutions, are exacerbating the problems quite dramatically.[318] As always, it is very helpful to apply a perspective from outside in order to realise to what extent we have become blind to the limitations of our political system, just how incapable we have grown of transcending our eurocentric perspective. These limitations become blatantly clear if we substitute our dominant perspective, centring on personal power, career, success and social status, with an equally possible one, which focuses on social competence, community values, trans-generational face-to-face communication, non-commodified life

---

[316] This insight led more than 30 British NGOs, the Real World Coalition, to publish a book urging 'official politics' to start noticing what is happening on the ground (Jacobs, 1996).

[317] Compare the graph in *Factor Four*, 1997, 262 as well as Vine Deloria jr., 1978, 119-120.

[318] See Paul Virilio, 'Der Medien-Putsch', in *Lettre International*, (1994) 25, 30-31 and Postman, 1987. Also note that today it would be virtually impossible for someone like Roosevelt, bound to a wheelchair, to become president of the US, quite irrespective of how qualified s/he might be for the job.

and efforts to encourage understanding (Bowers, 1995; see also Steiner, 1998, 311). Here is a Native American perspective on Euro-American democracy:

> Amongst the white folk people who want to gain power for themselves compete with one another. These people have to travel through the country to gain support of the people over whom they want to govern. And due to the way they do that they will only have a majority of people on their side. This means that all they have to do is to promise this majority something, undoubtedly mostly at the expense of the minority. I don't know whether I could dream up a worse system. Just imagine: to compete for something which in reality is an honour, containing an immense responsibility. An honour which can only come into being if a community trusts somebody. And yet there are people who turn such things into competitions. Of course, the method used for that is determined by economic reasons. In the traditional system however – and this is true for every Indian nation I know – the spiritual and social leaders will be chosen by the overwhelming majority of a tribe, in most cases even unanimously. The candidates are put forward and if anybody is unhappy with them, then somebody else will be proposed. Thus there is no minority which feels cheated since their candidate has fallen through. In this sense the one chosen in that way is really representative of the people who have elected him.[319]

But there are two further limitations to 'liberal' democracy as currently practised. As long as the relatively democratic countries actually depend on and exploit undemocratic practices abroad we have a system of "imperial democracy" where 'democratic' states are surrounded "with a wall of military force and social discrimination, particularly against those in other states whose labor is exploited to provide the economic base for the leisure and liberty of the democratic citizens." (Lummis, 1996, 136) If that strategy is pursued long enough in a globalised context, it will in fact hit home:

> Union busting and cheap labor under military regimes abroad become union busting and wage cuts under the regime of the "democratic" free market at home. In short, yes, the attempt to establish democratic empire produces precisely what we have now, a situation in which radical democracy is subversive in the "democratic" countries. (...) For the radical democrat, imperial democracy is no longer a possibility. Lest it corrupt its own spirit, the struggle for democracy must be not the struggle only for a democratic country but for a democratic world. (Lummis, 1996, 137-138)

So we can only seriously talk about having a democracy if our entire system is democratically based, and doesn't benefit from other people's or nature's exploitation.

---

[319] Johnny Mohawk quoted in Claus Biegert, 'Akwesasne – wo das Rebhuhn balzt', in *Akwesasne. Indianische Texte aus dem Widerstand* (1978) (Munich, Trikont), 15-22, here 19-20.

But there is another dimension to the question of global democracy. In the age of globalisation more and more issues cannot be dealt with at a national level any more, be it climate change, resource plundering or migration of refugees, employees and corporations. These start to seriously limit the self-determination of individual states,[320] and need to be solved by transparent, international and democratic co-operation between countries.

The progress myth: Science and Technology

> Much as feudal rule was secured and legitimated through religious faith, the domination of capital is legitimated through faith in technology and the productive forces, and through the religion of progress. (Bennholdt-Thomsen/Mies, 1999, 176)

> Modern man is defined by progress. His self-esteem is rooted in it and it is his deepest justification for the ruthlessness he displays towards his fellow men and nature. (José María Sbert)[321]

> History teaches us that the promises of salvation of modern science and technology are always followed by sobering-up. (Hartmut Böhme)[322]

> Like a drug habit, technological addiction provides an experience of short-lived euphoria, followed by the need to acquire a more powerful fix as soon as possible. (Bowers, 2000, 177)

> Technology is inherently political. (Bowers, 2000, 192)

There is no doubt that one of the most powerful driving forces behind our unsustainable current system – besides the growth and development illusions, but intricately linked to them – is the progress myth; i.e. the idea that Euro-American science and technology, in all their variations, are by definition always a boon to humankind and that they will always solve all our problems if we just wait long enough:

> This tendency to see science and technology as having a transcendent rather than an instrumental value sometimes takes on an almost religious quality and makes it difficult to have a scientific discussion about science. Statements such as that we should leap into the scientific future, presented as self-evident, are themselves without scientific (i.e., empirical) foundation. The *scientific* future means whatever "comes next" in the logic of scientific advance. To say that this must neces-

---

[320] See Alex Demirovic, 'Ökologische Krise und Demokratie', in *Demokratie radikal*, 1992, 62-71, here 63-65.

[321] Sbert, 'Progress', in *The Development Dictionary*, 1992, 192-205, here 195.

[322] Böhme, 'Wer sagt, was Leben ist? Die Provokation der Biowissenschaften und die Aufgaben der Kulturwissenschaften', in *Die Zeit*, No. 49, 30.11.2000, 41-42, here 41.

sarily be good for human beings and their world is not to make a scientific statement but is rather to make a profession of faith in science. As for what "comes
next" technologically, nuclear holocaust or the invention of some ro-
botic/biotechnological cyborg that renders human life "obsolete" are as easy to
predict as any other scenario. (Lummis, 1996, 104)

We have known since the beginning of the Industrial Revolution, but at the very
least since Rachel Carson's *Silent Spring* and the disaster of nuclear energy that
the price we pay for this kind of 'progress' very often outweighs the benefits and
advantages, particularly if you look at it in a sustainability framework, namely not
externalising the costs and negative consequences onto others. Yet despite at least
three decades of sustained critique of this myth, on philosophical, social, political
and ecological grounds, it seems that it is so deeply ingrained in the popular mind,
nicely helped along by a constant flood of ads and media brainwash, that it holds
the same power as ever, as we shall see especially when talking about information
technology and the internet.[323]

The ideology of arrogance: **homo scientificus**

The dominant image of the scientist – believed in by many scientists almost as
much as by large sections of the public – depicts (almost invariably) a man who
will solve all the remaining mysteries, who is the pinnacle of knowledge and a
penetrating mind. Additionally, to connect it to the previous chapter, in the minds
of many Southern leaders, such as Nehru, the myth of development and science
collapsed into one.[324] However, there is a stark contrast at work here.

   On the one hand this arrogance, which starts from the image that man is
God and knows everything, has been ludicrously inefficient[325] and botched if you
compare it to nature's performance. It has also produced a long and steady stream
of mistakes, wrong assumptions and miscalculations with devastating consequences. Rather than starting from the assumption that "Nature knows best"
(Commoner's third "law" of ecology) because "ecosystems (...) are extremely
intricate and are unlikely to be improved by random tinkering" (Carter, 1999, 21-
22), we think that we are so fantastically clever that we can do *everything*, and
better, of course. Yet if you compare the results of human technology with

---

[323] This even goes so far that very sophisticated critics of science, such as Nowotny, still accept its
underlying notion of redemption, its doctrine of salvation and its progress myth (see Nowotny,
2000a). This reminds me of Chomsky's analysis of mainstream media, where he shows that 'conservative' and 'liberal' media, at first glance divided by an unbridgeable gap, in fact share the same
relatively narrow base of fundamental convictions (see Chomsky, 1989).

[324] Claude Alvares, 'Science', in *The Development Dictionary*, 1992, 219-232, here 225.

[325] One of the most obvious examples is the automobile where only 1 per cent of the overall energy is actually used to move the person in it, all the rest is lost in moving the vehicle, heat loss,
friction etc. (see Vester, 1997, 127 and *Natural Capitalism*, 2000, 24).

ecosystems, it doesn't end favourably for us: "On the whole [human technologies] are poor imitations (...) which try to mimic the principle, but with a far worse degree of efficiency, an abysmal energy rating and a ridiculously primitive organisation if compared to nature." The reasons for this are on the one hand that "nature had much more time than we have to develop organisms to perfect maturity, testing them many thousands of times by trial and error" (Vester, 1997, 220), namely 4.5 billion years; on the other hand, natural design starts from and is integrated into the laws of sustainable systems. Rather than trying to emulate these systems with "biomimicry" (*Natural Capitalism*, 2000, 14-16) which would by necessity guarantee sustainability since their design and structure would be compatible with ecosystem laws, we are still doing the opposite: We are "superimposing human structures" onto nature, producing a "still immensely primitive artificial world which is adverse to nature". This, in effect, is regress, "even though there are still a number of people who have the guts to call it progress" (Vester, 1997, 221-222). Just take the example of the one nation always branded about as the most successful one, and therefore our role model:

> All the same, the industrial system of the United States cannot subsist on internal resources alone and has therefore had to extend its tentacles right around the globe to secure its raw material supplies. For the 5.6 per cent of the world population which live in the United States requires something of the order of forty per cent of the world's primary resources to keep going. (Schumacher, 1993, 96)

Yet on the other hand, in terms of impact, both positive and negative, the scientific intervention, i.e. the driving force behind industrial systems, has completely transformed a lot of nature: it has given humankind very large spoons indeed to dip into the planetary chocolate cake, to speak with Wackernagel.[326] It has long been understood why Euro-American science has been so 'successful' and the reason for this provides us with a direct understanding of the negative consequences this has brought upon us. The Cartesian world view, which dissects mind from body and allows us to take up a position as the distanced observer of nature, able to put nature as an object before us, to cut it up and deconstruct it into its constituent parts, has enabled the modern scientific revolutions to take place. The reductionist approach of the natural sciences, digging ever deeper into the parts, yielding with that an understanding of how things are constructed and built, has brought us many of the most astonishing advances in medicine, biology, physics and chemistry. It has also led to an ever more specialised scientific enquiry where biologists often don't understand each other's work any more because one concentrates on marine biology and the other is a biotechnologist.

---

[326] Mathis Wackernagel (1997), *Ecological Footprints of Nations Report and Slide Show* (Toronto, ICLEI [http://www.iclei.org/iclei/ecofoot.htm]).

Yet all these advances in understanding, which they undoubtedly are, have been bought at a price. Apart from the fact that Euro-American science's claimed 'objectivity' has been shown to conceptually favour industrial processes over natural processes, which tilted it towards unsustainability from the very start,[327] all these advances are flawed even more fundamentally because they start from the premise that men shall control nature and that nature exclusively "exists for the convenience of man" (Carson, 1991, 257). Hannah Arendt has pinpointed the problem:

> But it could be that we, who are earth-bound creatures and have begun to act as though we were dwellers of the universe, will forever be unable to understand, that is, to think and speak about the things which nevertheless we are able to do. (...) If it should turn out to be true that knowledge (in the modern sense of know-how) and thought have parted company for good, then we would indeed become the helpless slaves, not so much of our machines as of our know-how, thoughtless creatures at the mercy of every gadget which is technologically possible, no matter how murderous it is. (1958, 3)[328]

What Arendt is saying in essence is that scientific knowledge has proliferated at an enormous rate and constantly produces technical applications based on it, but we are not able any more to understand *what* in effect we are doing and setting free in the world, let alone take responsibility for it. "Think and speak about the things which we are able to do": this stands for complex understanding. There is a large gap, Arendt is claiming, between our cleverness in producing the most astonishing technical gadgets and our ability to fully grasp not just how you produce such a thing, but what the consequences of its production are: how does it interact with nature, how does it change the human-human relationship, what does it do to the social fabric, is it compatible with our desired political system (and so on)? All these questions, of course, would need to be answered in a sustainability framework (see p. 73), yet usually the only one that is posed and answered with yes is: is it technically possible?[329]

---

[327] Claude Alvares, 'Science', in *The Development Dictionary*, 1992, 219-232, here 222-224. Historically this comes as no surprise, given the fact that industrialism and Euro-American science are twin brothers, dependent on each other from the start.

[328] See, for a more recent formulation of the same problem: "Nowadays, there is much talk about the technology gap between industrialised and developing countries. There is much less talk about the fundamentally more dramatic *intellectual* gap within our *own* society, namely the fact that (...) our mental-psychological constitution has not kept pace with the technological development." (Vester, 1997, 469)

[329] See Günther Anders who has pointed out that we are caught in the belief that "we are permitted to, no: that we should, no: that we must, do what we can do" (Anders, 'Vorwort zur 5. Auflage', in Anders, 1988, VII).

The bulk of evidence to prove that this reductionist approach, which screens out any consequences, costs and impacts, is effectively alienating us from our life-support system and ourselves, is mounting by the day. From DDT to CFC and PVC, from nuclear energy to genetic engineering, from large dams[330] to the private car: almost all scientific breakthroughs leading to technical applications were originally viewed as solely beneficial to humankind until it turned out that they have negative consequences. The main reason why this happens is that we rush, every time, into technical applications without ever stopping to perform a life cycle analysis, an overall impact assessment to see whether the claims made by the vested interests actually do match up to what we should expect. Understanding as much leads to the "conclusion that the only future for a series of once celebrated triumphs of scientific-technological progress lies in renunciation", particularly with regard to "the use of atomic energy, the chlorine industry, most aspects of synthesizing chemistry, reliance on the automobile, and industrialized and chemicalized agriculture".[331]

Long before the current debate on genetic engineering, Frederic Vester made clear what is at stake, to take just one example:

> Since we have the tendency to technically implement every little partial new insight, before we have even attempted to think through the context and consequences (instead of starting all of our interventions in the environment only from already known interconnections), [genetics] in particular poses the threat that we switch the points for the future so that there is no turning back. (Vester, 1997, 159)

But why are people so drawn to and fascinated by this reductionist scientific enterprise which is so blatantly violating the precautionary principle – leaving aside the drive to conquest, to victory in the "battle with nature" which, when winning, humankind is bound to lose (Schumacher, 1993, 3)? Vandana Shiva has identified in the following example – where biotechnology can stand for much of the dominant scientific world view – the underlying fundamental reason for this need to control nature via knowledge and technology, namely biophobia:[332]

> A world-view that defines pollination as "theft by bees" [quote from Cargill] and claims that biodiversity "steals" sunshine [quote from Monsanto] is a world-view

---

[330] See *Dams and Development: A New Framework for Decision-Making. The Report of the World Commission on Dams* (2000) (London, Earthscan) [http://www.dams.org]; Lori Pottinger, 'Dammed if you do', in *The Ecologist*, 31 (2001) 1, 50-51; and the concrete experience of nation states/ dam builders oppressing local people: LS Aravinda, 'People's Knowledge in a Paperless Society. Organizing against the Narmada Dam', in *Z Magazine*, 14 (2001) 1, 44-49.

[331] Otto Ullrich, 'Technology', in *The Development Dictionary*, 1992, 275-287, here 281.

[332] See Gregory Cajete, 'Reclaiming Biophilia. Lessons from Indigenous Peoples', in *Ecological Education in Action* (1999), 189-206, here 190; and Orr, 1994, 136.

which itself aims at stealing nature's harvest by replacing open, pollinated varieties with hybrids and sterile seeds, and at destroying biodiverse flora with herbicides such as Monsanto's Roundup. The threat posed to the Monarch butterfly by genetically engineered bt. crops is just one example of the ecological poverty created by the new biotechnologies. (...) When giant corporations view small peasants and bees as thieves, and through trade rules and new technologies seek the right to exterminate them, humanity has reached a dangerous threshold. The imperative to stamp out the smallest insect, the smallest plant, the smallest peasant comes from a deep fear – the fear of everything that is alive and free. And this deep insecurity and fear is unleashing violence against all people and all species. (Shiva, 2000, 19)

Alvares has further elaborated *why* science has to hate tribal people, bees and silkworms: they "all process the resources of the forest at ambient temperatures, and hence without the polluting side-effects of waste heat and effluent associated with big industrial processes" with "high energy input".[333] When it comes to resource efficiency and zero emission production, science and technology cannot but end up with an inferiority complex compared to nature. But in addition to this fear/hate of nature which provokes the urge to control or even destroy it, there is another reason for the destructiveness of scientific and technological 'progress': illiteracy. And that, it has to be said, is true for the natural sciences just as much as for the humanities.

It is often astonishing how naive and ignorant major biologists, chemists, physicists and engineers are when it comes to social and political matters. Their historical, political and social understanding is often at best *very* limited and blindfolded by eurocentric assumptions (see Bowers, 1997, 49-50). That is *not* taking the moral high ground; this does happen equally with social scientists and humanities scholars such as me, when we try to reach an understanding of biology or physics. Sokal/Bricmont's (1998) close look at postmodern philosophers dabbling in quantum physics, mathematics and chaos theory should be a stark warning for 'soft scientists'. This illiteracy should make us alert to the necessity of *real* transdisciplinary dialogue. We therefore do not need the kind of colonising attitude, itself pure single-minded arrogance, of, say, evolutionary psychology or sociobiology, which wants to impose one single scientific model (fundamentalist Darwinism) onto all human understanding (and, besides, has a political agenda which legitimises the status quo and prevents change). As *Alas, Poor Darwin* (2000) shows, this would not be scientific progress, but a step back in understanding our human condition. Hilary Rose's chapter 'Colonising the Social Sciences?' (*Alas, Poor Darwin*, 2000, 106-128) demonstrates convincingly the blindness of Darwin's theory to its own contemporary context, which in fact shaped it considerably. Rose draws the following general conclusion:

---

[333] Claude Alvares, 'Science', in *The Development Dictionary*, 1992, 219-232, here 223.

After Chernobyl, BSE and GM food the public is no longer content to accept the reassurance of industrial and governmental scientists. Their view is seen as partial in two senses. First, because their reductionist training hinders them from seeing and understanding the whole human ecosystem, although it is within this that risk takes place; and, second, because neoliberal governments and industry are more interested in wealth creation than in quality of life. (ibid. 116)

This implies two things. The first is that scientists (and those transforming their findings into technologies), just as much as anybody else, have to adhere to the precautionary principle (see p. 40). Secondly, to break through the limitations of the reductionist approach – which by necessity has to produce results which might be entirely sensible in the relevant specialist sphere, but which are bound to have unforeseen consequences in other spheres – we need a new, inter- or transdisciplinary approach to science: "Coherent and imaginative interdisciplinary thinking which refuses to be limited by the inherited academic division of labour is urgently needed."[334] The most important point to make in this context is that we need to reorientate scientific investigation. Rather than starting from very special, limited problems, it should start from complex real-life problems. Attempting to solve those leads automatically to transdisciplinary approaches and transcends the boundaries of specialist knowledge.

Let me come back to the precautionary principle. If we want to achieve a sustainable society, then no part of society can pursue its fancies irrespective of the whole. This means that science and technology have to fit into a sustainability framework in terms of impact and they have to start assuming responsibility for their actions. The prudence of the precautionary principle is based on two insights: Firstly, our ignorance. I have quoted above how little we effectively know about the biosphere (see footnote 119 on p. 61). This, rather than encouraging us to ride roughshod over everything, should lead us to the following conclusion: "We should judge every scrap of biodiversity as priceless while we learn to use it and come to understand what it means to humanity. We should not knowingly allow any species or race to go extinct." (Wilson, 1992, 351) The consequence of this is that we have to use whatever we need to use with the utmost care, but also that we need to carefully evaluate what we plan to do in terms of impact:

> Besides societal and human values, there is also an intrinsic value to be attached to life itself and the life supporting systems. This means that even if something is judged as valuable within society, its costs in terms of negative effects on those systems could be so large as to make the total balance negative.[335]

---

[334] Ted Benton, 'Social Causes and Natural Relations', in *Alas Poor Darwin*, 2000, 206-224, here 222.

[335] Holmberg *et al.*, 'Socio-Ecological Principles for a Sustainable Society', in *Getting Down to Earth*, 1996, 17-48, here 39. A very good example of this intervention into a very complex system

Wilson as a biologist and Dürr as a physicist are highly critical of claims of omnipotence by scientists who essentially allege that we can easily survive in a totally man-made world, with the natural resource base destroyed: there is simply no scientific evidence whatever to uphold such *science fictional* wishes (see Wilson, 1992, 347-348 and Dürr, 2000, 109). I just would like to recall Vester's contention, when faced with this claim that the scientists will come up with the saving solution in the last instance: Come up with what? "New thermodynamic laws", "a shortening of the half-life of radioactive isotopes", "a change of gravity" or

> even a new mathematics and new system laws which decree that 2 and 2 is only 3 and that life can exist without integration and interaction, without structure and communication, without matter and energy exchange, without evolution and diversity of species? (Vester, 1997, 487)

It is time to face up to Gregory Cajete's insight that the first barrier we have to overcome on the way to a sustainable society is "modern society's psychology of denial and its unwillingness to admit that science and technology cannot solve the ecological crisis."[336] The reason for this is obvious because "faith in progress is faith in a purely intellectual, mathematical, scientific knowledge 'liberated' of all moral constraint and ethical context".[337] Yet reality is complex, built up from various spheres – we have differentiated between ecology, empowerment, equity, economy and equipment – and no single sphere can in isolation provide the solution. Any scientific-technological solution, however clever and congenial it might be, has to be implemented by humans in a social context at a given time in a certain cultural, economic and political context.

But if this is so, then science and technology have to become sensitive to these contexts. And that means that any action which is irreversible or produces a large impact becomes more and more questionable. As we have said above, a properly applied precautionary principle would mean prevention of any negative impact rather than (expensive) rectification; it would also mean the reversal of the burden of proof (see p. 75). Before any new devices and technologies are implemented, they would have to prove their positive overall balance, rather than that we would need proof of their negative impacts. All this points to the necessity of the revival of the notion of responsibility on the part of science and technology.

---

without considering the consequences is the impact modern medicine and hygiene had on population growth. While before death and birth rates kept each other in check, the sudden drop in infants' death and the prolongation of the average lifespan had tremendous and unforeseen impacts on the entire society (see Vester, 1997, 173-174 and *The Lugano Report*, 1999).

[336] Gregory Cajete, 'Reclaiming Biophilia. Lessons from Indigenous Peoples', in *Ecological Education in Action*, 1999, 189-206, here 192.

[337] José María Sbert, 'Progress', in *The Development Dictionary*, 1992, 192-205, here 200.

There is no hiding behind the neutrality of science; if scientists want to act within the human community then they have to bear the responsibility for their deeds.

> It is obvious that the responsibility of a scientist grows in equal measure with the escalating might and precision of the instruments that he leaves to an ignorant society and its elected representatives. (...) Since the consequences are on principle unforeseeable responsibility demands in reality that he doesn't allow such situations to develop in the first place. (...) The ability to act responsibly depends not only on the expert's knowledge of his field but also on judiciousness, on the ability to see his own knowledge and his own skills in the context of the wider reality and then to draw the necessary conclusions for his own perspective from this topological overview. Scientists today are highly specialised; they only rarely possess the intellectual breadth which would allow them to act responsibly, given the openness of the future. (...) This is why concerned scientists should never cease to demand that society provides institutions within which, in collaboration with the public, open and free discussions about all the contentious issues with regard to sustainability are held. (Dürr, 2000, 193-194)

This has rather far-reaching consequences for any sensible EfS, as we shall see in chapter 3. Suffice it to say here that an assessment of the problems created by a misunderstood notion of the freedom of research leads to the conclusion that, as in the other spheres, we need to arrive at an acknowledgement of limits with regard to science. If scientists cannot guarantee both the reversibility and the beneficial nature of their inventions then they (and society) will have to learn to say no. If we want to pride ourselves on being the only *thinking* species and a free one at that, then we should be highly critical of any inevitability claims. If scientists, technocrats or politicians tell us that we *cannot* but go with the newest technical fads, frequently alluding to the unstoppable train of progress, then we should be very clear in our minds that if we give in to this emotional blackmail we sacrifice at the same time our freedom to determine our own lives *and* our ability to stand back and adequately assess any proposed path of 'development' according to its merits and faults. Only such an overall assessment should set us going down a certain route, not threats and bribery, sticks and carrots.

This implies that science and technology – and any other profession or activity, for that matter – ought to become self-reflective, i.e. start to reflect upon their own history, their ideological preconditions, their dependency on certain economic and/or political structures,[338] their impact, and upon the relationship between what they claim and what they deliver.

---

[338] It is clear, for example, that "technologies are social products, not the result of some inevitable chain reaction in which a scientific discovery leads inexorably to a particular technological innovation" (*Fool's Gold*, 2000, 72).

When [science] starts to look into the social background in the context of research on sustainability, it will sooner or later hit upon itself. Then it will have to face up to the fact that a lot of our current problems are a consequence of scientific-technological progress (and of course its commercialisation). This requires – apart from an assessment of the role of science in a sustainable society by an external perspective and by society as a whole – a thorough self-reflection within science. How can we avoid in future that science itself always is among those who cause the problems? (...) The central question therefore is: How can we make sure that science itself – including the science of sustainability – is able to meet the requirements of sustainability? (Steiner, 1998, 310-311)

To put it in a nutshell: "Ecology doesn't need to become a science, but science needs to become ecological" because the "ecological crisis is very directly also a crisis of science".[339] Defila/ Di Giulio have attempted to define what the parameters of a responsible science would have to be if it wants to bring this crisis to an end:

Science is environmentally responsible if it works in an interdisciplinary way, reflects upon its assumptions and consequences and if it offers matching education. In addition it is also assumed that environmental responsibility in this sense cannot be delegated to individual university departments and faculties or to special degree schemes (such as environmental institutes, ecology/ environmental science degree schemes), but has to be the duty of science as a whole.[340]

But there is one other fundamental aspect to be considered. As Ullrich has aptly stated, also with a view to the eurocentric notion that the South will find the path of sustainable development with the help of technology transfer from the North:

It requires a lengthy search to find anywhere in this gigantic mountain of industrial processes and products examples that are not part of the system of externalizing techniques of plunder and which might be recommended without reservation to the Third World. (...) The scientific civilization of the West has scarcely any technologies on offer truly suited for the future – that is, humane and appropriate over the long term to nature.[341]

---

[339] Philipp Egger, 'Neue Umweltlehrgänge an den Universitäten; gesellschaftliche Notwendigkeiten und inneruniversitäre Möglichkeiten', in *Umweltbildung in Theorie und Praxis*, 1998, 157-165, here 161.

[340] Rico Defila and Antonietta Di Giulio, 'Was ist die spezifische Umweltverantwortung der Wissenschaft?', in *Umweltproblem Mensch*, 1996, 483-505, here 496.

[341] Otto Ullrich, 'Technology', in *The Development Dictionary*, 1992, 275-287, here 281-283.

The anti-democratic scientific-technological complex

> The institutionalization of knowledge leads to a more general
> and degrading delusion. It makes people dependent on having
> their knowledge produced for them. It leads to a paralysis of
> the moral and political imagination. (Illich, 1973, 85-86)

Today, it is very difficult to realise the full extent to which our lives are determined by the scientific-technological complex. We get sudden glimpses of it when there is a fuel shortage as in autumn 2000 or when a storm knocks out the electricity supply. But apart from these brief glimpses, we suppress this dependency as thoroughly as we have suppressed the knowledge about our ideological indoctrination with regard to the totalitarian nature of our economic system. Only when we look back into history (see, for example, Llewellyn, 1991) or immerse ourselves in the lives of the despised poor people of this world, do we realise how completely even our imagination has been poisoned by the ideology that the newest, fastest, slickest and fanciest new technologies are a) always the best and b) crucially necessary to survival. Against this, we have seen with Lummis (see p. 56) that "most of the technologies that a human being really needs to live an orderly, comfortable, and healthy life are ancient" (1996, 105) and that we can easily prioritise food, clothing and shelter over robots, computers and heart transplantation. In this respect, it is again immensely helpful to adopt the Gandhian perspective elaborated above (see p. 55) in order to realise the full impact and extent of our technological 'civilisation'. Those at the receiving end and *not yet* brainwashed into the progress myth feel it most sharply: "If one attempts to live close to the peasants or within the bosom of nature, modern science is perceived differently: as vicious, arrogant, politically powerful, wasteful, violent, unmindful of other ways."[342]

The inherently and deeply antidemocratic nature of our Euro-American scientific-technological project can be seen on three levels: firstly structurally, because it is an exclusive concept, delegitimising all other forms of knowledge; secondly, in terms of implementation, because it is neither developed nor introduced democratically; and thirdly, because it is remote from life, but close to totalitarian-military concepts.

Before you throw up your hands in horror at such a Luddite interpretation, I would like to state that this is not intended as an undifferentiated indictment of science and technology as such, but of a blind acceptance of a faith in one particular kind of science and the willingness to look away from the facts relating to the dependence of this science and technology not only on the laws of nature but also on identifiable economic and political power structures. It is an invitation to look at the full picture, rather than following the official PR which sells you five per cent of the picture as the whole story. Perhaps one of the most striking exam-

---

[342] Claude Alvares, 'Science', in *The Development Dictionary*, 1992, 219-232, here 232.

ples of this, particularly because it focuses on the most advanced military technology, were the laser-guided bombs used against Iraq in the 1991 Gulf War. Whenever I asked anyone afterwards about what they remembered from this war (and I have systematically done this with my Media Course students for the last ten years) they answered: "the laser-guided bombs which hit their target with clinical precision". Full scores there for the military propaganda of the Pentagon which did everything in its power to convey this message (in other times the procedures used were called censorship). Zero score for the truth, though, since after the war this myth of technological perfection and clean war was utterly destroyed: laser-guided bombs accounted for just 7 per cent of the total bombs dropped, the remaining 93 per cent being conventional bombs dropped by B-52s in indiscriminate carpet bombing, just as in the Vietnam war. More interestingly, 10 per cent of the laser-guided bombs missed their target.[343]

The totalitarian exclusion of all other forms of knowledge

To concerned scientists it has always been clear that science, if at all, can only give answers in specific and limited areas. Especially in the context of sustainability, we certainly need science to provide us with detailed and accurate understanding of ecosystems or the workings of living organisms. Yet it is equally clear that such knowledge and understanding only plays a certain part in the overall context of sustainable solutions. As the "Century of Development" has shown with numerous failed technological 'aid' projects, even the best technical solutions will backfire if they are not embedded in and driven by democratic control, and the real needs and empowerment of those affected.

This insight, however, that science and technology are just one and not even the most important part in a complex mixture of factors, is not reflected either in the way scientific-technological progress has been embraced *or* in the way it is inherent in the belief-structure of modern science. As far as the former is concerned we have already noted that for most political leaders of the past century, and irrespective of whether in the North or South, there has been a simple and automatic equation between scientific-technological progress, economic growth, 'development' and the well-being of people. Postman has an interesting psychological theory why this is so. He claims that by now we cannot believe any more "that history itself is moving inexorably toward a golden age." But "the idea that *we* must make our own future, bend history to our own will" is so frightening and such a "psychic burden" that, in order to compensate,

we have held on to the idea of progress but in a form that no eighteenth-century philosopher or early-nineteenth-century heir of the Enlightenment would have embraced – could possibly have embraced: the idea that technological innovation

---

[343] John R. MacArthur, 'Operation Wüstenmaulkorb', in *Die Zeit*, No. 9, 26.2.1993, 44.

is *synonymous* with moral, social, and psychic progress. It is as if the question of what makes us better is too heavy, too complex – even too absurd – for us to address. We have solved it by becoming reductionists; we will leave the matter to our machinery. (Postman, 1999, 41)

Rather than facing up to the reality that no technological gadget and no cloned robot will solve the complex questions of reality for us – except by taking any remaining autonomy away from us – we prefer to be lulled into the illusion that machines can do what we can't (see below on computers (IT) and education). This, then, leads to rather strange problem-solving strategies: "Upon recognizing a problem caused by technology, technologists naturally want to layer on still more technology." (Stoll, 2000, 139)

In terms of the latter, namely its conceptual design, Euro-American scientific methodology is exclusivist, since it claims that its methodology guarantees that it establishes the ultimate and *only* truth about a matter. Turned around, this means that all other forms of knowledge are disregarded and devalued.[344] Alvares talks in this context of "science's dictatorship" and notes the fact that most intellectuals in most nations have "succumbed to the totalitarian temptation of science". This totalitarian nature lies in the following:

> While science itself advanced its knowledge by dissent, by the clash of hypotheses, it summarily dismissed dissent from outside the scientific imperium regarding either its content or its methods and mode of rationality. The non-negotiability of scientific assumptions, methods and knowledge became a powerful myth elaborately constructed over several centuries.[345]

If you take this together with the abstractness of science and its obsession with technological fixes, it is clear that it produces a very limited picture of reality and therefore we shouldn't be surprised if it fails people and their real needs. Angayuqaq Oscar Kawagley and Ray Barnhardt tell a very interesting story in this context. Some scientists of the State Department of Fish and Game and the Department of Natural Resources wanted to do research in the Minto Flats, Alaska. They met with elders of the native people living there and basically regarded all this as a one-way process. These scientists assumed that only they knew how to acquire knowledge and that they would have to impart this knowledge to the ignorant natives. There were five scientists with different specialisations, all going about their ways with different methods, and ignorant of each others' approaches.

---

[344] As happens when soil scientists ignore or belittle the holistic knowledge of farmers about their soil (see Patricia Fry, 'Wie Bauern und Bäuerinnen Bodenfruchtbarkeit sehen: Ein Vergleich mit naturwissenschaftlichen Sichtweisen', in *Shifting Boundaries of the Real: Making the Invisible Visible* (2000), ed. by Helga Nowotny and Martina Weiss (Zurich, vdf Hochschulverlag), 157-176).

[345] Claude Alvares, 'Science', in *The Development Dictionary*, 1992, 219-232, here 229.

They were then completely dumb-struck when they were confronted by one elder of the natives, Peter John, who could provide them more or less offhand with most of the information they wanted to find out in the first place, and could indicate where and why they would run into trouble with the proposed methodology and their elaborate technical equipment; all this on the basis of accumulated knowledge over generations and an intimate knowledge of the area through long-term first-hand experience. In the end, it turned out that the ignorance was somewhere else than anticipated:

> While the scientists with their specialized knowledge and elaborate tools were well intentioned, the gulf between their compartmentalized, limited-time-frame view of the world and the holistic, multigenerational perspective of Peter John appeared insurmountable. The fish and game people couldn't see beyond their constituent areas of expertise to connect with what the elders were trying to tell them, though the Minto people had a quite sophisticated understanding of what the fish and game specialists were talking about.[346]

This leads to another crucial point about which we tend to delude ourselves. We always, as with the progress myth, assume that only what we consider positive – say, increased convenience through private transport – is happening. But, of course, the side effects and negative impacts, the other side of the coin, are happening as well. And this is also true for science (and, as we shall see, for Euro-American-style education): Science does not by default increase the overall stock of knowledge, because it devalues "ordinary people's epistemologic rights":

> It is an illusion to think that modern science expanded possibilities for real knowledge. In actual fact, it made knowledge scarce. It over-extended certain frontiers, eliminated and blocked others. Thus it actually narrowed down the possibilities for enriching knowledge available to human experience.[347]

The imposition of technological 'solutions' and the destruction of independent culture/ Ideology of neutrality

Because of this totalitarian, universalist assumption of modern science, its claim to be equally valid, irrespective of historical and cultural differences, and because of its universal acceptance by decision makers, it became a globalised empire, driving economic globalisation, with all the normal consequences of empires:

> All imperiums are intolerant and breed violence. The arrogance of science concerning its epistemology led it actively to replace alternatives with its own,

---

[346] Angayuqaq Oscar Kawagley and Ray Barnhardt, 'Education Indigenous to Place. Western Science Meets Native Reality', in *Ecological Education in Action*, 1999, 117-140, here 125.
[347] Claude Alvares, 'Science', in *The Development Dictionary*, 1992, 219-232, here 230.

superimposing on nature new and artificial processes. Naturally, the exercise provoked endless and endemic violence and suffering as the perceptions of modern
science sat clumsily and inappropriately on natural systems.[348]

It is quite obvious that the dominant paradigm of science/technology is anti-democratic. Shiva has distinguished between two scientific paradigms to highlight
the difference:

> In the dominant paradigm, technology is seen as being above society both in its
> structure and its evolution, in its offering technological fixes, and in its techno-
> logical determinism. It is seen as a source of solutions to problems that lie in soci-
> ety, and is rarely perceived as a source of new social problems. Its course is
> viewed as being self-determined. In periods of rapid technological transforma-
> tion, it is assumed that society and people must adjust to that change, instead of
> technological change adjusting to the social values of equity, sustainability and
> participation.
>
> There is, however, another perspective which treats technological change as a
> process that is shaped by and serves the priorities of whomever controls it. In this
> perspective, a narrow social base of technological choice excludes human con-
> cerns and public participation. The interests of that base are protected in the name
> of sustaining an inherently progressive and socially neutral technology. On the
> other hand, a broader social base protects human rights and the environment by
> widening the circle of control beyond the current small group. (Shiva, 1991, 231-
> 232)

It is fundamental to understand that distinction. The dominant paradigm operates
as any good ideology: it hides the real interests behind a discourse of legitimisa-
tion. The view that science and technology are neutral, universally applicable and
by definition 'correct' means that people have to surrender all self-determination
and autonomy to this impersonal process of infinite progress: TINA, or we have
no choice but to follow the holy path.

The second perspective – and this is borne out by the history of science
and numerous examples – holds that there is not just one, unchangeable, necessary
path of technological development, but that this path is only entered upon because
it is the path that favours the powers-that-be. It is not at all the case that scientists
freely roam the world of things to research and then develop their theories, insights and so on in an ideal world without any pulls and pressures. The reason
why certain technologies are favoured over others and are then sold to the public
as the only viable ones is closely connected to the interests of power. There is no
doubt that big multinational corporations, using massive industrial processes, are
very prone indeed to power concentration and domination of the relevant markets.
That is why such technologies as industrial agriculture, nuclear power, large

---

[348] Ibid.

dams, genetic engineering, petrochemicals, and the automobile have been favoured over devolved, decentralised processes such as organic agriculture or local
power generation through renewables. We have already seen in our critique of the
myth of global capitalism that it is a crucial ingredient of capitalist power to make
people dependent on the system, both as employees and as consumers. As soon as
people start to grow their own food and produce their own energy, corporate control over them is lost.

That clearly is an issue of democracy. Lummis has shown that this notion
of the "seemingly inescapable logic" of science and machines and the idea that
they are somehow above the political discourse is at the heart of the problem:

> If something deeply affects the order of our collective life and we are taught that
> we have no choice about accepting it when in fact we do, that is a problem for
> democracy. In other words, the doctrine that machines should never be judged
> and chosen by political criteria is itself antidemocratic. (Lummis, 1996, 79)

At root here is the difference between a technocratic society – where scientists,
engineers, politicians and their corporate sponsors decide what our problems are
and with which technology we are going to deal with them – and a democratic society where people determine their fate. The most recent example of this is genetically modified food. Without the slightest concern for the sociology of hunger, the
problems of industrial, monoculture agriculture, the biotech corporations and the
research-funding institutions decided that this was a profitable avenue to pursue.
There was no consultation with the public, which eventually led to the disaster of
genetically modified food in Europe, where the public is much less accustomed
than in the US to just gulp down what the technocrats prescribe for them.

If we are serious about sustainability, and that means empowerment and
self-determination, we have to bring science and technology, just like the corporations, under real democratic control. In other words, science and technology then
would need to address the most urgent and pressing problems of those in need, not
those with the biggest purse. This means, for example, that preventive medicine
would move centre-stage,[349] not – as is presently the case – the most expensive
and technologically advanced medical treatments for a very small rich clientele in
rich countries. It would mean that those affected by a technology would not just,
as current jargon has it, be 'consulted' during research and development, but
would actually drive and finally decide about the if and how and when of any
'technological progress'. As Joy says: "In any normal terms, and by any normal
standards, a corporation should not be permitted to toy with fundamentals like
genes, nanotechnology and the like without the full endorsement of those whose
lives will be affected should anything misfire." ('Discomfort and Joy', 2000, 38)

---

[349] See Nino Künzli, 'Endogenisierung von umweltwissenschaftlichen Elementen: nachhaltige
Krise in der Medizin?', in *Interdisziplinäre Themen in Fachstudiengängen* (2000), 33-37, here 34f.

This, though, means that we would need to revamp entirely the way science and technology are financed. For research funding, scientists are more and more at the mercy of the very corporations which want to turn the results of their research into palpable profits. Study after study shows that scientists on the pay roll of private corporations skew or are forced to skew their results in order to make them fit the objectives set by the corporations.[350] If scientists, rarely enough, refuse to comply, the research is often censored or withheld from the public (Klein, 2001, 99-101). This situation is not primarily the fault of the scientists, since there is tremendous pressure in terms of financial dependence and/or career prospects on them. It has in the first instance to do with the fact that corporate scientists are not there to discover the truth, but are part of a mechanism to print money for their employers. Such is the pressure of the capitalist system that it produces all sorts of technologies just to make corporations richer, rather than society concentrating on what it would need to come to terms with its most serious problems.

It seems to me that we cannot but accept the conclusions arrived at by Arpad Pusztai after his dismissal instigated by state and corporate power – and quite irrespective of where we stand on the validity of his research on GM-potatoes:

> By accepting money from an industry which has aggressively set out to dominate many aspects of life and society, science and scientists are becoming servants of multinational concerns whose motives are at best questionable and at worst positively detrimental. The alternative seems to me to be clear: we must help the public to understand that if they want independent scientific advice in today's complex world, they will have to pay for it, somehow, from the public purse. This will then release scientists from their servitude to "big business". Furthermore, scientists must be, and be seen to be, transparent. They must publicly declare all financial and other interests, just as MPs are obliged to do. Only in this way can we begin to win back the public trust we have often deservedly lost.[351]

I think we should be very clear about the consequences of this. There is plenty of evidence to suggest that Lummis is right when he states:

> Choose a technology and you choose the politics – the order of work – that comes with it. Choose mass consumption and you choose mass production and a managed order of work. Choose the big factory and you choose managerial oligarchy and social inequality. And again, there is a sharp difference between the inequal-

---

[350] See Jennifer Ferrara, 'Revolving Doors: Monsanto and the Regulators', in *The Ecologist. Special Issue: The Monsanto Files*, 28 (1998) 5, 280-286, and Steven Gorelick, 'Hiding damaging information from the public', in ibid. 301.

[351] Arpad Pusztai, 'Academic Freedom: Is it dying out?', in *The Ecologist*, 30 (2000) 2, 26-29, here 29.

ity separating manager and worker and that separating master and apprentice. (Lummis, 1996, 98)

We have grown blind with regard to the institutional, political, social and structural consequences that come with certain types of technology. We generally tend to assume that any new technological development or device comes on its own, without strings attached, so to speak. But that, of course, is far from true. Any technology, as Otto Ullrich writes, forces its "laws upon society in such a way that cultural self-definition and autonomy cannot be maintained for long." This is so because of "a little noted characteristic" of technology: with it

> typically comes an infrastructural network of technical, social and psychological conditions, without which the machines and products do not work. For an automobile to be truly used, one needs a technological infrastructure composed of networks of streets with petrol stations, refineries, oil wells, workshops, insurance, police and ambulance services, lawyers, automobile factories, warehouses for spare parts, and much more besides. And, on the psycho-social side, one needs people who will conform to all the installations and facilities and institutions and who can function within them. And so one needs driving lessons, training for children in crossing streets, conscientious petrol station and garage repair owners, and in general, the expert and diligent industrial worker, which in turn means schooling, disciplining, and yet more schooling.[352]

This fact that "a technology mediates human experience through its selection/ amplification and reduction characteristics" (Bowers, 1995, 79) is also the most hideous aspect of the "creeping cultural imperialism", the totalitarian character of Euro-American scientific-technological global rule: it brings with it a ruthless destruction of social, cultural and political native structures, something which is usually not noticed until it is too late:

> Through technological "development aid" more euphemistically called technical assistance, from the industrialized countries, they receive "Trojan machines" (to use Robert Jungk's phrase), which conquer their culture and society from within. They are forced gradually to absorb an alien industrial work ethic, to subordinate themselves completely to unaccustomed time rhythms, to value objective relations higher than human relations, to experience increasing stress and to regard it as normal, and to accept jobs without regard to motivation or meaning.[353]

This again points to the fact that indeed we need a sustainability impact assessment for any new technology before we embark on it, one which evaluates not

---

[352] Otto Ullrich, 'Technology', in *The Development Dictionary*, 1992, 275-287, here 284-285; see also, using the concrete example of Ladakh, Norberg-Hodge, 2000, 142.

[353] Otto Ullrich, 'Technology', in *The Development Dictionary*, 1992, 275-287, here 285.

just positive, but also negative impacts, and in all dimensions. Postman, though, has pointed out that this is not attractive to the powers-that-be. He fears that even the "most intelligent entrepreneurs" cannot be expected to ask the question of the overall value of a new technology to society, since "they are, after all, dazzled by the opportunities emerging from the exploitation of new technologies, and they are consumed with strategies for maximizing profits. As a consequence, they do not give much thought to large-scale cultural effects." (Postman, 1999, 50)

This relates back to my general thesis that we need a reinvention of our political sphere if we are to adequately deal with things like new technologies. We have to become clear in our minds that new technologies, like new ideologies or new dominant paradigms in science, shift the discourse:

> What is at issue are the changes that might occur in our psychic habits, our social relations and, most certainly, our political institutions, especially electoral politics. Nothing is more obvious than that a new technology changes the structure of discourse. It does so by encouraging certain uses of the intellect, by favoring certain definitions of intelligence, and by demanding a certain kind of content. (Postman, 1999, 51)[354]

The military bias and remoteness from real life

The trouble is that scientific-technological development, far from being neutral, has always been in the service of particular interests, most notoriously, right up to the present day, the interests of destruction.[355] Already in 1923 Bertrand Russell complained about three "harmful effects of science": "increase in the total production of commodities, increase in the destructive potential of the war machine, and the mechanization and trivialization of cultural activities".[356] There is an intricate link between science and destruction (military/war): for most of the "Century of Development" scientific endeavours were "concentrated in the main (in money and personnel) on increasing the war machine's productivity in killing" and up to recently "the largest part of Western technological assistance has comprised these destructive weapons" (see also footnote 285 on p. 154): "The effect of all this highly modern technology in these lands can be described unambiguously – it in-

---

[354] Postman has shown this for the political consequences of the technology TV: "Ronald Reagan, for example, could not have been president were it not for the bias of television. This is a man who rarely spoke precisely and never eloquently (except perhaps when reading a speech written by someone else). And yet he was called The Great Communicator. Why? Because he was magic on television. (...) The point is that we must consider whether or not (or to what degree) the bias of a new medium is relevant to the qualities we require of a politician." (Postman, 1999, 51-52)

[355] One particularly pertinent example was the use of IBM Hollerith machines to streamline the deportation of the Jews and the operation of concentration camps in Nazi Germany (see Edwin Black (2001), *IBM and the Holocaust: The Strategic Alliance Between Nazi Germany and America's Most Powerful Corporation* (New York, Crown)).

[356] Quoted in Otto Ullrich, 'Technology', in *The Development Dictionary*, 1992, 275-287, here 277.

creases hunger and misery, it hinders independent development, and it secures corrupt regimes against popular revolutions."[357]

Drawing on her experience in Ladakh Helena Norberg-Hodge has shown that the kind of technology which would be really geared to local people's needs and their geographical, social and ecological reality is the exact opposite of Euro-American technology:

> Truly appropriate technology would be far less costly than "high" technology – not just in purely economic terms but, very importantly, in its impact on society and the environment. It would be born of research in specific social and geographical settings, and be tailored to them, rather than vice versa. As anyone who has been close to the land knows, variations in wind, water, sun, soil and temperature are significant even within very short distances. Just as brick-making in Ladakh varies from region to region, depending on the type of mud available, so small-scale installations adapted to local conditions are required if we are to make optimum use of available resources. This would entail a listening, intimate knowledge of nature – a very different approach from the heavy-handed ways of industrial society. (2000, 164)

If we take this recognition together with what we know about the structure and objectives of the globalised capitalist economy, it seems that "the development of modern technology" cannot

> be even conceived without social injustice in the industrial heartlands and between them and the colonies. Technological development without capitalist, imperialist exploitation, and especially without the military technology that is the driving force, would certainly not generate progress in the conventional sense of the term. (Bennholdt-Thomsen/Mies, 1999, 114)

It seems undeniable that the development and usage of technology would look entirely different if it were directed at conservation and survival, i.e. sustainability. For as it stands, it is removed from the real needs of people and tightly locked into the capitalist cycle of 'useless desire production': "Technological progress provides the majority of people with gadgets they cannot afford and deprives them of the simpler tools they need." (Illich, 1971a, 59)

In order to change this, we need to become aware of just how much scientific theory and the corresponding technologies construct and determine our reality, rather than serve our needs. Donald MacKenzie has very convincingly shown how this happens in the case of modern 'high' finance. The entire current global financial markets as they dominate economic life (i.e. 97 per cent speculation, 3 per cent directly related to real production and trade) would not exist without the

---

[357] Ibid.

economic theory and the accompanying mathematical equation by Black-Scholes-Merton (for which the last two received the Nobel Price for economics in 1997) and the power of modern computers to make the formula useable. When the three scientists first published their formula in 1973 the market in speculative capital was virtually zero and their formula was proved wrong in 40 per cent of the empirical cases to which it was applied. Now speculative trading amounts to US$87.9 trillion per annum and most of it conforms to the equation. In other words, if you get enough people to believe in your ideology, with all its assumptions, preconditions, and the world view it conveys, the world turns into what you predicted: "self-fulfilling prophecy" we call it with Robert K. Merton, ironically the father of the Robert C. Merton of Black-Scholes-Merton. Or as MacKenzie sums it up: "finance theory describes not a state of nature but a world of human activity, of beliefs and of institutions. Markets, despite their thing-like character, their global reach, and their huge volumes, remain social constructs".[358]

To give a detailed and concrete example of this process, of the implications of an unquestioning acceptance of the formula 'progress equals technology' I will devote a special subchapter to information technology (IT).

Information technology against sustainability

> In truth the threat of the new electronics to independence could be greater in the late 20th century than even colonialism was. (Anthony Smith)[359]

> Human development, it turns out, really can't be reduced to information processing. (*Fool's Gold*, 2000, 11)

It seems as if the scientific-technological complex is capable of coughing up a new myth production machine whenever the old ones finally start to disintegrate. After the progress myth slowly but surely began to be dismantled in the 1970s and 1980s, primarily due to the failure of large-scale nuclear, chemical, automobile and agricultural industries, along came the saviour: the computer, and with it information technology (IT) or Information and Communication Technology (ICT), as it is now called in Orwellian newspeak.

In the 1990s IT, which Nürnberger claims is the driving force behind globalisation (1999, 9), has grown to be the rescuer from all sorts of problems. Even such an otherwise credible organisation as the *Forum for the Future* (UK) introduces its outline for a proposed e-lab with a quasi-religious quotation by Charles Leadbeater who compares the so-called New Economy to the Holy Grail

---

[358] Donald MacKenzie (2000), 'Fear in the Markets', in *London Review of Books*, 22 (2000) 8, 31-32, here 32. [http://www.lrb.co.uk/v22/n08/mack2208.htm]
[359] Smith (1980), *Geopolitics of Information: How Western Culture Dominates the World* (New York, Oxford University Press), 176.

and claims it will "combine economic growth with sustainable development". The *Forum* then continues: "Information technology, telecommunications and e-commerce can generate real gains in environmental productivity, social cohesion and quality of life."[360] Everybody involved in education knows that over the last ten years all government commissions, funding bodies and education ministers automatically assumed that throwing money at IT in schools and universities will improve standards and make our kids and students cleverer. Linking up the poorest countries to the internet and inundating them with computers is seen as the solution to the 'development problem'. In other words, IT is the strongest reincarnation of an unreflected progress, growth and salvation myth that we have had in a long while.[361]

Yet, as soon as you start to look closely at the claims made by IT proponents they start to crumble very quickly. We just need to look at them in the light of the sustainability parameters we have developed in chapter 1, and it becomes clear why the claims *cannot* be accurate.[362] Vester has, however, pointed out why it is so difficult for us to understand this; it is the same reason why we generally have so much difficulty in grasping the implications of scientific change (as elaborated above, p. 179): "The development of computers has brought about a technical revolution so fast that our understanding of the intellectual and psychological consequences couldn't keep up at all." (Vester, 1997, 57; see also Bowers, 1997, 53-54)

IT revolution accelerates plundering of the earth

The claim that the IT revolution will solve our ecological problems rests on the assumptions that if we can do all our communication via email we will save paper, if we can do all our business dialogues via video conferencing we won't need to fly around the globe that much any more, if we can do our work and shopping online from home, work and leisure travel will decrease dramatically. All this should result in a considerable fall in resource consumption.

These ideas, in turn, are driven, on a deeper level, by the old dreams of progress and technology which will turn our life into a leisurely paradise where we don't have to lift a finger. It is very striking that the discussions about the "end of utopia" have never reached IT. As far as computers and the internet are concerned the frenzy of what-you-can-do-you-must-do is still in full swing, com-

---

[360] *Forum for the Future*, 'e-lab: sustainable solutions for the new economy', outline (London, March 2001), 1.

[361] For a collection of some of the absurdly fantastical predictions of computer advocates (and an account of how they collapsed) see Stoll, 2000, 113-120.

[362] One aspect usually forgotten is the tremendously high and climbing costs of participating in IT, including the numerous costly failed implementations of IT solutions (see Bowers, 2000, 16-17).

pletely oblivious to any social or environmental consequences:[363] "Undoubtedly enthusiasm for technology is on the rise again".[364] And science, producing technology as its applications, still is, as we saw above, the utopian discourse providing the ideology to legitimise unlimited, uninterrupted progress.[365] To corroborate this claim you just have to look at any computer magazine. All you will find are advertising pleas by the computer industry disguised as editorial content with the – obviously unsustainable – aim of shortening the lifecycle of new computers so that this growth sector continues to be one. Every new software version, every new processor, every new development in printers or any other peripherals is sold as so revolutionary and so absolutely essential that you cannot under any circumstances let it pass you by.[366] In vain will you search for any lifecycle analysis of the incredibly large "ecological backpacks" of computers (see below).[367]

But the serious book market is hardly an improvement. The recently published *Where's IT Going* (Pearson/Winter, 2000) is a classic example. The book which purports to look at the future of IT and its impact on society, is just a techno-freak's fantasy (see, for example: "The whole environment will seek to serve you and to shape itself to you wherever you go." [ibid. 8]), written by IT specialists who quite apparently have never ever thought beyond a computer screen. It is a sad example of the ecological illiteracy and social ignorance Orr comments upon, and this, I'm sure, passes for the knowledge elite. Not a word about the environmental impact of the proposed vision and the relationship between 5 billion computers (for all of us, isn't it?!) and the carrying capacity of this Earth.

For all such claims and fantasies are in fact based on three fallacies:

---

[363] See Giaco Schiesser, 'Wir fahren auf der Autobahn. Mit Multimedia auf dem Weg ins ökologische Zeitalter?', in *Die WochenZeitung*, No. 31, 2.8.1996, Dossier Multimedia & Ökologie, 6-7.

[364] Particularly as far as medicine, genetics and biotechnology are concerned (see Franz Mechsner, 'Der lange Arm der Apparate', in *GeoExtra: Das 21. Jahrhundert: Faszination Zukunft*, (1995) 1, 45-55, here 46). But even if you look for example at a memo by the German postal union and the media union on future developments it is soaked with unquestioned utopian dreams and technocratic illusions (see 'Die Kommunikation der Zukunft spielt sich in Datennetzen ab', in *Frankfurter Rundschau*, No. 200, 29.8.1995, 16).

[365] See Helga Nowotny, 'Science and Utopia: On the Social Ordering of the Future', in *Nineteen Eighty-Four*, 1984, 3-17, here 8-9, 13. See also, on the issue of diffusing social conflicts with science utopias: Burkhart Lutz, 'Das Ende der Wachstumsmechanik als gesellschaftliche Herausforderung', in *Utopien – Die Möglichkeit des Unmöglichen*, 1989, 9-20.

[366] A random pick: 'New Power for your PC', cover story, in *PC World*, October 2001 [http://www.pcworld.com/magazine/index/0,iss,1909,00.asp] or 'The Next Quantum Leap in Computing: 64 Bits: Are you ready?', cover story, in *PC Magazine*, 20 (2001) 16 [http://www.pcmag.com/current_issue/0,3026,i%253D1143,00.asp].

[367] See Rolf Jucker, 'Fiends of the earth', in *PC Magazine*, 6 (1997) 6, 19.

1. 'Computers don't use any resources'

We have seen when discussing Life Cycle Analysis that the production of computers is a very resource and energy intensive business (see p. 89). From a sustainability point of view they are utterly flawed: they use around 50 per cent of all the energy consumed over their lifetime during production (grey energy), they have a very short lifespan, their production process shifts masses of material, uses thousands of litres of water[368] and scarce raw materials; PCs are produced in a way that makes it difficult to remanufacture and recycle them, their backward compatibility is very bad (the US and the Russian army, it turned out at the time of the so-called Year-2000-bug, have a considerable number of weapon systems controlled by computers and programming languages which are not replicable any more; considerable amounts of data produced with earlier computers cannot be read and deciphered any more today); their energy consumption is high (particularly monitors) and inefficient. This leads Sachs to the following conclusion:

> Quite a few champions of the information society proclaim that electronic impulses, travelling at the speed of light, will finally square the circle: simultaneity and ubiquity can be achieved without any cost to nature. They are, more likely than not, mistaken. To be sure, the data highway can be travelled without noise and exhaust fumes, but the electronic networks require quite a lot of equipment. (...) In particular, numerous components require the use of an array of high-grade minerals that can be obtained only through major mining operations and energy-intensive transformation processes. As it turns out, between 8 and 18 tons of energy and materials – calculated over the entire lifecycle – are consumed by the fabrication of one computer. When one considers that the production of an average car requires about 25 tons, it is clear that the ecological optimism surrounding the on-line future is misplaced. On the contrary, there is no reason to believe that mass computerization will weigh drastically less heavily on nature than mass motorization. (Sachs, 1999, 192-193)

So, to equip every human being on Earth with a PC, and a networked one at that, as the digital doctrines of salvations of IT gurus such as Negroponte would have it,[369] would straightaway deplete various scarce metals, trigger an unsatisfiable

---

[368] "The manufacture of computer wafers, used in the production of computer chips, uses up to 18 million litres of water per day. Globally, the industry uses 1.5 trillion litres of water and produces 300 billion litres of wastewater every year." (*Corporate Europe Observer*, (October 2000) 7, 23)

[369] The fictional aspect but also the arrogance of these IT revolutionaries who impose the world view of the rich global upper-class onto the rest of the world is quite obvious in the following passage from Negroponte's cult bestseller. Just notice at the start that apparently in some sense, which eludes me, "bits" can combat hunger: "Bits are not edible; in that sense they cannot stop hunger. (...) But being digital, nevertheless, does give much cause for optimism. (...) Digital technology can be a natural force drawing people into greater world harmony. (...) But more than anything, my optimism comes from the empowering nature of being digital. The access, the mobility, and the ability to effect change are what will make the future so different from the present. The infor-

demand on energy provision and cause the already frail waste management of the earth to collapse pretty much instantly. Quite apart from the fact that the voices advocating the use of IT "grow weak when it comes to the profound responsibilities we all have in using these powerful machines for the benefit of humanity rather than simply exploiting them for our own personal profit or pleasure" (*Fool's Gold*, 2000, 71).

Yet there is another, an equity aspect to this. Far from being a tool to free up the Southern countries and narrow the gap between rich and poor, it is already leading to an increase in this gap and to a consolidation of US global dominance:

- In January 2000 there were 242 million internet users worldwide (4 per cent of world population) half of whom lived in the US or Canada (120 million), with another 110 million in Europe (70 million) and Asia/Pacific (40 million). In other words, 95 per cent of users lived in highly industrialised countries. By 2005 between 350 and 765 million users are expected (5.8 or 12.7 per cent of world population).
- Of the 1.72 million people online in Africa (0.2 per cent of total population), 1.62 million live in South Africa.[370]
- Around two-thirds of all the internet host computers worldwide are sitting in the US.
- American domain names do not need a country code, as opposed to those in any other country in the world.
- The dominant language is American English.
- 80 per cent of the world population don't have reliable phone lines, without which access is very difficult. On the other hand 75 per cent of all phones are concentrated in just 8 of the highly industrialised countries.
- More than half of the population on Earth has never made a phone call. Of those 49 countries which have less than one phone line per 100 people 35 are to be found in sub-Saharan Africa.
- A third of the world population has no access to electricity, let alone computers. The possibilities to access an internet host computer in a highly industrialised country are 8,000 times higher than in a poor country.

---

mation superhighway may be mostly hype today, but it is an understatement about tomorrow. It will exist beyond people's wildest predictions. (...) My optimism is not fuelled by an anticipated invention or discovery. Finding a cure for cancer and AIDS, finding an acceptable way to control population, or inventing a machine that can breathe our air and drink our oceans and excrete unpolluted forms of each are dreams that may or may not come about. Being digital is different. We are not waiting on any invention. It is here. It is now." (Nicholas Negroponte (1995), *Being digital* (New York, Knopf), 228-231 [http://www.obs-us.com/obs/english/books/nn/ch19epi.htm]). See also the excellent analysis of cyberspace as replacement of God: Hartmut Böhme, 'Die technische Form Gottes', in *Neue Zürcher Zeitung*, No. 86, 13./14.4.1996, 69.

[370] Figures above are taken from CommerceNet Research Center, 'Worldwide Internet Population', accessed 15.12.2001 [http://www.commerce.net/research/stats/wwstats.html].

- At the moment there is an acute danger that the internet-boom in developing countries is diverting vital funds from more important projects, such as provision of safe water, hygiene or basic medical provision.[371]

So the transformation towards an IT society is a typical example of the externalisation of costs, hailing a solution as an environmentally friendly one, just because you forget to do the full accounting:

> A look at the computer branch further along shows just how much high-tech industry lives off the new ecological division of labour. In the case of 22 computer companies in the industrialized countries, more than half of their (mostly toxic) microchip production is located in developing countries. Does this not show in outline the future restructuring of the world economy? The software economies of the North will pride themselves on their plans for a cleaner environment, while the newly industrialized economies will do the manufacturing and contend with classical forms of water, air and soil pollution, and the poorer primary economies will do the extracting and undermine the subsistence basis of the third of humanity that lives directly from nature. (Sachs, 1999, 152)

So the IT revolution is clearly unsustainable not just because of ecological limits, but also because it violates the principles of empowerment, equity and economic independence. The computer industry, dominated largely by a few multinational corporations (Microsoft, Intel, Apple etc.),[372] is just the newest in a long line of Euro-American technologies deepening and perpetuating colonial exploitation. In other words, the internet becomes a new commercial sphere where recent developments have made sure that "it is a global norm that the infrastructure which forms cyberspace is in private hands", so that it inevitably follows "that the internet is no longer an instrument of scientific research but a planetary advertising board for capitalist corporations".[373] In fact, we have to realise "that computer technology represents the digital phase of the Industrial Revolution; that is, it perpetuates the primary goal of transforming more aspects of everyday life into

---

[371] Figures for remaining bullet points taken from: Duncan Pruett with James Deane, *'The Internet and Poverty: real help or real hype?*, Panos Briefing, (April 1998) 28 [http://www.oneworld.org/panos/briefing/interpov.htm]. It is interesting to note that even Bill Gates has realised now that "the world's poorest 2bn people desperately need healthcare, not laptops" (quoted in Edward Helmore, Robin McKie, 'Computers cannot cure the world's ills, admits Gates', in *Guardian Weekly*, 9-15.11.2000, 12).

[372] Ask, for example, any manufacturer of notebook computers about the origin of the parts (disk drive, motherboard, monitor etc.): they will tell you that there is just a handful of producers worldwide, and whether you buy a Toshiba, Dell or IBM notebook, you will pretty much have the same machine (information supplied by the senior director of Ergo Computing UK in a personal conversation, June 1999).

[373] Dan Schiller, 'Wer besitzt und wer verkauft die neuen Territorien des Cyberspace?', in *Le Monde diplomatique [deutsch]*, 2 (1996) 5, 4-5.

commodities that can be manufactured and sold, now on a global basis." (Bowers, 2000, 40)

Bennholdt-Thomsen/Mies sum this up by looking at IT from a Southern perspective. As it turns out, for the most important problems we face today, as judged from the Gandhian perspective of the poorest human being, IT is utterly superfluous:

- It contaminates the environment to a high degree.[374]
- It is a technology designed for military purposes, and we doubt whether it can ever free itself from this original logic (Weizenbaum).
- Who produces it and for whom?
- We do not believe in the possibility of decentralised application of micro-processors; their production and sales are highly monopolistic; their use always depends upon centralised supply (of energy, cables, etcetera).
- There is any amount of evidence that it does not make subsistence work any easier.
- So who is supposed to use it? Who does the work of providing on a daily basis? Or are there no children to be cared for in this model?
- It is not necessary. There are other subsistence technologies, tried and tested for hundreds or thousands of years, which are adapted to the environment, to community structures and cultural specificities. (1999, 180)

2. *'You can keep an online society running without energy usage'*
Yet this is not even half the story, as Sachs points out: "Those who hail the rising information and service society as environment-friendly often overlook the fact that these sectors can grow only on top of the industrial sector and in close symbiosis with it." (Sachs, 1999, 40) We have seen above, discussing the progress and development myths, what that means. You cannot run a highly industrial society without a massive infrastructure which is resource and energy demanding both to construct and to run. To keep an IT and online society running you need an electricity grid to meet constant demand, you need fibre optic, satellite or similar data transmission networks connecting all the users; you need not just PCs, but servers, data storage facilities, adequately air conditioned and cooled/heated rooms to store the equipment, offices, factories to produce all the equipment, ser-

---

[374] Particularly problematic are the motherboards, the 'heart' of PCs: "Printed Circuit Boards contain heavy metals such as Antimony, Silver, Chromium, Zinc, Lead, Tin and Copper. According to some estimates there is hardly any other product for which the sum of the environmental impacts of raw material extraction, industrial refining and production, use and disposal is so extensive as for printed circuit boards." (CARE conference, Vienna, 1994, quoted in: Silicon Valley Toxics Coalition, 'Just say no to e-waste: background document on hazards and waste from computers', Clean Computers Campaign [http://www.svtc.org/cleancc/pubs/sayno.htm]). For an inventory of chemicals used in semiconductor manufacturing, see http://www.svtc.org/hightech_prod/desktop.htm.

vice networks; it produces transport for deliveries, maintenance and so forth and so on.

> From an ecological point of view, electronic communication is assuredly less wasteful of resources than is physical transport. Yet one should not underestimate the additional strain that the construction and maintenance of a digital infrastructure place upon the earth's resources. High-quality materials used in hardware and peripherals are obtained through numerous refining processes that impose a large (and often toxic) extra burden on the environment, cables of all kinds use a lot of material, and satellites and relay stations also cannot be had without a drain on the environment. (Sachs, 1999, 145)

Let me just give two figures from Germany to illustrate this: "Despite ongoing technical improvements even today around 30 per cent of primary energy in the conversion sector (in generation of electricity and district heating, in refineries, and in delivery of natural gas) is 'lost'", in the form of "unused waste heat" (*Greening the North*, 1998, 57, 58 [figure 4.8]). And 6.4 per cent of total consumption of primary energy is used for "the erection and maintenance of buildings", before any heating, cooling and energy usage (*Greening the North*, 1998, 58 + figure 4.9).

And far from inducing a fall in energy consumption, the sectors with an intensive reliance on IT such as the service sector, in particular banking and insurance but also the university sector, have seen a massive increase in electricity usage in recent years.[375]

---

[375] "Computers are one of the fastest growing users of electricity in North America today. (...) According to the U.S. Environmental Protection Agency (EPA), computers account for 5 percent of worldwide commercial consumption. If nothing is done, this figure is expected to jump to 10 percent, which translates to 70 billion kilowatt hours per year, by the year 2000. (...) Unfortunately, the University of Toronto's (St. George Campus) computer use and energy consumption patterns reflect the general global trends. In the last ten years, electrical consumption per gross square foot floor (GSF) on the campus has increased on average 2 percentage every year." (Alice Chan, Wendy Siu, Deanna Li and Fiona Wong (1997), *The Role of Computers in Energy Consumption on Campus* (University of Toronto) [http://www.cquest.toronto.edu/env/env421h/energy/computers.html]) And further: "Here are just a few realities of today's energy-hungry new economy: Silicon Valley companies consume as much power as small steel mills – and their requirements are growing over 7 percent per building, per year; a personal computer and its peripherals typically boost power consumption in homes by about 5 percent per year; the Web's invisible infrastructure consumes at least twice as much power as desktop hardware; although the amount of power it takes to create, process or transmit a single bit is cut in half about every 18 months, the number of bits in play is doubling much faster." (Peter Huber and Mark Mills, 'Got a Computer? More Power to You', in *Wall Street Journal*, 7.9.2000)

3. '*Efficiency revolutions lead to a fall in resource usage*'
The most notorious mistake made by the self-proclaimed IT revolutionaries (and many others, as we have seen above, p. 57) is that they look at just one narrow aspect and hope that any resource efficiency improvements gained there will lead to an overall reduction. Yet this is displaying ignorance about the "dynamics of growth", which, if "not slowed down", will eat up any "achievements of rationalization" in "the next round of growth": "In fact, what really matters is the overall physical scale of the economy with respect to nature, not only the efficient allocation of resources." (Sachs, 1999, 41) If you care to look at the overall picture, it is paradoxically "often precisely the economic gains from improved technical efficiency that increase the rate of resource throughput":

> For decades, efficiency has been the driving force behind competition and growth – per unit gains have fuelled new rounds of expansion. (...) Efficiency gains on the micro level are therefore – over time – likely to be eaten up by growth in volume on the macro level. (...) There is no logical connection between statements of relative efficiency and statements of absolute scale, but it is in the end the absolute scale of resource consumption that matters. (Sachs, 1999, 183-184)

On past experience, what we have to expect from a full-scale global IT revolution is precisely *not* a decrease in energy and resource consumption, a fall in physical travel etc., but a massive expansion of these sectors:

> Whatever the many prophets of the information age merrily predict, electronic networking will in the long term probably generate more physical travel than it replaces. Anyone who has established close contact with distant places via electronic media will sooner or later want to seal the contact face to face. In any event, the main effect is a positive feedback between electronic and physical transport systems: globalization itself means transport and still more transport. (Sachs, 1999, 145-146)

At least for the first decades of the IT revolution we know for a fact that this is true. The transport sector is by far the fastest growing sector in resource consumption. Since the 1960s road traffic in Germany has risen by some 700 per cent and the growing sector of "leisure" travel accounts for over 50 per cent of energy consumption of all spare time activities (*Greening the North*, 1998, 58, 59). Air travel worldwide is growing faster than almost any other sector at 5 per cent per annum.[376]

---

[376] Theophil Bucher-König, 'Wird Fliegen zu einer Frage des Gewissens? Die Beurteilung des Flugverkehrs auf Grundlage der Nachhaltigkeit', in *Reisen & Umwelt. Informationsmagazin von SSR Travel*, (September 1999) 7, 16-19, here 18.

So we can see once more that unless we try to look at the full picture and take into account all the externalised costs, the lifecycle, energy, and materials consumption and all the knock-on effects of a restructuring of the social, political and economic sphere towards an IT society, we will be fooling ourselves when it comes to a realistic assessment of the likely benefits and costs.

Information revolution is not knowledge revolution

> You cannot know until you have had time to learn, and impatience will gain nothing but confusion. (Llewellyn, 1991, 259-260)

> We move the problem of learning and of cognition nicely into the blind spot of our intellectual vision if we confuse vehicles for potential information with information itself. We do the same when we confuse data for potential decision with decision itself. (Illich, 1973, 86)

> Computer-mediated data and information strengthens the modern belief that objectivity and fact are separate from values. (Bowers, 2000, 72)

> My point is that it's possible to do perfectly well without any computers or high-tech teaching devices. For many – children and adults alike – zero computing is perfectly acceptable and even singularly desirable. (Stoll, 2000, 71)

The hype about the IT revolution and the internet is also dogged by another fundamental flaw. It is always and by definition assumed that more information is good, and that it automatically leads to better understanding, more knowledge, and deeper insights. Nothing could be further from the truth because this assumption is based on mistaking the mere mass of data for knowledge. Yet data does not easily translate into knowledge. And this is why we might live in an information society, but certainly not in a knowledge society; on the contrary, knowledge "as wisdom, understanding, insight" "is more and more lacking". If there *is* knowledge it is "knowledge as power, a tool to dominate, either nature or 'the others'" (Dürr, 2000, 117-118; see also Stoll, 2000, 185-186). In fact, it seems that we are "becoming both more clever and less intelligent", as borne out by the fact that we are "able to perform amazing technological feats while being unable to solve [our] most basic public problems": "As Exhibit A, consider our phenomenal and growing computer capabilities side by side with our decaying inner cities, insensate violence, various addictions, rising public debt, and the destruction of nature all around us." (Orr, 1994, 51)

Data alone doesn't help here, for if you want to gain knowledge from mere data you need time to reflect, you need previous knowledge in order to evaluate, contextualise and integrate new data into existing knowledge, you need the ability to filter data:

Statements about facts – that is, information – can be wrong, and often are. Thus, to say that we live in an unprecedented age of information is merely to say that we have available more statements about the world than we have ever had. This means, among other things, that we have available more *erroneous* statements than we have ever had. Has anyone been discussing the matter of how we can distinguish between what is true and what is false? Aside from schools, which are supposed to attend to the matter but largely ignore it, is there any institution or medium that is concerned with the problem of misinformation? Those who speak enthusiastically of the great volume of statements about the world available on the Internet do not usually address how we may distinguish the true from the false. By its nature, the Internet can have no interest in such a distinction. It is not a "truth" medium; it is an information medium. (...) Here, I am addressing a problem no culture has faced before – the problem of what to do with too much information. (Postman, 1999, 91)

The more data we have, the less time we have to evaluate it, check its sources, or assess its relevance. And in the midst of this data overload and the surrounding hype we lose sight of the fact that one of the most crucial abilities for survival is "not to notice *irrelevant* information in the first place or to suppress it efficiently" (Dürr, 2000, 227). So we clearly do not just need any amount of extra information; the crucial point is a reduction of speed to generate time to process information, to turn it into usable knowledge. And for that we need to learn the art of complex thinking, which no amount of data overload will ever teach us. We have seen in chapter 1 that in order to be able to meet the challenge of sustainability we need a veritable change in the way we look at the world, at processes and at how we evaluate them. And we have also seen that there is quite enough information gathered already – admittedly with some notable exceptions in specific areas – in order to gauge with sufficient precision what will lead to an increase in sustainability and what will lead to the opposite. Yet, as Vester has noted, the IT revolution, with its reliance on a specific technology and its increase in speed, will not give us the breathing space in order to achieve this transformation; on the contrary it has the tendency to lock us in old ways of thinking which are completely inadequate for coping with complex reality (as we have already seen, the IT 'revolution' catapults us back into an age of quasi-religious technocratic faith which we thought we had long left behind us) (see Vester, 1997, 94).[377]

And it is not just a problem of the mind. As we should expect on the basis of the precautionary principle and its demand to keep impact as low as possible so as to minimise the chances of large-scale unforeseen effects and to keep open the chance of reversing developments, any decision on a particular type of technology – and IT is certainly no exception – leads to a physical inscription of this technol-

---

[377] Vester inadvertently proves this point himself when he falls back into an unreflected IT utopia (see 1997, 114f.).

ogy on the earth, into the daily living patterns of people, into the communication and hardware infrastructure, and thus determines to a large extent what is possible and what isn't: it literally "cements the organisation of human movements" and behaviour (see Vester, 1997, 116-117). But it is not just this 'physical' aspect which is problematic. Bowers has shown the specific "cultural assumptions embedded in software programs" which "influence culture" almost always in unsustainable ways (2000, 125, 127).[378]

Now the trouble is that almost anybody in the educational sector seems to believe in the most naive claims of the IT 'revolution'. The last decade has certainly not seen a commitment to real learning on the basis of what we know from a long experience of work in the classroom, rather the contrary. Instead of changing the curriculum to facilitate complex, experience-based real learning, in real contexts, there is a relentless drive to do exactly the opposite, namely to invest very heavily into what is called *virtual* learning. While the money available to pay teachers, to increase the time spent on crucial issues and to lower the average class sizes constantly goes down, there are literally billions of pounds or dollars spent on computers, e-learning, open and distance learning, the virtual university and whatever else it might be called. All government agencies, funding bodies and the media[379] seem to be utterly convinced that computers in classrooms automatically improve the quality of learning and make students, by some magic, extremely clever: all you have to do is to place them in front of a computer monitor (see Bowers, 1995, 75-91 and 2000, 111-139).

The trouble is, as always with such claims to bring about a perfect utopia, that nobody so far has gone to the trouble of trying to prove that computers in fact *do* enhance deeper and broader learning.

> 30 years of research on educational technology has produced just one clear link between computers and children's learning. Drill-and-practice programs appear to improve scores modestly – though not as much or as cheaply as one-on-one tutoring – on some standardized tests in narrow skill areas, notes Larry Cuban of Stanford University. "Other than that," says Cuban, former president of the American Educational Research Association, "there is no clear, commanding body of evidence that students' sustained use of multimedia machines, the Internet, word processing, spreadsheets, and other popular applications has any impact on academic achievement." (*Fool's Gold*, 2000, 3)

And the U.S. National Science Board goes on:

---

[378] Bowers shows how even "programs widely regarded as having the most educational merit" are in fact saturated with "cultural assumptions and values" which are unsustainable, limit the child's/ student's imagination and learning and are therefore reinforcing eco-illiteracy (2000, 128-139).

[379] An exemplary article of that sort is Tim O'Shea, 'University of future takes shape in cyberspace', in *Guardian Weekly*, 2-8.11.2000, 24.

The fundamental dilemma of computer-based instruction and other IT-based educational technologies is that their cost effectiveness compared to other forms of
instruction – for example, smaller class sizes, self-paced learning, peer teaching,
small group learning, innovative curricula, and in-class tutors – has never been
proven. (Science & Engineering Indicators, 1998, quoted in *Fool's Gold*, 2000,
95)

Quite to the contrary. Firstly, computers, if not carefully used, "pose serious
health hazards" to users, including "repetitive stress injuries, eyestrain, obesity,
social isolation, and, for some, long-term physical, emotional, or intellectual de-
velopmental damage", which led the US Surgeon General to warn that our children "are the most sedentary generation ever." (*Fool's Gold*, 2000, 3)[380]

But computers are also positively detrimental for the formation of a
competent new generation that is capable of dealing with our complex reality.
Studies show that far from being highly motivating to students they positively
bore girls, they "stunt imaginative thinking" and they "isolate children and students emotionally and physically, from direct experience of the natural world"
(*Fool's Gold*, 2000, 4). Computers in fact might produce exactly the incompetence
which perpetuates unsustainability: the US National Science Board, for one, "reported in 1998 that prolonged exposure to computing environments may create
'individuals incapable of dealing with the messiness of reality, the needs of community building, and the demands of personal commitments.'" (*Fool's Gold*, 2000,
4) And far from producing the shift from teacher-centred to student-centred
learning, as claimed by the IT-advocates, it is clear to anybody who ever looked
into an IT lab at a university or school that "the actual shift is to computer-cen-
tered, not student-centered, education." (*Fool's Gold*, 2000, 29)

To sum up, as Postman notes, "there certainly does not exist compelling
evidence that any manifestation of computer technology can do for children what
good, well-paid, unburdened teachers can do" (1999, 46-47; see Stoll, 2000, 85-
89),[381] yet on the other hand there is a lot of evidence – certainly borne out by the
personal experience of those people in higher education I asked about this – that
the depth and breadth of knowledge of the average student at university has
sharply declined during the same period that computers have started to appear in
classrooms and at universities (see box 14). I don't want to construct a monocausal
link here since the influence of the mass media, particularly TV, and the disinte-

---

[380] For a full list of physical, emotional, social, intellectual and moral hazards of IT use, particularly, but not exclusively, in early childhood, see *Fool's Gold*, 2000, 39.

[381] An experiment in Austin, Texas, has yielded the same results: 16 primary schools received an
additional US$300,000 per year for five years to improve standards. The only two schools where
standards had been raised by the end of the trial period were also the only ones that had *not* invested the money in computers, but in teacher training, improved co-operation with parents and
curriculum redesign (see Thomas Homberger, 'Computer verbessern die Schulqualität nicht', in
Neue Zürcher Zeitung, No. 71, 26.3.2001, 41).

gration of family and social culture are bound to have at least as much impact. But it is certain that computers did nothing to change the downward trend. It is astonishing that anybody could fall for the claims in the first place.[382] One would just have to look towards the most experienced practitioners in the distance learning field, in Britain the Open University (OU). Every year, students at the OU say that they would like to have more face-to-face tutorials, and satisfaction rates for traditional teaching methods are consistently higher than for computer assisted learning.[383] Other research also shows that the most effective ingredient in any education is a "close relationship with caring adults" or tutors (*Fool's Gold*, 2000, 4). In fact, we do find for effective education what we have already found for strategies for meaningful lives, namely that both are "remarkably low-tech": The most important human rather than technological tools for this include (and, even though it refers to younger children, the passage can be applied to students and adult learners as well):

> good nutrition, safe housing, and high-quality health care for every child – especially the one in five now growing up in poverty. They also include consistent love and nurturing for every child; active, imaginative play; a close relationship to the rest of the living world; the arts; handcrafts and hands-on lessons of every kind; and lastly time – plenty of time for children to be children. (*Fool's Gold*, 2000, 95)

Yet despite all this "U.S. public schools have spent more than $27 billion on computer technology and related expenses in the last five years". The question obviously arises, why? If you look at the composition of the advisory boards urging such techno mania there is no surprise any more, at least in the US. They hardly ever consist of teachers on the ground or educational experts, but all too often of exponents of the hardware and software multinational giants such as Microsoft and all the big computer producers which are set to earn inordinate amounts of money if all schools, all universities and colleges adopt their products and are locked into a cycle of replacing both hard- and software every few years (see *Fool's Gold*, 2000, 80-84): "Wiring and computerizing America's schools is an urgent priority – not for children, but for high-tech companies that need to constantly expand their market." (ibid. 84)

---

[382] Rather surprisingly in view of the long history of failed educational technology, notes Larry Cuban, Professor of education at Stanford University: "education policymakers have careered from one new technology to the next – lantern slides, tape recorders, movies, radios, overhead projectors, reading kits, language laboratories, televisions, computers, multimedia, and now the Internet – sure each time that they have discovered educational gold. Eventually, the glimmer always fades, and we find ourselves holding a lump of pyrite – fool's gold." (*Fool's Gold*, 2000, 97)
[383] Diana Laurillard (1998), *Technology Strategy for Academic Advantage* (Milton Keynes, OU), 7 and 13.

Box 14: IT does not improve standards of knowledge
I will use technology when I judge it to be in my favor to do so. I resist being used *by* it. In some cases I may have a moral objection. But in most instances, my objection is practical, and reason tells me to measure the results from that point of view. Reason also advises me to urge others to do the same. An example: When I began teaching at NYU [New York University], the available instruments of thought and teaching were primitive. Faculty and students could talk, could read, and could write. Their writing was done the way I am writing this chapter – with a pen and pad. Some used a typewriter, but it was not required. Conversations were almost always about ideas, rarely about the technologies used to communicate. After all, what can you say about a pen except that you've run out of ink? I do remember a conversation about whether a yellow pad was better than a white pad. But it didn't last very long, and was inconclusive. No one had heard of word processors, e-mail, the Internet, or voice mail. Occasionally, a teacher would show a movie, but you needed a technician to run the projector and the film always broke.

NYU now has much of the equipment included in the phrase "high tech." And so, an eighteenth-century dinosaur is entitled to ask, Are things better? I cannot make any judgments on the transformations, if any, technology has brought to the hard sciences. I am told they are impressive, but I know nothing about this. As for the social sciences, humanities, and social studies, here is what I have observed: The books professors write aren't any better than they used to be; their ideas are slightly less interesting; their conversations definitely less engaging; their teaching about the same. As for students, their writing is worse, and editing is an alien concept to them. Their talking is about the same, with perhaps a slight decline in grammatical propriety. I am told that they have more access to information, but if you ask them in what year American independence was proclaimed, most of them do not know, and surprisingly few can tell you which planet is the third from the sun. All in all, the advance in thought and teaching is about zero, with maybe a two- or three-yard loss.

It gives me no special pleasure to report these observations, because my university spends huge sums of money on high-tech equipment. Strangely, many professors seem to prefer that money be spent on technology instead of on salary-increases. I don't know why this is so, but here is one possibility: There are always some professors who have run out of ideas, or didn't have any to begin with, and by spending their time talking about how their computers work, they can get by without their deficiency being noticed. I don't mean to be unkind, but I can think of no other reason.

Neil Postman (1999), *Building a Bridge to the Eighteenth Century. How the Past Can Improve Our Future* (New York: Knopf), pp. 55-57.

Information revolution decreases freedom

The IT revolution turns people into ever more totally dependent customers of large corporations, rather than increasing their freedom. From the software and hardware, to energy and network connections and to institutions running the show, we are ever more intricately caught in a web of dependencies from which there is, within the IT world, no escape.

But this also goes for the bold claims that the internet will in fact reinvent democracy and thereby fulfil the age-old dreams of self-determination and real democracy. Norberto Bobbio treats such claims with caution since, so he claims, the explosion of electronically available information benefits the "democracy of the governing", not the "democracy of the governed":

> No ancient despot, no absolutist monarch of early modern times, even if he sur-
> rounded himself with thousands of spies, ever had the faintest chance to amass all
> the information about his subjects which today the most democratic of all gov-
> ernments can extract from electronic data processing. The old question which
> runs through the entire history of political thought, namely "who guards the
> guards?", can now be rephrased: "who controls the controllers?" If it isn't possible
> to find an adequate answer to this question, then democracy is doomed.[384]

Nor should we underestimate the pitfalls of an "electronically transmitted people's will" because it will benefit those who can shout loudest, whip up emotional feelings before any electronic vote and so on:

> A virtual democracy is a non-existing democracy. Direct democracy has always
> been thought of as a democracy of dialogue: decisions are reached through talk-
> ing with one another, through listening to the ideas of others and through elabo-
> rating one's own. If this process is reduced to pressing a button on a remote
> control, we don't have democracy, simply an expression of will. The immediate,
> concrete interactivity loses its content and transforms itself into a dangerous mul-
> tiplier of ignorance.[385]

Yet there is another aspect to this debate about freedom. Information technology is "not culturally neutral"; it is an expression of a very specific set of values and ideas about the world and how to deal with and represent it, so that "the introduction of computers into other cultures undermines a cultural group's ability to maintain the integrity of its traditions and avoid becoming technological satellites of the West." (Bowers, 2000, 18) Just like the other dominant paradigms of the Euro-American ideology ('free market', 'liberal democracy' etc.), it is a colonial tool. In so-called developed countries we might not recognize the specific, reductionist cultural assumptions encoded in information technology because we share these assumptions, but "members of other cultures are aware that when they use computers they must adapt themselves to radically different patterns of thought and deep culturally bound ways of knowing" (Bowers, 2000, 22). In other words, "computers are, in

---

[384] Norberto Bobbio, 'Die Zukunft der Demokratie', in *Sie bewegt sich doch*, 1993, 57-76, here 68.

[385] Giovanni Sartori, Department of Political Science, Columbia University, speaking on *Panorama*, 23.7.1994, quoted in Riccardo Stagliano, 'Der elektronisch übermittelte Volkswille', in *Le Monde diplomatique [deutsch]*, 2 (1996) 5, 8-9. On the impossibility of a "computo-cracy" (direct democracy via internet or similar) see also: Bobbio, 'Die Zukunft der Demokratie', in *Sie bewegt sich doch*, 1999, 65.

effect, simply amplifiers of Western traditions", and that, of course, is true for the use of computers in education as well (Bowers, 1995, 83-89, here 84).[386]

For all these reasons there is scarcely any hope that the "empowering quality" of the "digital revolution" which Negroponte mentions, will come to fruition.

---

[386] For an extensive list of "cultural amplification and reduction characteristics" of computers, see Bowers, 1997, 56-59.

b.   Legitimising Destruction

> Power concedes nothing without a demand. It never did and it
> never will. (Frederick Douglass, quoted in Zinn, 1996, 179)

Through the juxtaposition of the parameters of a sustainable society in chapter 1 and our current unsustainable practice in chapter 2.a we have seen how large indeed the gap is between what we ought to be doing and what we are doing. I am writing this at a time when George W. Bush, the president of the largest, most wasteful and destructive polluter on Earth, the US, has just single-handedly decided to destroy ten years of negotiations by not submitting the Kyoto-protocol to Congress.[387] His predecessor Clinton prepared the ground for this when the US-delegation, with massive support from corporate lobbies,[388] killed off any implementation of the Kyoto-protocol at the Hague conference in November 2000.[389] Maybe this is a very telling sign, because it does instruct us as to where we have to look for reasons for the non-implementation of change towards sustainability.

We have seen in chapter 1 that the sustainability agenda is on everybody's lips, embedded in numerous government strategy papers and displayed in full-colour print in countless annual reports of TNCs, not to speak of the thousands of NGOs allegedly fighting for its implementation. Yet in chapter 2.a we have seen that on all levels, not just the environmental one, we are far from the necessary social, political and economic structures which would facilitate sustainability.

So we have to ask what accounts for this gap. We have already identified a number of reasons above. Sustainability is simply not conducive to the profit motive, so the entrenched business interests all over the world – whether in their lobbying circles such as the ERT (European Roundtable of Industrialists) or their business-state collaborations such as the OECD or TABD (Transatlantic Business Dialogue) or in their global institutionalised forms such as the WTO, IMF or the World Bank – do whatever is in their power to put a hold on whatever might, in their short-term view, threaten the bottom line. We have also seen that both the institutionalised liberal democracy and the basic structures of the capitalist economic system are actively harmful when it comes to self-determination of the people and/or respect for our life-support system Earth and other creatures. At the

---

[387] Andreas Missbach, 'Wo Kühe waren, werden Fische sein: George W. Bush erledigt das Klimaprotokoll von Kioto', in *Die WochenZeitung*, No. 14, 5.4.2001, 1-2.

[388] Corporate Europe Observatory (CEO) (2000), *Greenhouse Market Mania. UN climate talks corrupted by corporate pseudo-solutions. A CEO Issue Briefing* (Amsterdam, CEO).

[389] See the international press, for example: Maike Rademaker, 'Hart gerungen, nichts erreicht', in *die tageszeitung*, 27.11.2000, 4; Madeleine Bunting, 'The hot air balloon', in *The Guardian*, 27.11.2000. Hermann Scheer has pointed out the absurdity of the entire debate about emission trading because in the last analysis it is an attempt to let the main culprit, the fossil fuel economy, off the hook. The perversion is well captured in US finance minister Summers view that the South is "scandalously underpolluted" (Hermann Scheer, 'Die Ironie des Scheiterns', in *die tageszeitung*, 30.11.2000, 12).

same time, "virtually everywhere there is a pattern of denial" (Orr, 1999a, 220-221), a worldwide denial of knowledge which would clarify the relationship between the powers-that-be and the systematic destruction of both nature and people and their dignity and freedom. The consequence "is a poverty of wisdom", "reinforced by a fundamental ignorance of ecology and the basics of global change" (Orr, 1999a, 221).

Yet there are further reasons. Schumacher has identified a central one: the myths of consumerism, scientific-technological progress and economic growth are so appealing precisely because they are based on simplification, ignorance and with it arrogance: "The strength of the idea of private enterprise lies in its terrifying simplicity. It suggests that the totality of life can be reduced to one aspect – profits." (Schumacher, 1993, 215) And if you have the choice between a hard time learning of and facing up to the real workings of the world, and blissful ignorance bathed in a constant diversion from the real issues through consumption, there is little wonder that the business interests have such an easy ride.

Schumacher stresses as well, as seen in chapter 1.c, that a sustainable society is by no means difficult to conceive or impossible to achieve, but because the consumer society feeds very cleverly into a dependency loop, nurturing human inclinations like passivity, laziness and resistance to changing preconceived ideas, people end up "like a drug addict who, no matter how miserable he may feel, finds it extremely difficult to get off the hook." (Schumacher, 1993, 125-126)

The attempt to achieve sustainability has indeed all the odds stacked against it. One would have to give up dear and long-indoctrinated ideas such as the ones elaborated in chapter 2.a; one has to face a steep learning curve to come to terms with reality, rather than ideology; one has to understand the parameters of sustainability and learn, for example, that *not doing* certain things is often the most viable option;[390] one will face misery, incredible tales of abuse, exploitation and human arrogance; and almost all institutions, the establishment in North and South, are structured against the task. Subcommandante Marcos has described the uphill struggle:

> The task of progressive thinkers – which is to remain sceptically hopeful – is not easy. They have understood how things work and, *noblesse oblige*, they must reveal what they know, dissect the information and pass their findings to others. But to do this they must also confront neoliberal dogma, which is backed by media, banks, corporations, army and police. Since we live in a visual age progressive thinkers, to their considerable disadvantage, have only words with which to fight the power of the image.[391]

---

[390] See Gerhard Scherhorn, 'Revision des Gebrauchs', in *Ökointelligentes Produzieren und Konsumieren*, 1997, 47.

[391] Subcomandante Marcos, 'Do not forget ideas are also weapons', in *Le Monde diplomatique [English]*, (2000) 10, 10-11, here 11.

It is a long-term, intensive, human-centred struggle which acknowledges that all our activities and thinking are intricately linked with the past, shaped by it. And only by knowing this past – such an utterly anachronistic idea at a time when every new technological whim is supposed to wipe out the past and reinvent the world – can we come to terms with the present reality:

> Solidarity, compassion and resistance against injustice need spirituality in order to work long-term and with intensity. A rule of such spirituality is: know your own origins and traditions of liberty! You need to know a lot to escape hopelessness and cynicism. We are also responsible for our courage. But this is nourished by stories of success. You need to know that human beings once have escaped slavery in order to believe that it is possible to escape as well. You need to know what the dead whom we call ours have dreamt so that their dreams can sharpen our dreams and our conscience. Solidarity doesn't simply come into being at the moment when it is needed. You have to educate yourself in love, and education is a long-term endeavour.[392]

Unsustainability has infected our present down to the very bone; it has infected all our institutions, our way of thinking, our language, our political and social structures, our economic system. Over much of this we have no direct influence. Yet there are two conveyor belts of unsustainable thinking and action where we are all directly implicated and to which we are exposed, on a daily basis: the media and education. So before I can finally sketch out an education for sustainability (EfS), I will have to deal with these two systems of disinformation which are crucial tools legitimising the current global destruction. I want to deal with the media first, since mass media such as TV have outgrown the traditional educational system both in terms of influence and length of exposure (see Postman, 1987).

The denial of knowledge: media disinformation and corporate propaganda

Why is it then that the term sustainability is freely branded about, clogs up the search machine on the internet, yet whenever you probe deeper, people are thoroughly unfamiliar with the concept, a large proportion have not even heard of the term and can certainly not describe in any meaningful way what it should mean?[393] Why is it that the concept of sustainability as elaborated in chapter 1 could so easily be reduced to three factors (ecology, economy, society), excluding the crucial political sphere, without anybody noticing, never mind objecting? Why is it that whenever I ask my media studies students, either none or at most 2 of 15

[392] Dorothee Sölle and Fulbert Steffensky, 'Nie ganz verscharrt', in *die solitaz*, 21./22.10.2000, 34.
[393] See Plant, 1998, 41 and Fritz Reusswig, 'Die ökologische Bedeutung der Lebensstilforschung', in *Umweltbildung und Umweltbewußtsein*, 1998, 91-101, here 99.

students ever know what the acronyms MAI, IMF, WTO, ERT or even OECD stand for, let alone could elaborate on what these agreements and institutions actually do and how they impact on their lives?[394] Why is it that we do not know anything about the most noble traditions of human struggle, the anarchist, libertarian and socialist struggles at the end of the 19th and in the early 20th century, about the Spanish Republic, all incidents when people engaged in mutual aid, organised their lives in a self-determined, democratic way and were time and again crushed by business, trade unions, governments and armies (see Marshall, 1993 and Zinn, 1996)? Why do we always only honour the deeds of destruction and exploitation and not the blacks who stood up against slavery, the women who stood up against machos, the Native Americans who stood up against the white man, the anarchists and workers who stood up against corporate power, the anti-war people who stood up against the Pentagon and US government, the indigenous peoples who stood up and still stand up against (neo-)colonialism?

Bill McKibben explains the blindness which is created by the cultural imperialism of corporate (American) media:

> We [US] send out so much information that we receive very little ourselves; we get no sustained looks at other ways of life. It's as if we had a telephone that we could only talk into. And, at least from an environmental standpoint, *we* are the ones desperately in need of missionaries that could show us how to live closer to the ground. (...) Increasingly, there *are* no other cultures and economies that might suggest different approaches to things; there is only one system, ours. (...) In the end, argues Rensselaer economist Sabine O'Hara, we have endangered "socio-diversity" nearly as completely as biodiversity. (McKibben, 1997, 53-54)

But hang on. Isn't it absurd and wilful slander to insinuate that – in the age of the internet, of hundreds of digital TV channels, of thousands of newspapers world wide, of hundreds of thousands of new books published each year – we are in fact short of information and knowledge, indeed that existing knowledge is deliberately hidden from us?

In the face of the current problems and the sustainability challenge we do need, as we have seen when discussing the shortcomings of 'liberal' democracy, competent, not just fully informed but also prudent, sensible and far-sighted citizens, who are capable of dealing with the data avalanches let loose on them. Yet it is unlikely that any number of digital TV channels or in fact computers on any desk will achieve this. What we will promote with this explosion of data and telecommunications technologies, though, is the acceptance of the underlying ideol-

---

[394] A recent ICM survey found that 64 per cent of the British public do not know what GATS, the Treaty of Nice or the TRIPS Agreement are ('The Great British Environmental Survey 2001', in *The Ecologist*, 31 (2001) 4, 33-39, here 38).

ogy which is transforming the world into a "global market place".[395] If we look at the war over market shares for information and communication technology in the last few years, a war which has led to the mergers of Time Warner with CNN, Disney Corporation with ABC, of News Corporation and MCI, of Microsoft with General Electric and in early 2001 of AOL with Time Warner, we get a glimpse of what is at stake here. The prime concern is not to provide an international network and information system which would allow all citizens of this earth to access as cheaply and as conveniently as possible the most important information and knowledge which would enhance their fight for self-determination and democratic control. Instead we are witnessing "the globalization of the Western ideology of consumerism" (Karliner, 1997, 177), a new round of global colonialism which seeks to perfect its control over the world's consumers, making them "operational" for the world's largest advertising agencies and producers of consumer goods. In short, not enlightenment, but commodification and "monoculture of the mind" (Vandana Shiva) is the aim. And since colonialism always needs a colonial power, we are not surprised to learn that "the audio visual and film industry has grown into the main export of the US, the most important foreign exchange earner".[396] It would be rather astonishing if the further development of the communication networks, video on demand over the internet and other gimmicks would change any of this.

There is no historical evidence that the ruling powers would ever concede anything without a demand, that they would voluntarily give up any of their power. So it is clear that the business world and global power elites are not sitting around passively to wait until somebody might challenge their authority and power. With massively increased technological help since the invention of newspapers, the radio and TV, they use any propaganda tool available to them to influence public opinion.[397] I do apologise if I sound like an old-style class-war activ-

---

[395] Riccardo Petrella, 'Die Gefahren einer Techno-Utopie', in *Le Monde diplomatique [deutsch]*, 2 (1996) 5, 5.

[396] Ignacio Ramonet, 'Wer sind die Citoyen des Cyberspace? Medienkonzentration und Pressefreiheit', in *Die WochenZeitung*, No. 49, 6.12.1996, 24. The unidirectional flooding of the world's media channels by US products and the accompanying stifling of local and regional production can be seen not only in the South, but also in Europe. Today more than half the programmes, series and films shown on European TV are made in the US, with a much higher rate on private TV stations. With the help of the WTO the trade ministry in Washington will try any trick to stop anything which gets in the way of this global cultural steamroller and producer of ignorance. According to this logic everybody trying to support local or regional culture contravenes free trade (see Ramonet, ibid.). A good ally in the production of this world wide smokescreen which brings the blessings of 'Dallas' and Disney into the remotest corners of this planet is Rupert Murdoch, whose Star TV can theoretically be received by 2.5 billion people in Asia (see Alan Rusbridger, 'The moghul invasion', in *The Guardian Weekend*, 9.4.1994, 6-10, 43, here 8; and Rusbridger, 'The global visage', in *The Guardian*, 24.7.1995, 16-17).

[397] For a collection of the dirty tricks in the books of corporate PR to 'massage' public opinion for millions of dollars, see Stauber/Rampton, 1995.

ist, but there is simply too much evidence to suggest that this indoctrination has been taking place for a long time, and *not* as a conspiracy theory, driven by some evil, ugly, fat capitalist bosses meeting in some secret board room. Chomsky/ Herman have shown that what we witness are inbuilt institutional necessities of any unjust system: corporate media and PR are part of the strategy to retain power at all costs, to spread the ideology that the interests of those in power are the interests of all in society, and to hide the fact that this is ideology rather than accurate description of the facts (see Chomsky/Herman, 1994; Chomsky, 1989).

We have already referred to the most prominent recent example of this, which became known as corporate "greenwash". In its most extreme form it is an attempt at brainwashing the public into believing that the biggest environmental destructors and polluters[398] are actually the most successful heroes fighting for a sustainable world (see p. 27). For many years now the world's oil, car and chemical TNCs, together with industrial associations and utilities, have been using long-term advertising campaigns in order to influence and determine the debate about environmental problems, to infiltrate the environmental movement and to present themselves to the public as the "greenest of the green". This "ecopornography" (Jerry Mander) gobbled up more than US$1 billion in 1997 and was described by *PR Watch* as "the PR equivalent of a prolonged carpet bombing campaign" (Karliner, 1997, 170). And it is clear that it works: Chevron's "People Do" campaign, running for more than ten years, costing between $5 and $10 million a year and 'accommodating' concern for green issues through, for example, the butterfly "preserve" at the company's El Segundo refinery (costing the company $5,000 a year), has led to an increase in pump sales of 10 per cent in total and of 22 per cent amongst "potentially antagonistic, socially concerned types" (Karliner, 1997, 172-174).

It is to the credit of Alex Carey to have shown in detail how the global business interests have quickly recognised that in democratic societies the manipulation of the opinions of citizens, i.e. the "manufacture of consent", is of utmost importance if you want to retain power. In totalitarian regimes the rulers can resort to violent oppression, but this avenue is closed off in democracies where in theory the people are free to dispose of the business oligarchy ruling them. In order to prevent this and to portray the narrow interests of the elite as the general interests of society, emotionally charged terms like "America", "democracy", "progress", or "freedom" are equated with the current economic structure, i.e. "market economy" (Carey, 1997, 75-84). We have seen in chapter 2.a how successful this indoctrination has been, so that these interpretations of the terms are in fact considered *common sense* and almost universally accepted. As we saw before, Carey has summarised this fact as follows: "It is arguable that the success of business propaganda in persuading us,

---

[398] For an example with regard to the oil industry see Suzanne Simon, 'Texaco's Ecological Terrorism of the Ecuadorian Amazon', in *Z Magazine*, 13 (2000) 10, 52-57.

for so long, that we are free from propaganda is one of the most significant propaganda achievements of the twentieth century." (1997, 21) This concerted and long-term attack on the awareness of citizens by business propaganda, through advertising and corporate media, has led to the consequence that our perception of truth has been fundamentally shifted (witness also the discourse of postmodernity): it is not important any more whether something is true or false, but only whether something, anything, can be presented as credible (see Carey, 1997, 83).

Particularly the concept of "virtual reality" aims at diverting people's interests even more from 'real reality' and the existing problems that might affect them. It creates 'perfect' technical dream worlds, entirely uncontaminated by the real world. Jörg-Uwe Albig has described this escape from reality with regard to the theme parks which are created everywhere: "According to statistics more and more people spend their holidays not in real countries any more, but in the artificial geography of theme parks", with grave consequences:

> The stunt worlds with their monstrous blatancy do not strengthen trust in reality, but trust in the media. In theme parks, in the artificial paradises of shopping malls and in the digital hallucinations of virtual reality, the world appears exactly as we know it from the ads. The Grand Canyon in "Port Aventura" looks like the one in the Marlboro ad; the pyramid inside the Luxor hotel in Las Vegas is a copy of a Camel one. (...) The TV stations have long been inundated with applications for the vacant flat whenever somebody dies in "Lindenstrasse" [a popular German soap] and in a drawing competition in kindergartens, every third child coloured the cows in purple [as in a German chocolate ad].[399]

But this is only imitating a long-standing trend in audiovisual and print media where soap operas, sports coverage, game shows, ads, "reality TV" and much more allow the viewers to escape into a fictional realm which has precious little to do with daily life and the pressing problems of the world: "One of the functions that things like professional sports play in our society and others is to offer an area to deflect people's attention from things that matter; so that the people in power can do what matters without public interference."[400]

I am not claiming that this is the only job the media do or that this is the only way to use media. Understood as tools to uncover secret deals, to enable citizens to comprehend issues by providing them with the necessary sources, to keep a check on the political, social and economic powers, they are an indispensable tool for sustainable change and in fact people's power anywhere. Yet this implies that they are independent from corporate and state control, free from financial

---

[399] Jörg-Uwe Albig, 'Im Sog der Illusionen', in *GeoExtra: Das 21. Jahrhundert: Faszination Zukunft*, (1995) 1, 143-156, here 146 and 156.

[400] Chomsky in James Peck, 'Interview with Noam Chomsky', in Chomsky, 1992, 1-55, here 36.

dependence on advertising and not controlled by the cultural elite.[401] And this is less and less the case. Most print media are dependent for 70 to 80 per cent of their revenue on advertising.[402] The newest fad in Europe – free newspapers with an abysmal editorial content, almost exclusively reliant on news agency material or poorly researched texts[403] – is 100 per cent advertising financed, as are almost all private TV stations.[404] And for citizens, it is increasingly the same story: access to information is dependent on cash, education, social status and time, so that the theoretical right to information, as guaranteed by many a constitution, looks different in reality:

> But freedom of that sort [access to information in principle], though important for the privileged, is socially rather meaningless. For the mass of the population of the United States, there was no possibility, in the real world, to gain access to [independent information about the secret war in Laos], let alone to comprehend its significance. The distribution of power and privilege effectively limits the access to information and the ability to escape the framework of doctrine imposed by ideological institutions: the mass media, the journals of opinion, the schools and universities. The same is true in every domain. In principle, we have a variety of important rights under the law. But we also know just how much these mean, in practice, to people who are unable to purchase them.[405]

This fact should, one would have thought, oblige those sectors of society who have the time, money and education to use commodified and other information, to put these advantages to good use for society as a whole rather than to perpetuate their own power:

---

[401] The cultural elite, like any other elite, tends to do everything they can to retain their power. It is "the norm for the educated classes" that they "are typically the most profoundly indoctrinated and in a deep sense the most ignorant group, the victims as well as the purveyors of the doctrines of the faith." (Noam Chomsky, 'The Manufacture of Consent (1984)', in Chomsky, 1992, 121-136, here 126) See also Bourdieu, 1992, 755.

[402] Hermann Meyn (1996), *Massenmedien in der Bundesrepublik Deutschland* (Berlin, Edition Colloquium [=Zur Politik und Zeitgeschichte; 24]), 94. This dependence often has serious consequences: when the most influential German news magazine, *Der Spiegel*, started a car-critical editorial stance after the first oil price hike in 1973, all the big car manufacturers withheld their ads and *Der Spiegel* quickly changed the editorial line again (regarding the influence of advertisers on media see Chomsky, 1989, 8).

[403] Just like the first online-only newspaper in Germany (see Jutta Hess, 'Das kleine Informative mit der Maus', in *die tageszeitung*, 30.11.2000, 17).

[404] Hermann Meyn (1996), *Massenmedien in der Bundesrepublik Deutschland* (Berlin, Edition Colloquium [=Zur Politik und Zeitgeschichte; 24]), 166.

[405] Noam Chomsky, 'Equality. Language Development, Human Intelligence, and Social Organization (1976)', in Chomsky, 1992, 183-202, here 189.

> For a privileged minority, Western democracy provides the leisure, the facilities, and the training to seek the truth lying hidden behind the veil of distortion and misrepresentation, ideology, and class interest through which the events of current history are presented to us. The responsibilities of intellectuals, then, are much deeper than what Macdonald calls the "responsibility of peoples", given the unique privileges that intellectuals enjoy.[406]

Yet the trouble is that it is very difficult to get to the bottom of almost any matter because of the power constellations. The key myths upholding the power of money and special interests have been so effectively planted, repeated and disseminated that when we encounter the truth we are unlikely to believe it. George Monbiot has described this very aptly after yet another set of studies proved that traditional farming techniques (planting several breeds in one field) "resulted in spectacular increases in yield" (18 per cent more rice per acre, without using pesticides) and a 94 per cent decrease of the most devastating fungus, rice blast, when compared with modern rice-growing methods (planting a single, hi-tech variety across hundreds of hectares). Why is it, then, that we still believe that our industrial agribusiness practices are better and more advanced?

> All this requires an unrelenting propaganda war against the tried and tested techniques of traditional farming, as the big companies and their scientists dismiss them as unproductive, unsophisticated and unsafe. The truth, so effectively suppressed that it is now almost impossible to believe, is that organic farming is the key to feeding the world.[407]

How does this constant disinformation work? There are various aspects to normal 'information' provision through mainstream media which together work to perpetuate the current structures and to counteract the ability of people to get to know the real world.

The first level, which again stresses why the media are at the same time so important and potentially such a threat to democracy, consists in the fact that today we are utterly dependent on media as our 'window to the world'. Whereas in previous times the village or town was the world and it was quite possible to get first-hand experiences relating to the most important events in life, in 'developed' countries this has become more or less impossible. We partake in the world through the glasses of the media, i.e. we experience most things in a mediated way.

---

[406] Noam Chomsky, 'The Responsibility of Intellectuals (1966)', in Chomsky, 1992, 59-82, here 60. See also, in similar vein: Schumacher, 1993, 172-173.

[407] George Monbiot, 'Biotech has bamboozled us all', in *The Guardian*, 24.8.2000.

The medium of television is a privatized ritual because without the encounter with real embodied events we depend on this medium as a means of relating to a wider world. It is estimated that in the typical North American family, each individual will have been spoken to by television figures more than they have been spoken to by each other. (O'Sullivan, 1999, 26-27)

This has crucial implications. We all know that there is an unbridgeable difference between going to a football match ourselves, experiencing the event through all our senses, but from a particular point of view (which we choose, within limits), and watching the same game on TV where we have different points of views, but they are chosen by the camera-crews and most sensory data is filtered out:

> The subjective experience of observing and reflecting on something that appears on the screen as objective, distant, and separate from self is very different (...) from orally based interactions, in which memory and the five senses attune themselves to the context of the experience and which involve physical and mental participation in the reciprocal patterns of ongoing community life. (Bowers, 2000, 34-35)

In other words, quite contrary to the popular belief that "pictures can't lie",[408] it is clear that the world presented to us in the media is a constructed world. The journalists and camera-crews select 'news-worthy' information, filter other information out, decide on the point-of-view, the importance, the length and depth of coverage. They act, in other words, as gatekeepers who decide which information we get and which is kept from our view.[409] This also means, as can be shown for any number of topics,[410] that coverage is influenced by the dominant values of society and the media corporations, and by of the prevalent myths of society (such as 'growth', 'progress', 'free market'). Chomsky has shown that individual journalists can and do frequently break through these limitations, but that the institutional structures are such that this is unlikely to happen – not least because it is

---

[408] This reason is routinely given by those of my media course students who argue that TV is a more reliable and truthful medium than newspapers.

[409] Ferguson details information on media campaigns in the US to manipulate public opinion by not talking about certain issues, misrepresenting facts, not printing certain polls, etc. (Ferguson, 1995, 389-390).

[410] One of the most drastic examples, and also one of the best researched, is the massive manipulation, distortion and outright falsification of the truth during the Gulf War of 1991 (see Chomsky, 'Afterword', 1992a, 407-440; John R. MacArthur (1993), *Second front – Censorship and Propaganda in the Gulf War*, foreword by Ben Bagdikian (Berkeley; London, University of California Press); Philip M. Taylor (1992), *War and the Media: Propaganda and Persuasion in the Gulf War* (Manchester, Manchester University Press, 1992); *Triumph of the image: the media's war in the Persian Gulf – a global perspective* (1992), ed. by Hamid Mowlana, George Gerbner and Herbert I. Schiller (Boulder, Westview Press [=Critical studies in communication and in the cultural industries])).

detrimental to one's career within such an institution – so that the dominant ideology and preconceptions are perpetuated as a matter of course. In other words, "the media serve the interests of state and corporate power, which are closely interlinked, framing their reporting and analysis in a manner supportive of established privilege and limiting debate and discussion accordingly" (Chomsky, 1989, 10).

This almost complete dependence on media, at least in the industrialised world, makes it ever more important that the work of the media is as transparent as possible. We need to know who's funding the particular media outlet; we need to know the sources, the context, the background. This is more and more important, as Stauber/Rampton have shown, because the journalists, under ever more pressure to be fast and efficient, yet with less and less time and resources, are more than ever inclined to use perfectly pre-prepared "feature articles" and "video news releases" (VNRs) produced by corporations or PR agencies. These products are then presented to the public as if the newspaper or TV station had produced them, yet they are thinly disguised PR campaigns for corporate interests (see Stauber/Rampton, 1995, 184-186).

Neil Postman has shown that mass media like TV and radio are, due to their structure, not really suited to the transparency just mentioned, since they decontextualise everything and desensitise people from asking "why":

> If I were asked to say what is the worst thing about television news or radio news, I would say that it is just this: that there is no reason offered for why the information is there; no background; no connectedness to anything *else; no* point of view; no sense of what the audience is supposed to do with the information. It is as if the word "because" is entirely absent from rite grammar of broadcast journalism. We are presented with a world of "and"s, not "because"s. This happened, *and* then this happened, *and* then something else happened. As things stand now, at least in America, television and radio are media for information junkies, not for people interested in "because"s. (1999, 94; see also Postman, 1987)

The other level of disinformation is the subtle but constant initiation into the rites of consumers. We have made above the distinction between an enlightened media which aims to serve the people in their quest for adequate and truthful (as much as possible, at least) information and one which treats media as just another commodity, ruled not by an attempt at truth but according to the profit maxim, with dramatic consequences for a democratic public sphere:

> It is now a reality all over the globe that a significant portion of our lives is mediated by an instrument designed to advertise and sell commodities; that instrument being the mass media technology called television. With the advent of commercial television there has been a steady decline in public life. (...) It is important to note here the aims of TV productions *vis-à-vis* the viewer. The commercial interests that sponsor TV programming proceed on the assumption that there is a

population of viewers who have to be formed and conformed (i.e. a passive public) to commercial and consumption values. This connection does not add up to a
sense of a public life. (O'Sullivan, 1999, 26-27)

And this initiation into the ideology of consumerism starts at an early age, training
the kids with unrelenting impact (and dramatic health consequences on top of the
mental ones):

> Not surprisingly, children's snack selection and food requests – mostly for sweet-
> ened cereals, candy, and salty snack foods – parallel the products advertised on
> their favorite television shows and influence their families' purchasing habits.
> American children watch an average of 21 to 28 hours of television each week.
> By high school graduation, the typical teen will have spent 12,000 hours in
> school and watched between 15,000 and 18,000 hours of television. With up to
> 21 ads per hour in TV programming for children, 16% of children's viewing time
> is actually devoted to advertising. The average American child sees over 10,000
> commercials a year, a large proportion of which are devoted to selling food prod-
> ucts, many high in calories, cholesterol, sugar, and salt, but low in nutrients. (...)
> Television watching is a likely and significant contributor to the declining health
> of American youth: One study found that obesity risks among adolescents in-
> creased with higher average hours of television viewing. In part, this higher risk
> derives from munching snack foods and drinking soft drinks during the sedentary
> hours in front of the tube. The trend towards more inactive entertainment is exac-
> erbated by the increasing popularity of computers and video games. (Cobb *et al.*,
> 1999, 36)[411]

So the future consumers are trained into passivity, unhealthy eating habits and intellectual submissiveness by the mass media (see box 15 for TV's levelling out at
the lowest common denominator). No wonder can they not make sense of slightly
more elaborated concepts such as sustainability.

Box 15: TV's dumbing down effect
To begin with, television is essentially nonlinguistic; it presents information mostly in visual images. Although human speech is heard on television, and sometimes assumes importance, people
mostly watch television. And what they watch are rapidly changing visual images – as many as
1,200 different shots every hour. The average length of a shot on network television is 3.5 seconds; the average for a commercial is 2.5 seconds. This requires very little analytic decoding. In
America, television watching is almost wholly a matter of pattern recognition. What I am saying is
that the symbolic form of television does not require any special instruction or learning. (...)
Watching television requires no skills and develops no skills. That is why there is no such thing as
remedial television-watching. That is also why you are no better today at watching television than
you were five years ago, or ten. And that is also why there is no such thing, in reality, as children's

---

[411] "There is very little the culture wants to do for children except to make them into consumers."
(Postman, 1999, 125)

programming. Everything is for everybody. So far as symbolic form is concerned, *ER* is as sophisticated or as simple to grasp as *Sesame Street*. Unlike books, which vary greatly in syntactical and lexical complexity and which may be scaled according to the ability of the reader, television presents information in a form that is undifferentiated in its accessibility. And that is why adults and children tend to watch the same programs. (...) It requires no instruction to grasp its form, and it does not segregate its audience. (...) For reasons that have partly to do with the accessibility of its symbolic form, and partly to do with its commercial base, television promotes as desirable many of the attitudes that we associate with childishness – for example, an obsessive need for immediate gratification, a lack of concern for consequences, an almost promiscuous preoccupation with consumption.

Neil Postman (1999), *Building a Bridge to the Eighteenth Century. How the Past Can Improve Our Future* (New York: Knopf), pp. 191-193.

On a further level it is clear that concerted PR campaigns and media efforts do, to a large degree, shape public opinion. This can easily be shown historically: see the change of public anti-war opinion in the US into war-fever after a sustained campaign before World War I [Carey, 1997], or the predominantly pro-union/anti-business stance of a majority of Americans into a business-friendly one between World War I and II [Zinn, 1996]). But there are also plenty of examples today. Kelley and Geer have shown that "voters' decisions on the smallish list of considerations that actually appear to move them are importantly affected by campaigns, the media and other influences" (quoted in Ferguson, 1995, 392). And Burson Marsteller, the world's largest PR company and corporate whitewasher, itself a TNC, has constantly 'massaged' public opinion: it was already responsible for the crisis management after Bhopal, the Exxon Valdez oil spill and instrumental "in keeping discussion of TNCs off the Rio de Janeiro Summit agenda" (Madeley, 1999, 165). More recently it has run the concerted campaign by the European biotechnology corporations to get genetically modified food accepted by the public. Burson Marsteller's advice to these companies, which want a high return on their research investment irrespective of the consequences for humankind and nature, was clear: blur people's vision. For example, they urged biotech companies "to tell stories, rather than conduct objective debates"[412] and to concentrate on "symbols eliciting hope, satisfaction, caring and self-esteem".[413] Occasionally, the relevant people are quite frank about their aims: "The point of PR, says a Mobil Oil Company executive 'is getting people to behave the way you hope they will behave by persuading them that it is ultimately in their interest to do so.'" (Madeley, 1999, 165)

But we also have to be aware that the tentacles of this comprehensive influence by corporate interests are *not* limited to the mass media by any means.

---

[412] Susan Boos, 'Das globale Grünwaschen: Die PR-Strategien der Gentech-Lobby', in *Die Wochen-Zeitung*, No. 29, 18.7.1997, 3.

[413] John Vidal and George Monbiot, 'Dollar power of bio firms', in *The Guardian*, 16.12.1997, 4.

In the last few years many other means have been developed which have the same aim of brainwashing people into believing that "what's good for business is good for you", such as the donation of IT and other equipment by companies to schools or the sponsorship of universities and entire research disciplines by corporate donations.

A last point: pseudo-information or as it is sometimes called 'plurality of information' is one of the most persistent problems with today's media and it stems largely from what Postman has termed decontextualisation. You can pick almost any news item on TV or in the print media and show how it subtly nudges our perception into specific directions, how vague it is, how impossible it often is to verify claims within it, how it deprives us of any tools to assess the relevance and validity of the claims made (see box 16). It stems from the confusion between truth and opinion (see p. 166) – much worsened by the postmodern dogma that everything is equally valid. Any opinion, however absurd and demonstrably false it might be, is allocated equal time or space and therefore importance as reports that are soundly researched, offer in-depth analysis and give viewers or readers the chance to verify for themselves what they have been presented with. Very helpful for this strategy to increase confusion is the attempt to instil uncertainty into the public. It is well known and happily exploited by corporate PR that if people are unsure about an issue, they will not act. This strategy has been amply exploited in connection with environmental issues, especially global warming.[414]

Box 16: Constructing the world: an example

One of the clearest examples in recent media history of the construction of a suitable image of the world, rather than an adequate representation of what is happening, was the portrayal of the international protests in Seattle against the WTO, in Washington against the IMF (1999), in Prague against the World Bank/IMF (2000) and in Davos against the World Economic Forum (WEF) (2001). The world's mainstream media have unequivocally portrayed the protesters as misguided, violent rioters and thus been able to avoid talking about the real issues that brought the protesters there in the first place: "Each of the protesting individuals and social action groups share a common disdain for institutionalised power structures that service the corporate elites of the world at the expense of working people and the environment." Quite apart from the fact that the violence was mostly either a media construct or a police provocation, facing up to the real issue of how to create a world that "respects the environment and minimizes domination in any form" would shake the foundations of injustice and exploitation on which current power is based. No wonder the masters of the world were scared and have orchestrated a disinformation campaign (see Peter Phillips, Director, Project Censored, Sonoma State University, 'Mainstream Corporate Media Dismisses Democracy' [http://www.indymedia.org/display.php3?article_id=4507]).

Yet let me give you a concrete example of a mainstream article which shows what I mean by disinformation, eco-illiteracy and distortion of the truth: On 15 September 2000 *The Independent* (UK) ran a story by Fred Pearce entitled "Rainforests – who needs them?" Here is an extract:

---

[414] Sharon Beder, 'Corporate Hijacking of the Greenhouse Debate', in *The Ecologist Special Issue: Climate Crisis*, 29 (1999) 2, 119-122, here 119.

"The romance of the rainforest has seduced scientists into inflated notions of the forests' economic value to their inhabitants, according to a new study. Whisper it quietly, but far from being a cornucopia of riches, rainforests may often be fit only for chopping down. If the world wants to keep the forests, it will have to pay – and handsomely. This is the controversial conclusion of Ricardo Godoy, an anthropologist from Brandeis University in Waltham, Massachusetts. He has conducted the first detailed household inventory of the fruits of a rainforest. Godoy sent teams of his students into a remote region of eastern Honduras, in Central America, to catalogue what Indian villagers harvest from their forests and to ask what they fetch at market. (...) For more than a decade, biologists have taken it as an article of faith that most rainforests are a rich source of food, medicines and traditional building and craft materials. (...) But the suspicion has grown that some ecologists have been guilty of wishful thinking. As Godoy points out, few of the studies undertook actual inventories of harvested products. The researchers 'focused on what they saw as the potential value of the forests, rather than what was actually taken from the forest.' That sounded like bad economics as well as fanciful thinking. A small amount of a valuable rainforest fruit might fetch a high market price. But harvesting 10 times as much will more likely just flood the market and send prices tumbling. (...) This is rough news for environmentalists, but hardly surprising, according to Godoy. 'People in the rainforest are poor. If the forest produced high economic value to these people, they would not be poor.' Honduras is the second poorest country in the western hemisphere, and the Tawahka Indians among its poorest, most isolated, inhabitants."

The tone is already set at the beginning with loaded terms like "romance", "seduced" and "inflated", and then hammered home by the "whisper it quietly .... chopping down". It immediately suggests, without any evidence given – and it is also not forthcoming in the rest of the text – that up till now scientists in fact have been fools. Juxtaposed is, ironically, another scientist and we are now to believe what he says, without any context that would give us any reason why we should do so. Why, if the others are so easily fooled, should he be the messenger of the truth? Godoy, in fact, might be right but the text itself gives us no handle whatever to decide this. We only can see with worry that, *according to the text*, Godoy seems to have a very curious understanding of biodiversity and the workings of ecosystems. The article suggests that his thinking is unduly influenced by a notion entirely alien to biodiversity, namely economic profit within a market system. And for the presented notion of the rainforest people, exclusively defining them via the term "poor" and the absence of "economic value", read cash, is remarkable by any standards, but particularly for an anthropologist. It seems to be deeply steeped in the long discredited 'development model'. But the scientist is not to blame in the first instance. The text might misrepresent his views entirely, and that is precisely the problem. Rather than receiving reliable information, the critical reader is left either utterly confused (which, as we have seen, might be intended) or at least not wiser at all. Yet the text calls for a "radical rethink".

Education: part of the problem, not the solution

> The ecological crisis raises fundamental questions about the dominant culture's way of knowing, its moral values, and its way of understanding human/nature relationships. (Bowers, 1997, 65)

> However, a curriculum committed to social critique is clearly at odds with dominant political and social practices and consumer-led economic systems. Moreover (...) prevailing educational systems shut out the most effective teaching strategies, those of personal experience and dynamic learning, and appear to prohibit the attainment of the outcomes of informed and politically active citizenry. (Plant, 1998, 96-97)

> If western civilisation is in a state of permanent crisis, it is not far-fetched to suggest that there may be something wrong with its education. (Schumacher, 1993, 60)

> Modern education conditions us for "consumer consciousness". (Cajete, 1999, 193)

When I argue with Plant that current educational practice on all levels is one of the most important factors in legitimising and perpetuating the unsustainability of the present, that in fact the dominant paradigm of education "makes little or no acknowledgement of the sustainability challenge",[415] this flies in the face of common sense. That Euro-American education is by definition liberating has been part of the 'development model' and it is certainly a deeply entrenched core conviction of the discourse of environmental education and education for sustainability. Any number of official declarations and UNESCO documents abound with claims that education is the saviour to all sorts of ills, from "underdevelopment" to "overpopulation". And the EfS discourse is no exception: "Education is the primary agent of transformation towards sustainable development", we read in UNESCO's manifest *Educating for a Sustainable Future* (1997, 36, §118) and in a recent German textbook on EfS it says: "the only meaningful answer to the worldwide problems of the uncontrollable excesses of human civilisation is a pedagogical answer" (*Umweltbildung im 20. Jahrhundert*, 2001, 2).

This talk of education as a guarantor of salvation is wrong for two main reasons. Firstly, it is plain that today the "shadow curriculum", to borrow David Orr's phrase,[416] of the economic Master Science and the accompanying consumer culture combined with the mass media, in particular TV (see p. 215ff.), is the primary educator (see also Illich, 1971, 33). The shaping forces of our world are *not* primarily the teachers and lecturers (however much they might like this to be the case), but the stockbrokers, industrialists and mass entertainers. So the almost

---

[415] Stephen Sterling, 'Education in Change', in *Education for Sustainability*, 1996, 18-39, here 21.
[416] Before him, Illich talked of the "hidden curriculum" (1971, 2, esp. 32-33 and 1971a, 45).

exclusive stylisation of 'education' as the one social agent *responsible* for a transition to sustainability – note that *Agenda 21*, the Rio Conference blueprint for this transition, devotes an entire chapter (36: "Promoting Education, Public Awareness And Training") to education, yet not a line to the responsibilities of corporations – looks more and more like a convenient scapegoating exercise: first you claim that education will pull us out of the self-created mess, then you blame education if it is not happening, but the prime actors are all absolved from any responsibility and from any need to change their behaviour.

Yet this, of course, doesn't mean that education is not important at all and that it doesn't have its fair share in societal responsibilities. And here we find the second reason why this generalised talk about the liberating and positive qualities of education is wrong. David Orr has pinpointed the problem: "It is not education, but education of a certain kind, that will save us." (Orr, 1994, 8) In order to understand this we need to have a close look at what is wrong with the prevailing type of education. It will *not* save us; on the contrary, it is an important contributing factor to unsustainability. The problem can best be expressed in the following paradox: On the one hand, as I have demonstrated in the previous chapters, we are in dire need of enlightenment, information, education, knowledge and wisdom to turn the powers-that-be, the makers and shakers, the world's politicians, most of them 'highly educated' people, and the world's middle-classes into ecoliterate human beings, capable of changing their thoughts and actions into sustainable ones. Yet on the other hand there are literally millions, if not billions of people – 'primitive', 'uneducated', and 'underdeveloped' by Euro-American standards – who are actually very ecoliterate and live sustainably. They don't need 'our' education, on the contrary: we need to learn from them.

This determines the levels of our analysis. We have to find answers to the following questions: 1) Why are the best-educated people (in the conventional sense) also the ones whose thoughts and actions are the most unsustainable? And 2) Is our Euro-American educational model really so beneficial and so superior to other forms of learning?

Euro-American-style education produces, rather than overcomes unsustainability

To my mind, one of the most important aspects of David Orr's work has been his consistent identification of Euro-American-style education as *a crucial perpetuator and producer* of unsustainability, rather than a saviour from it. It is quite clearly *not* the poor peasant farmer in Thailand or the subsistence farmer in the Andes who are the problem; it is us, the 'well educated' elites of the world who wreak havoc on Earth, who destroy the biosphere, who invent ever new technologies, chemicals and processes which accelerate this destruction, rather than ending it. Orr says it with all the necessary clarity and frankness:

It is worth noting that this [destruction of the world] is not the work of ignorant people. Rather, it is largely the result of work by people with BAs, BSs, LLBs, MBAs, and PhDs. Elie Wiesel once made the same point, noting that the design-ers and perpetrators of Auschwitz, Dachau, and Buchenwald – the Holocaust – were the heirs of Kant and Goethe, widely thought to be the best educated people on earth. But their education did not serve as an adequate barrier to barbarity. What was wrong with their education? In Wiesel's (1990) words, "It emphasized theories instead of values, concepts rather than human beings, abstraction rather than consciousness, answers instead of questions, ideology and efficiency rather than conscience." I believe that the same could be said of our education. (...) It is a matter of no small consequence that the only people who have lived sustainably on the planet for any length of time could not read, or like the Amish do not make a fetish of reading. My point is simply that education is no guarantee of decency, prudence, or wisdom. More of the same kind of education will only compound our problems. (Orr, 1994, 7-8)

Rather it [destruction of our world] is the work of people who, in Gary Snyder's words, "make unimaginably large sums of money, people impeccably groomed, excellently educated at the best universities – male and female alike – eating fine foods and reading classy literature, while orchestrating the investment and legis-lation that ruin the world". (ibid. 17)

So plainly the focus has to be on the factors in our current educational regime which produce these results, as opposed to ecoliteracy. On that basis we can then proceed in chapter 3 to outline what EfS would have to entail. I would like to dis-tinguish several of these factors, which mirror to a large extent what Elie Wiesel has pointed out above.

1. Mass education as historical necessity for industrialisation and consumerism

Historically there is no question that the mass education system, with its compul-sory school attendance by all children, was not set up to empower and liberate the children and growing adults, but was a necessary tool to produce the frame of mind needed by the industrial system: it was (and of course still is) a massive feat of social engineering. It trains pupils into modes of subservience to authority, it equips them with the skills which allow their most effective utilisation in the in-dustrial system and the economy more generally. It was and is the conveyor belt without which the nation state, its bureaucracy, industrialisation and colonialism would never have been possible. Right from the start, despite the pedagogical rhetoric of the enlightenment, of Rousseau or Pestalozzi, mass education in the Euro-American sense was always a tool to consolidate the power of the rulers, rather than a project of empowerment (yet there is no reason why it couldn't be such a project, see chapter 3). Edmund O'Sullivan has made clear that this is still the case today, since the current educational system

serves the needs of our present dysfunctional industrial system. Our present educational institutions which are in line with and feeding into industrialism, nation-
alism, competitive transnationalism, individualism and patriarchy must be
fundamentally called into question. (O'Sullivan, 1999, 7)

And while originally the prime task was making the masses operational for the
industrial system, the factories and state offices (see Lummis, 1996, 52-53), it is
now the prime function of schooling, besides the mass media, to 'educate' consumers: "Our educational institutions have been apologists for the industrial society and they are part of a broad hegemonic process for consumer dream structures.
Optimism and belief in the consumer industrial society are still a part of everyday
education." (O'Sullivan, 1999, 43; see also Illich, 1971, 74)

Far from being a subversive cesspit, as some conservative critics would
make you believe, the educational system, if you judge it from the smooth running
and expansion of Euro-American consumer capitalism, is oiling this system very
well or at least not producing any sand to obstruct it.

This can best be seen in the way that economic logic has swept the educational system in the last decades, in particular the employability discourse. When
my university recently rewrote its 'Learning and Teaching Strategy', the word employability was implanted into every second sentence. You will not find, in this
strategy, any reference to educational standards that might have to be upheld irrespective of demands made by the economic sphere. O'Sullivan has noted that this
is not an isolated case:

> Today, formal educational institutions are being enlisted to prepare the next gen-
> eration for the needs of the global marketplace. We are beginning to see this phe-
> nomenon across the curriculum, with the possible exception of education in the
> early pre-school and elementary school years. When we examine the policy goals
> for education in the secondary and post-secondary school system, we now see a
> predominance of globalization language. The deep penetration of the global mar-
> ket mentality is overshadowing the older language of nation-state loyalty.
> (O'Sullivan, 1999, 33)

The only purpose of university education these days, it seems, is to relieve the
economy of the task of producing totally malleable receptacles of 'transferable
skills', utterly devoid of any contextualised knowledge, social responsibility, or
ethical understanding of their place in life and the biosphere. The promotion of
economics to Master Science, the reduction of all spheres to economic imperatives turns human beings into commodities, just like everything else. And this is
reflected in the educational system:

> The biggest problem with the current educational system is that education and
> upbringing is increasingly instrumentalised for the production of "human re-

sources". The education of the personality loses more and more importance. (...) Just like any other material or immaterial resource, human beings are viewed as commodities which are available, always and everywhere. They possess neither civil nor other rights of a political, cultural or social nature, because costs are the only factors determining their exploitation. The resource human being owes its right to existence and income solely to its profitability, competitiveness and efficiency. At all times it has to prove how useful it is – so that the "right to work" is replaced by the obligation to prove one's "usability".[417]

Quite obviously, the notion of furthering a balanced understanding of the ecological, political, social and economic parameters and principles of a sustainable society as developed in chapter 1.c have no place whatsoever in such a conception of education and of human beings.

2. *Underlying biophobia of our educational values*

Unsustainability is founded in deep cultural beliefs or "root metaphors", as Chet Bowers calls them, which shape much of our perception and understanding (1995, 31) and which drive us to destroy what we depend on. Orr has labelled this set of beliefs "biophobia" and identified its constituent parts: 1. that the world is not alive, and therefore "not worthy of respect"; 2. a distancing of nature by transforming animals and plants into machines, with the help of the Cartesian mechanistic world view: "no obligations or pity are owed to machines"; 3. a belief in "hard data" which replaced any "remaining sympathy we had for nature"; 4. the combination of power, cash and knowledge to transform the world on a large scale; 5. the ideology of unlimited growth; 6. "the sophisticated cultivation of dissatisfaction which could be converted into mass consumption"; 7. the transformation of politics into "the pursuit of material self-interest" which disempowers citizens (Orr, 1994, 133-134).

The concerted attack on the traditional wisdom of people who knew that they were part of the biosphere and had to live in respect for it – still to be found in indigenous cultures – led to a complete alienation of people from themselves as living beings and from animals, plants and nature in general. This "earth alienation" was aptly captured by Hannah Arendt quite a while ago:

Without actually standing where Archimedes wished to stand (...) still bound to the earth through the human condition, we have found a way to act on the earth and within terrestrial nature as though we dispose of it from outside, from the Archimedean point, and even at the risk of endangering the natural life process

---

[417] Riccardo Petrella, 'Humanressourcen für den Weltmarkt. Fallstricke der Erziehungspolitik', in *Le monde diplomatique [deutsch]*, 6 (2000) 11, 22.

we expose the earth to universal, cosmic forces alien to nature's household. (Arendt, 1958, 238)

Orr goes even further. He claims that biophobia is actually a necessary precondition of our current system, which also explains why it is by definition unsustainable:

> For our politics to work as they now do, a large number of people must not like any nature that cannot be repackaged and sold back to them. They must be ecologically illiterate and ecologically incompetent, and they must believe that this is not only inevitable but desirable. Furthermore, they must be ignorant of the basis of their dependency. They must come to see their bondage as freedom and their discontents as commercially solvable problems. (Orr, 1994, 136)

This view is shared by Cajete. In fact, biophobia might be the defining principle of industrialism and capitalism alike, which is why any EfS will have to focus on averting it:

> Because biophobia underlies aspects of the prevailing mindset of modernism, it influences the "hidden curriculum" of modern Western education. Indeed, the evolution of biophobia as expressed in the attempt to control and subdue nature has its own unique historical progression in Western religious, philosophical, artistic, and academic traditions. Biophobia also underpins the epistemological orientation of most Western governmental, economic, religious, and educational institutions.[418]

The fundamental corrective to biophobia is the insight that the earth doesn't need us, but that we cannot survive without the earth (Wackernagel/Rees, 1996, 115). Once we start to acknowledge the "profound significance of indigenous knowledge" (O'Sullivan, 1999, 100) and its insight that there is no environment (in the sense of different from us, opposed to us, outside us), but only a being part of the world, it becomes clear that "the missing element [in our education] is the relation of humans to the other-than-human components of the world that we live in".[419] The Cartesian world view with its contemporary effect of externalising most of the costs of our lifestyles onto other people and nature, has made us blind to the fact, amply borne out by even the most rudimentary knowledge of how both our bodies and minds work (see Rose, 1998), that there is no life without the world around us, the air, nutrients etc. And this, of course, has implications not just for our physical well-being:

---

[418] Gregory Cajete, 'Reclaiming Biophilia. Lessons from Indigenous Peoples', in *Ecological Education in Action* (1999), 189-206, here 190.
[419] Thomas Berry, 'Foreword', in O'Sullivan, 1999, xi-xv, here xi.

We cannot live simply with ourselves. Our inner world is a response to the outer world. Without the wonder and majesty and beauty of the outer world we have no developed inner world. As all those living beings around us perish, then we perish within. In a sense we lose our souls. We lose our imagination, our emotional range; we even lose our intellectual development. We cannot survive in our human order of being without the entire range of natural phenomena that surround us. [420]

3. Abstractness of educational discourse: neither place nor responsibility

Our current educational practice emphasises "theories instead of values, concepts rather than human beings, abstraction rather than consciousness, answers instead of questions, ideology and efficiency rather than conscience." (Elie Wiesel, quoted in Orr, 1994, 8). This means in essence that it is an abstract discourse, more dedicated to itself as a discourse, rather than to a serving attitude towards real life in an actual, concrete, historical location:

> As commonly practiced, education has little to do with its specific setting or locality. The typical campus is regarded mostly as a place where learning occurs, but is, itself, believed to be the source of no useful learning. It is intended, rather, to be convenient, efficient, or aesthetically pleasing, but not instructional. It neither requires nor facilitates competence or mindfulness. By that standard, the same education could happen as well in California or in Kazakhstan, or on Mars, for that matter. (Orr, 1999, 229)

It can be clearly seen that such an education can not do much for local people to empower themselves, to arm them with the most useful knowledge about their own situation, their surroundings, their embeddedness into ecosystems, their dependence from them and the ways within which they can rely on them and be nurtured by them. Instead, such a mind set produces quite logically the hordes of international business apparatchiks, working for TNCs, the World Bank or NGOs, and descending onto local livelihoods about which they know nothing, but also care nothing, since there is no necessity for a sense of responsibility: if the company decides to relocate, that temporary place of work loses all significance.

4. Education versus wisdom

Our notion of education is very intimately bound up with all the other myths of Euro-American ideology (linear progress, unlimited growth, development as improvement). We constantly, if mostly unconsciously, start from the assumption that education (of our sort) is always good, always better than what was before.

---

[420] Ibid.

Chet Bowers has shown that this is hardly surprising since most educators, even those considering themselves progressive and critical, are caught in these myths and thus perpetuate them in their teaching. The most dominant "root metaphors" which perpetuate the idea that new is always better are the notion of progress and the Euro-American concept of individual creativity and intelligence (see Bowers, 1995, 41-74, 92-134; 1997, 174-197). To transcend this we have to link the discussion back to our distinction between mere knowledge and wisdom, between cleverness and intelligence (see p. 205). Postman forcefully makes the point that we are lacking not information or data, not even knowledge, but wisdom. The crucial point is

> the difference between mere opinion and wisdom. It is also the difference between dogmatism and education. Any fool can have an opinion; to know what one needs to know to have an opinion is wisdom; which is another way of saying that wisdom means knowing what questions to ask about knowledge. (Postman, 1999, 96)

So, contrary to the hype about the 'information society' and the 'information highway',

> the problem to be solved in the twenty-first century is not how to move information, not the engineering of information. We solved that problem long ago. The problem is how to transform information into knowledge, and how to transform knowledge into wisdom. If we can solve that problem, all the rest will take care of itself. (Postman, 1999, 98; see also Orr, 1994, 11; Dürr, 2000, 117-118)

This is more fundamental than we tend to allow, for the production of data, the filling of brains with bits of disconnected items of information, the dissemination of 'educational standards' of reading, writing and maths occupies time, space and energy. This is why, despite the capacity of our individual brains to deal with an astonishing amount of information and knowledge, for society as a whole education and knowledge production seems more like a zero-sum game:

> What can be said truthfully is that some knowledge is increasing while other kinds of knowledge are being lost. For example, David Ehrenfeld has pointed out that biology departments no longer hire faculty in such areas as systematics, taxonomy, or ornithology. In other words, important knowledge is being lost because of the recent overemphasis on molecular biology and genetic engineering, which are more lucrative but not more important areas of inquiry. (...) It is not just knowledge in certain areas that we are losing but also vernacular knowledge, by which I mean the knowledge that people have of their places. (...) All things considered, it is possible that we are becoming more ignorant of the things we must know to live well and sustainably on the earth. (Orr, 1994, 9-11; see Bowers, 2000, 54-60)

5. Education versus empowerment

> Everyone learns how to live outside school. We learn to speak,
> to think, to love, to feel, to play, to curse, to politick, and to
> work without interference from a teacher. (Illich, 1971, 28-29)

It is a specific Euro-American and historically relatively new idea that you cannot live life without a proper education. Only from the 17th century onwards did this idea develop and the more ground it gained, the more it devalued any other form of non-institutionalised knowledge (see Illich, 1971, 26-27). Yet institutionalised compulsory school education was not just any odd education, but a system of indoctrination based on specific values with specific goals. This context alone would suffice to question the ability of conventional education to further anything connected with human empowerment – and we know from our discussion of 'liberal democracy' that institutionalised forms of anything tend to counteract freedom and self-determination (see p. 165ff.). To put it provocatively: "For most men the right to learn is curtailed by the obligation to attend school." (Illich, 1971, xix) And it is not surprising that most people don't like it:

> [Most people] understand [education] as being made to go to a place called
> school, and there being made to learn something they don't much want to learn,
> under the threat that bad things will be done to them if they don't. Needless to
> say, most people don't much like this game, and stop playing as soon as they can.
> (Holt, 1991, 34)

That this is so, and necessarily so, is relatively easy to understand: If you start from the assumption, as our system of education does, that there are certain overarching goals which are not to be questioned (liberal democracy, market economy, growth, progress, development), supported by certain values (competitiveness, sacredness of property, pursuit of material self-interest), then it follows by necessity that educational institutions have but one job, namely to disseminate the values, information and knowledge necessary to achieve these goals and maintain these values; in other words, to act as an educational funnel (see Illich, 1971, xix-xx). Yet if you start from the assumption that education should empower human beings to acquire the necessary knowledge to take control of their lives, to balance self-determination with their responsibilities towards their communities and to respect the limits of the biosphere, then you see that the above institutions cannot succeed with this. In order to do this education would need to overcome the biophobia diagnosed above, the alienation from real life: and in radical democratic tradition it would have to bring about a sense of responsibility for one's education: "Only when a man [sic] recovers the sense of personal responsibility for what he learns and teaches can this spell be broken and the alienation of learning from living be overcome." (Illich, 1971a, 48; see also 1971, 24) This sense is clearly missing in a time when students sue their lecturers for having failed to educate them properly, as has happened in the US and in post-Thatcher Britain. I don't want to defend the lecturers' side for one minute, but it does expose a very odd idea of education, precisely the commodified notion of it: "I pay money for my goods, so I want the goods delivered". In other words, I as a student have nothing whatever to do with my education, except that I cough up some cash and get some commodified object in return. Empowerment?

But Ivan Illich has also pointed out another radical flaw in institutionalised, compulsory schooling. It is very much akin to the fundamental problem of the 'development model'. Just by instituting it you automatically declare one end of the scale as 'uneducated', 'underdeveloped', 'primitive' and the other end as 'educated', 'developed' and 'civilised'. This has various consequences which turn the notion that access to education for all furthers equality of opportunities into a joke: it leads to an abandoning of any traditional, craft or other, knowledge the poor might have; it locks them into a feeling of being inferior to 'better educated' people; it trains them to accept knowledge disseminated by institutions and therefore makes it easier to abuse them in any institutional context; it makes them dependent on schools and other institutions, since they cannot just get on and live their lives, they first have to get an education in order to be able to do it (or so it is claimed) (see Illich, 1971, 7-10).

> School makes alienation preparatory to life, thus depriving education of reality and work of creativity. School prepares for the alienating institutionalization of life by teaching the need to be taught. Once this lesson is learned, people lose their incentive to grow in independence. (Illich, 1971, 47)

And it is very clear that the modern educational system has not led to a more egalitarian and equitable society; on the contrary it has led to the establishment of new classes and casts of privileged 'priests' who hang on to their privileges of power with all their might:

> School has become the world religion of a modernized proletariat, and makes futile promises of salvation to the poor of the technological age. The nation-state has adopted it, drafting all citizens into a graded curriculum leading to sequential diplomas not unlike the initiation rituals and hierarchic promotions of former times. (Illich, 1971, 10).

Your position in the various hierarchies, itself not exactly an expression of a democratic society, depends up to this day very much on your formal education, on your diplomas and certificates. Yet we all know that the real abilities of someone have often very little to do with formal qualifications. Illich made, with this in mind, the radical, but entirely sensible suggestion to discount all this paper work as discrimination: "To make this [constitutional disestablishment of the monopoly of the school] effective, we need a law forbidding discrimination in hiring, voting, or admission to centers of learning based on previous attendance at some curriculum." (Illich, 1971, 11) That this should be so has two main reasons: Firstly, as indicated in the motto above, most of what we learn about life we do *not* learn in formalised, institutional education: "In fact, learning is the human activity which least needs manipulation by others. Most learning is not the result of instruction. It is rather the result of unhampered participation in a meaningful setting. Most people learn best by being 'with it'." (Illich, 1971, 39)

Secondly, it is very clear that the current educational system is *not* enabling people to nurture wisdom, to further sustainability. It positively obstructs the production of the necessary imagination (see Illich, 1971, 23). According to Vester and the Danish family therapist Jesper Juul, the current educational system not only prevents pupils and students from acquiring necessary knowledge about life on Earth, it is positively harmful because it disseminates knowledge of destruction (see Vester, 1997, 470; Valentin, 2000, 12).

Euro-American-style education as colonialisation of the mind

Bearing the above in mind it seems clear that Euro-American-style education cannot serve as model for the world. It is saturated with values, ideological concepts and institutional structures which are incompatible with a sustainable society. And this is true whether we look at the ecological, political, social or economic, let alone the scientific-technological aspects. The harm done by the relentless expansion of our educational system across the globe, in the wake of economic imperialism, is very hard to assess. But in terms of strategies for survival, of knowledge for sustainability it might well be catastrophic. We can only glimpse part of what we have lost, through sheer ignorance and arrogance, when we look at historical accounts of traditional, say village, knowledge and life (see, as just two examples, Llewellyn, 1991 or Yeşilöz, 1998 and 2000), or if we look at contemporary indigenous societies which have managed to retain at least some independence from Euro-American colonialisation (see Bennholdt-Thomsen/Mies, 1999 and Norberg-Hodge, 2000 for examples).

What we generally see if we bother to look closely is that education, at least in the current context, is something of a zero-sum game: One form of education will destroy the other: "It is important that woman teach our future generations what education really is [says a Ladakh village headman]. Now our children go to school, and learn to read and write. But there is education also in farming, nurturing, spinning, running a family." (quoted in Kingsnorth, 2000, 35-36) What is lost are "non-commodified aspects of community life" such as face-to-face and trans-generational communication, "elder knowledge", "ceremonies that are essential to the moral coherence and identity of the cultural group",[421] narratives,

---

[421] Chet Bowers, email conversation, 10.3.2001, but see also Bowers, 1995.

craft knowledge and much more. All this knowledge, which in a sustainability context is actually far more important than formal skills such as reading and writing (even though they might not necessarily be mutually exclusive), is exterminated by formal schooling which doesn't allow children to be part of the community any more. It cuts them off from traditional knowledge, therefore from the ability to survive on their own, turning them into dependants of the modern industrial system, from employment to consumer goods and social benefits. Only when we compare our life in highly industrialised societies with subsistence societies can we start to understand what we have *lost* in terms of self-determination. We are *entirely* dependent on the industrial machine; most of us would not know how to survive for just a week if the energy supply broke down, the supermarkets closed and traffic came to a standstill. And taking into account that our system is only to be had at the cost of overexploitation of nature and people far away, it might even dawn on us that our system is not desirable at all, since it is so weak and locked into dependency. This might make understandable why Illich talks of our educational system, which perpetuates the industrial society, as educational pollution, akin to environmental pollution (see Illich, 1971, 1).

Box 17: Learning from Ladakh: indigenous versus 'modern' education

[Modern education] isolates children from their culture and from nature, training them instead to become narrow specialists in a Westernized urban environment. This process is particularly striking in Ladakh, where modern schooling acts almost as a blindfold, preventing children from seeing the context in which they live. They leave school unable to use their own resources, unable to function in their own world. (...) For generation after generation, Ladakhis grew up learning how to provide themselves with clothing and shelter; how to make shoes out of yak skin and robes from the wool of sheep; how to build houses out of mud and stone. Education was location-specific and nurtured an intimate relationship with the living world. It gave children an intuitive awareness that allowed them, as they grew older, to use resources in an effective and sustainable way. None of that knowledge is provided in the modern school. Children are trained to become specialists in a technological, rather than an ecological, society. School is a place to forget traditional skills and, worse, to look down on them. (...) The basic curriculum [of Euro-American education in Ladakh] is a poor imitation of that taught in other parts of India, which itself is an imitation of British education. There is almost nothing Ladakhi about it. (...) Most of the skills Ladakhi children learn in school will never be of real use to them. They receive a poor version of an education appropriate for a New Yorker. They learn out of books written by people who have never set foot in Ladakh, who know nothing about growing barley at 12,000 feet or about making houses out of sun-dried bricks. (...) Modern education not only ignores local resources, but, worse still, makes Ladakhi children think of themselves and their culture as inferior. They are robbed of their self-esteem. Everything in school promotes the Western model and, as a direct consequence, makes them ashamed of their own traditions.

Helena Norberg-Hodge (2000), *Ancient Futures: Learning from Ladakh* (London, Rider Books), pp. 110-113.

We know that our educational system is not even a benefit for the poor in the North, so it would be devastating if it were transplanted into the South as a saviour (yet this is precisely what we are doing):

> It would be a disaster for the countries of the Third World if we were to offer them in a kind of exchange of methods our anaemic school and teaching approaches. It would be a barely rectifiable mistake to inflict these on people who have been spared the academic-intellectual terminology and therefore would have the chance to develop an education which leads to an immediate understanding of the numerous different aspects of reality without dismembering it. (Vester, 1997, 477)[422]

Our educational model is simply not desirable for a sustainable world because it is, essentially, a colonialisation of the mind. It is, often via the mentioned "shadow curriculum", akin to an indoctrination:

> Nations in the developing world fear the increase of cultural, economic and political imperialism by powerful developed nations, and the loss of language and traditional ways of life. (...) The ability of foreign broadcasters to beam poor-quality TV and satellite into countries regardless of the wishes of a country is a warning of the implications of unregulated education via the Internet. This can only offer variable quality in tandem with a threat to cultural integrity, which could be described as electronic cultural imperialism.[423]

That is nothing new in itself; the trouble is that this type of indoctrination accelerates the destruction of our life-support-system.

---

[422] For another radical critique of our educational approach and how it for long time has been failing the poor, see Agee/Evans, 1960, 289-315, esp. 290ff.
[423] David Bird and Brian Nicholson, 'A critique of the drive towards the globalization of higher education', in *ALT-J (Association for Learning Technology Journal)*, 6 (1998) 1, 6-12, here 9.

Universities as perpetuators of destruction

> The truth is that without significant precautions, education can equip people merely to be more effective vandals of the earth. (Orr, 1994, 5)

> Mainstream Western thinkers from Adam Smith to Freud and today's academics tend to universalize what is in fact Western or industrial experience. Explicitly or implicitly, they assume that the traits they describe are a manifestation of human nature, rather than a product of industrial culture. (Norberg-Hodge, 2000, 2)

All of what we have just said applies to university education, only more so. Yet universities all over the world have a very special obligation when it comes to furthering sustainability. There are three fundamental reasons for this.

The first reason stems from the fact that these institutions are primarily factories for the reproduction of existing power structures. University graduates are a minority, they are trained to serve the ruling minorities and they themselves become part of this ruling minority. Illich has pointed out how expensive university education in 'developed' countries is. This alone would imply a responsibility to put this privilege to good use for society as a whole:

> The university graduate has been schooled for selective service among the rich of the world. Whatever his or her claims of solidarity with the Third World, each American college graduate has had an education costing an amount five times greater than the median life income of half of humanity. (Illich, 1971, 34)

Yet it doesn't stop there. Through their overconsumption patterns and their life-styles academics all over the world set 'standards' which other people try to emulate: "In each country the amount of consumption by the college graduate sets the standard for all others; if they would be civilized people on or off the job, they will aspire to the style of life of college graduates." (Illich, 1971, 35) And since we know that these overconsumption patterns are positively obstructing a transition to sustainability, academics bear a considerable responsibility in this sense. If they are role models for society, for behaviour and attitude, they have a duty to change their lifestyles and make them compatible with sustainability. Yet today's "typical university graduate does not recognize that the consumer lifestyle is ecologically unsustainable." (Bowers, 2000, 56)

The second reason for a special obligation of universities goes even deeper than this. Academics are not just responsible for our unsustainable present through their role model function, but more profoundly since they are the ones who create and run our political and social institutions, who theoretically underpin and in fact run our capitalist economy, who are as scientists responsible for the technological direction our society takes, who dominate the mass media and are responsible for manipulation, propaganda and manufacturing consent, who educate the children and students and so forth. Academics initiated and sustained the 'development model', supported and furthered colonialism through their theories, destroyed and 'barbarised' indigenous peoples all over the world (see Forbes, 1981, 16, 94, 126).[424] As Orr, already quoted above, clearly said, "It is worth noting that this [destruction of the world] is not the work of ignorant people. Rather, it is largely the result of work by people with BAs, BSs, LLBs, MBAs, and PhDs." (1994, 7) Bowers has phrased it slightly differently, thereby highlighting the paradox we are facing: "The promotion of our highest values and prestigious forms of knowledge serve to increase the prospects of ecological collapse." (1997, 3) If any group in society has to face up to its deeds and start a rapid reorientation of its thinking, its inbuilt assumptions, its world view and its actions then it is the academics.

The Association of University Leaders for a Sustainable Future who issued the Talloires Declaration in 1990 shares this view. They acknowledge that universities are responsible for the leaders of tomorrow:

> Universities educate most of the people who develop and manage society's institutions. For this reason, universities bear profound responsibilities to increase the awareness, knowledge, technologies, and tools to create an environmentally sustainable future. (*Talloires Declaration*, 1995, 2; see also *The Essex Report*, 1995, 4)

The third reason why universities have a very special obligation to society in the just cited sense lies in the fact that they are afforded a very privileged status by society:

> Society has conveyed a special charter on institutions of higher learning. Within the United States, higher education institutions are allowed academic freedom and a tax-free status to receive public and private resources in exchange for their contribution to the health and well-being of society through the creation and dissemination of knowledge and values. These institutions have the mandate and potential to develop the intellectual and conceptual framework for achieving this goal. (...) They have the unique freedom to develop new ideas, comment on society, engage in bold experimentation, as well as contribute to the creation of new knowledge. (Cortese, 1999a)

The question, already partially answered of course, is whether universities live up to this threefold special responsibility they have *vice versa* society. The mentioned *Talloires Declaration* could be seen as a good start, but when we consider that it

---

[424] As a recent example see the abuses the – in the US – fêted ethnologist Napoleon Chagnon inflicted on the Yanomani people of the Amazon through his Euro-American arrogance (Patrick Tierney (2000), *Darkness in Eldorado. How Scientists and Journalists Devastated the Amazon* [New York, Norton]).

has been signed by around 220 university leaders worldwide (whereas there are around the same number of universities in Britain alone) and that a signature of a university leader does not necessarily mean concerted and decisive action on the ground (as we know in Britain from the institutional approval of environmental policies for universities, which very often are not worth the paper or web space they are published on) it is rather an abysmal result of a ten year project. In fact, it is rather exaggerated to claim that sustainability issues have more than a marginal importance in the global tertiary education sector (see also *Education and Learning for Sustainable Consumption*, 1999, 26).

Far from being leaders in the transition to sustainability universities are still leaders in pushing the world ever faster down the road to degradation and exploitation of people and nature. A typical example of the trend is one recent study from Switzerland. It compares university education provision in the US and Switzerland and develops strategies for the future. Not a word about sustainability and the responsibilities of the university sector for it, yet plenty of the same old neoliberal "make education competitive", a reiteration of many of the capitalist myths exposed above.[425] And all this despite the fact that we know that these strategies are positively damaging and will produce just more of the same, and faster. It is clear that

> designing a sustainable human future requires a paradigm shift toward a systemic perspective which encompasses the complex interdependence of individual, social, cultural, economic and political activities and the biosphere. This shift emphasizes collaboration and cooperation, while current higher education stresses individual learning and competition, producing managers ill-prepared for cooperative efforts. (*The Essex Report*, 1995, 6)

Let us look at some of the reasons why this "pattern of denial" is so strongly present at the university level.

1. The first reason should, after our previous discussions, not come as a surprise. Universities are part of our industrial society and therefore as strongly influenced by the underlying values as the rest of society. Cultural patterns mould the educational institutions through language in the form of "meta-narratives" and "root metaphors" (Bowers, 1995, 31) as much as through "the very fabric of bureaucracy, management, and committee structure" (Orr, 1999a, 221). Yet we have seen above that most, if not all of the currently dominant underlying values are detrimental to sustainability.

2. Precisely due to the restructuring of universities according to 'market values', "colleges and universities have become over-managed and under-led institu-

---

[425] See Jürg Steiner (2000), *Hochschulen im Wettbewerb: USA – Schweiz* (Zurich, Vontobel-Stiftung).

tions operating more and more like businesses with customers. College presidents increasingly regard themselves as CEOs whose chief mission is fundraising." (Orr, 1999a, 221) Because research budgets are more and more dependent on corporate funding and student numbers are more and more determined by dodgy league tables, 'economic viability' becomes the only *raison d'être* for the educational institutions. This transformation of students and education into commodities just like toothpaste has led to a dramatic decline in the quality of education. Nobody likes to admit it in public, but even the most cursory glance at some trends in the last few years shows that quality has to suffer. How do you expect the standards to improve if:

• you cut one and a half years off high school education as has happened in Switzerland in the last ten years, particularly with a view to the fact that understanding complexity, interdependence and the broad range of spheres implicated in EfS needs time and space.

• you put pressure on students to finish their studies ever faster, as happens in Germany and Switzerland at present. This leads to an ever more narrow core specialisation in most subjects which denies the necessary time and space to broader, inter- and transdisciplinary approaches. Just one example: Zurich University offers biologists a chance to study "ethics of biology" as a minor, yet this opportunity – even though essential when biotechnology raises the most fundamental ethical questions about science and research and their impact on society – is rarely taken up due to the pressures of the main subject.

• so-called 'transferable skills', clearly forced onto universities through the economic imperative, eat increasingly into teaching provision, again occupying time which should be spent on more vital topics.

• you replace experienced teachers with computer equipment.[426]

• students come from high schools to universities with ever lower qualifications. They are less and less capable of expressing a well-founded personal opinion, of establishing complex contextual explanations of things, often due to an almost complete lack of historical, cultural and political background knowledge.[427] We are just fooling ourselves with continuous talk about improving standards. We achieve it by a manipulation of the system. The personal and professional opinion of everybody I ever spoke to in the last five years in British German Studies is that we have a dramatic decline in standards. While students twenty years ago would be expected to have a

---

[426] In 1999, as if to prove the point, the cut in salary costs (through forced 'voluntary' retirements and redundancies) equalled exactly the spending on new IT equipment at the University of Wales Swansea (UK).

[427] How else do you explain that a second year university student in Britain can proclaim in all earnest that the Hitler-Stalin-pact was struck in 1968, and nobody in a class of 15 is the least surprised?

good command of German when they started their studies and were certainly expected to produce written and spoken German of a high accuracy to gain a good degree, we are today issuing degree certificates for students who can barely string a German sentence together and write translations into German where not one verb form, not one adjective ending is correct. And even the best ones cannot produce a text without mistakes in every sentence. Yet, at the same time we all claim in our reports for *Teaching Quality Assessment* and the like that our standards improve, that we add value etc. How do you do it? You lower the difficulty of say the final year translation year by year (shorter texts, easier language) and keep level or increase the number of good marks you award. I very much hope that this is an isolated exception, but looking at the public discourse by academics about any subject and the international literature, I fear it is not. And this is not even talking about the students' motivation and qualification to tackle challenges like sustainability.[428]

3. Rather than living up to the special status and putting the privileges of time for research, reflection and thought to good use for society, there is scarcely any intellectual or moral leadership coming from universities. "The result is a poverty of wisdom in high places reinforced by a fundamental ignorance of ecology and the basics of global change." (Orr, 1999a, 221) The more fundamental question, though, is how such leadership should develop on the basis of such ignorance. So, rather than being at the forefront of sustainable change, you will find that academics are mostly loyal defenders of the status quo, very reluctant, in case they "jeopardize their upward mobility", to criticise the most dramatic challenges to sustainability, "from free trade and the electronic global economy to the efforts underway to re-engineer the fabric of life". This timidness is, again, "reinforced by the financial and ideological dependence of colleges and universities". So much so, that the necessary debates are not generated by these institutions of "free inquiry": "The fact is that all-too-often disciplinary standards, professional loyalties, and words like 'rigor' are used to suppress debate about fundamental assumptions and paradigms." (Orr, 1999a, 221)

4. The notion of academic freedom is misunderstood in the individual-centred sense that any researcher is free to pursue their interests, wherever their fancy might take them. This idea has become a holy cow within the academic community and is vigorously defended, usually with reference to "progress" and

---

[428] "An annual survey of entering freshman indicates that 74.9% of incoming first-year students prefer being 'well-off' to developing a philosophy of life or improving their minds (New York Times, January 1, 1998). Interest in causes such as environment or racial justice has apparently declined sharply. Twenty years earlier the percentages were reversed. The study concludes that this is the most apathetic and apolitical generation surveyed since the poll began." (Orr, 1999a, 222)

the "good of humanity", as Bowers states (1997, 212). The proponents of this absolutist idea of academic freedom tend to forget two things. Firstly, there is, at present, no system in place which would check new findings, research results and 'breakthroughs' against the claims with which they are carried out and 'sold' to the public. As elaborated above, there wouldn't be a problem with new research findings if only those were implemented which indeed improved the quality of life of humankind within a sustainability framework. Secondly, it is usually ignored that at present academic freedom is not at all free from moral, social, political and economic pressures which determine what legitimate research is and what isn't. Asking for a redefinition of academic freedom within sustainability limits means therefore "not imposing a new regime of political correctness", but accepting the fact "that there is a need to radically change the moral guidelines that have always been integral, though largely implicit, in the exercise of academic freedom" (Bowers, 1997, 216).

So a mixture of ingrained social values, institutional structures, personal privilege[429] and power politics makes sure that the universities, with some noble exceptions, remain leaders in destruction rather than the opposite.

Box 18: 'Liberal university education': still part of the problem

To make the limitations of the current view of university education more palpable I would like to use an example, namely Martha Nussbaum's *Cultivating Humanity. A Classical Defense of Reform in Liberal Education* (1998). The intention is also to show how narrow the spectrum of conservative-liberal mainstream thinking on education is (see Chomsky, 1989) and how unfit this spectrum is to deal with the issue of sustainability.

Nussbaum's book – apparently important and influential in the States and claiming to be a reasonably representative survey of the state of university and college education in the US – does not discuss or even mention sustainability or ecological issues. Nussbaum states at the beginning: "But on the whole, higher education in America is in a healthy state. Never before have there been so many talented and committed young faculty so broadly dispersed in institutions of so many different kinds, thinking about difficult issues connecting education with citizenship" (1998, 2-3) and goes on to describe some of the rather lofty academic activities with which American faculty is occupied as good examples. On the basis of the discussion of the abysmal failure of the university system to face up to the challenge of the real world this doesn't seem to be too trustworthy an analysis. It actually serves to corroborate Orr's view that most of these "talented and committed young faculty" haven't got the faintest idea about the world they are living in. And Nussbaum herself gives away how governed she is by the new Master Science when she claims that one important reason for the need for multicultural education is that "our economy is to remain vital" (6).

---

[429] David Orr has very sarcastically but correctly described it thus in a conference speech: "All of us here tonight are part of this system. We are well paid. We have sabbaticals and time off to do research. We fly to exotic places to discuss how to save the world thereby adding to the problem of climatic change. Relative to the vast majority of people, we have a good thing going. And our standard of living and our enlightenment, too, demands that the theft continue [alluding to an Orwell quote that our standard of living is based on theft from the Third World]." (1999a, 222)

The problem with *Cultivating Humanity* is not so much that it is wrong, but that it is so limited in its perception, even though it argues for a critical re-evaluation of one's limits. I do think that it can serve as a basis for arguing what humanities can contribute usefully to EfS (namely to think critically, but in full awareness of the traditions that came before us (294); to develop compassionate imagination which can develop empathy, understanding and critical evaluation of traditions and thought systems of other cultures (through literature [chapter 3, 85-112]); connected with a deliberate attempt to understand as many different cultures as possible, to understand minorities, including women and other ethnicities; to understand that we are sitting in one boat, our world). Yet at the same time the text's limitations are obvious. EfS would need to be much more. We would not only need compassion for other human beings, but also for the natural world and the other species; we would not only need to be aware of all the other people on Earth, their suffering, but also of the suffering of nature; we need not only be aware of other cultures, be able to imagine different worlds, but we also need some sound knowledge, if we are to think critically and meaningfully about our human condition, about the limitations placed upon us by our nature, by the life-support system around us. We also need a far deeper, much less ivory-tower, understanding of democracy, economy, of the politics of our place, which for a start would make all the well-meaning humanistic ideals of Nussbaum a mockery in the face of existing power-relations and also in the face of the behaviour of those who are very well trained in all the nice things Nussbaum advocates. Possibly the biggest limitation in her argument is that she believes that acquiring an independent, critical mind is enough. Never mind, what these highly trained, critical thinkers then do, how they lead their life (see for example 293). In other words, all that she is saying would have to be reworked, embedded into the context of sustainability. O'Sullivan has nicely said about similar ventures as Nussbaum's: "Probably one of the most prominent omissions in the critical pedagogical approach to education (...) is its lack of attention to ecological issues. My major criticism is the pre-eminent emphasis on inter-human problems frequently to the detriment of the relations of humans to the wider biotic community and the natural world." (O'Sullivan, 1999, 63-64)

Some examples of Nussbaum's naivety:

- She seriously thinks that what she advocates will lay the foundation for "a conception of citizenship for the future" (8), without even thinking about sustainability issues (see also 9ff.).
- She can talk about the notion of "whole human being" (9) without ever talking about the human condition being grounded in nature.
- To speak with Orr, Nussbaum is in fact advocating to perpetuate the kind of education that has wrecked and is wrecking the planet. Since, evident even from her analysis, what she calls the ability to think critically is also the basis of the Industrial Revolution, this enquiring intellect, completely detached from traditions, from care for the earth and community, this critical inquiry is actively damaging to Earth and us, unless it is framed by sustainability.
- She is not free of prejudices herself and goes on to elaborate how fantastic the US democracy is in a rather naive, patriotic way (294), quite apart from a wholly idealised and entirely inappropriate view of Euro-American-style democracy (see for example 8f.).
- Being a "citizen of the world is often a lonely business. It is, in effect, a kind of exile" (83): A sentence like this reveals with dramatic clarity the eurocentric limitations of Nussbaum's views. It does not say much about any sensible concept of world citizenship, but a lot about our Euro-American notion of life with its disregard of community, place and embeddedness in nature.

Were the universities and other institutions of higher education to accept their responsibility and start to pay back what society has invested in them, the necessary "major shift in the thinking, values, and actions (...) in their relationship with the natural environment" (*The Essex Report*, 1995, 4) would have profound implications for all activities.

I will elaborate on the necessary changes in more detail in chapter 3, but would like to at least sketch them out here.

1. *Curriculum greening:* The one area which is probably the most important, yet least developed, is what is called curriculum greening, in other words a radical rethink both of how and what we teach.

2. *Institutional greening:* The second major area of change would have to be institutional greening:

> The university is a microcosm of the larger community, and the manner in which it carries out its daily activities is an important demonstration of ways to achieve environmentally responsible living. By practicing what it preaches, the university can both engage the students in understanding the institutional metabolism of materials and activities, and have them actively participate to minimize pollution and waste. (*Talloires Declaration*, 1995, 3)

This links in with our observation that universities serve as a role model to society and influence behavioural patterns. It also has to do with the "shadow curriculum" mentioned: it is the worst kind of pedagogy if lecturers preach sustainable lifestyles in their courses and both their personal lifestyles and the institution's everyday operations are anything but sustainable. It is a sore point that, with some notable exceptions, the discussion about institutional greening and "academic architecture" "as crystallized pedagogy" "that teaches as effectively as any course taught in them" (Orr, 1994, 113; see below p. 301) is largely screened out on the continent, even in such innovative (in terms of curriculum greening) institutions such as the IKAÖ in Berne.[430]

3. *Redirection of research:* If contemporary science and technology is primarily responsible and actively contributing to accelerating depletion of natural resources and destruction of the life-support system (see p. 176ff.), we urgently need a redirection of research. As argued above this redirection needs to be transdisciplinary in nature and democratic in process (see p. 73), in other words, society has to agree on this direction of research, rather than being the passive victim of its applications (as has happened with genetically modified food and is happening with much of the rest of biotechnology). Schumacher saw this in the early 1970s, identifying the need to distance research from the military-industrial-scientific complex:

> What needs the most careful consideration, however, is the *direction* of scientific research. We cannot leave this to the scientists alone. As Einstein him-

---

[430] Personal discussion with Prof. Ruth Kaufmann-Hayoz, Berne, 9.11.2000.

self said, "almost all scientists are economically completely dependent" and "the number of scientists who possess a sense of social responsibility is so small" that they cannot determine the direction of research. (...) That the direction should be towards non-violence rather than violence; towards a harmonious co-operation with nature rather than a warfare against nature; towards the noiseless, low-energy, elegant, and economical solutions normally applied in nature rather than the noisy, high-energy, brutal, wasteful, and clumsy solutions of our present-day sciences. (Schumacher, 1993, 116-117)

Unfortunately, "the vast majority of so-called research turned out in the modern university is essentially worthless. It does not result in any measurable benefit to anything or anybody." (Smith quoted in Orr, 1994, 10) This seems an utterly provocative statement at first, but if you compare it with an honest assessment in most academic disciplines and in the light of what Vester says about contemporary research, it seems rather accurate:

What are we to do with all these data [produced by scientific research]? Where are they going, who uses them, where are the results leading? Obviously not very far. New developments replace each other at an unprecedented speed, yet the thousands upon thousands of research results do nothing to diminish human suffering on the planet. Year by year there are more hungry people, year by year there are more illiterate people and still wars, oppression and economic instability. The research activities seem to be running in the wrong direction, furthering self-gratification rather than satisfying social needs. (Vester, 1997, 480)

I have no doubt that the constantly increasing pressure on academics to publish articles, research and conference papers as well as the dependence of research funding on these activities do not lead to an improved quality, but to a proliferation of useless publications, combined with a high degree of repetitiveness. We can call the results of this process "MPUs – Minimum Publishable Units":[431] because all academics know, if they are honest, that a good idea comes their way maybe every two, four or five years, yet the institutions and their career goals put pressure on them to publish various papers *every year*, they cut their better ideas up into various small units and parcel them into individual articles. This I can certainly corroborate for humanities research. Acting as an editor has led to a sobering disillusionment: the journals and conference volumes are clogged up by articles rewritten time and again, with a slight shift in emphasis; articles are hastily written, with little regard to logical consistency, let alone useable insights. Bowers talks of the pressure on "faculty members to devote time to research and writing on issues that are of-

---

[431] Steven Rose (1992), *The Making of Memory* (New York; London, Bantam Press), 300.

ten of minor significance" (1997, 230). Another structural feature is specialism. After a few years, when you are an expert in a certain field, it becomes ever more difficult to contribute really new and worthwhile insights to the debate. The tendency is to repeat and repackage previous insights into new 'ideas'. So the most logical thing would be to refocus one's research attention and develop a new specialism, broaden one's approach or work in transdisciplinary ways. Yet the way careers are built and institutionally honoured, the way one is viewed by the scientific community is forcing most people into the eternal republication of the same: you are seen as the expert on XYZ; nothing else is acceptable.

There is another aspect to the dubious nature of much of the research currently produced, and again, it has a lot to do with the "shadow curriculum". Due to the eco-illiteracy of the researchers, i.e. their ignorance of the wider implications of their research, the social, political, economic and ecological contexts within which it or its applications are operating, a lot of the research is blinded by its unconscious assumptions and cultural narrow-mindedness (see Bowers, 1995 and 1997). More often than not this amounts to "a justification" for the researcher's own society by "simply projecting one single type of existence on to all mankind, in generalising and universalising it as *that of the human being in essence*":

> However, this abstract human being can easily be identified: it is male, white, over eighteen years old, lives and works in an industrialised surrounding and thinks accordingly. (...) Hence cultural ignorance applies not only to others but to one's own society as well, because this way of thinking does not know its own culture either. A basic methodological principle therefore says that I can know about my own situation only by reflecting it in the other. (Bennholdt-Thomsen/Mies, 1999, 162)

In other words, research will only begin to move in a sustainable direction when it starts to become self-reflective with regard to the human condition and the limits set by the biosphere.

4. *Redefinition of professional excellence:* Fundamental change is also necessary with regard to what we see as successful careers, as excellence in a profession. All these notions need to be redefined in a sustainability framework:

> We are currently preparing to launch yet another of our periodic national crusades to improve education. (...) The answer now offered from high places is that we must equip our youths to compete in the world economy. The great fear is that we will not be able to produce as many automobiles, VCRs, digital TVs, or supercomputers as the Japanese or Europeans. In contrast, I worry that we *will* compete all too efficiently on an earth already seriously over-

stressed by the production of things economists count and too little production of things that are not easily countable such as well-loved children, good
cities, healthy forests, stable climate, healthy rural communities, sustainable
family farms, and diversity of all sorts. (Orr, 1994, 16)

It is as plain as it can be: we do not need more economists, brainwashed into
neoliberalism, we do not need more nuclear physicists, we do not need more
clever biotechnologists to solve our problems: "The planet does not need more
successful people", instead "it needs people who live well in their places. It
needs people of moral courage willing to join the fight to make the world hab-
itable and humane. And these qualities have little to do with success as our
culture has defined it." (Orr, 1994, 12) A successful career should be measured by the ecological footprint a person has left on Earth, the contribution
s/he has made to alleviate suffering, to halt ecological destruction, to improve
self-determination and equity of human beings, to improve the overall life of
the community and, to come back to Gandhi, the improvements s/he has made
to the lives of the poorest in the community, on Earth. Anything which has
counteracted the change to sustainability should not count for, but against a
sensible notion of 'success', let alone excellence or intelligence (see Bowers,
1995, 15). Interestingly, such a sustainable notion of excellence would be very
close to the old notion of wisdom.

Taken seriously, this would obviously have direct consequences on careers, on appointments in universities (but also in other institutions and businesses). Orr has outlined what kind of questions would drive the selection
process:

> All candidates for tenure appear before an institution-wide forum to answer
> questions such as the following:
> * Where does your field of knowledge fit in the larger landscape of learn-
>   ing?
> * Why is your particular expertise important? For what and for whom is it
>   important?
> * What are its wider ecological implications and how do these affect the
>   long-term human prospect?
> * Explain the ethical, social, and political implications of your scholarship.
>   (Orr, 1994, 102)

Society, it seems clear, doesn't need any more specialists, but broadminded experts for any fields who are at the same time 'sustainability experts', i.e. "ecoliterate". This also seems to be the need of those employers interested in a

transition to sustainability[432] and of students who consistently complain about the lack of sustainability themes.[433]

What is true for judging the excellence of individuals should also apply to institutions as a whole. We know that conventional, economically driven notions of excellence tend to "measure only the excellence with which some institutions do what should not be done" and "there is that unavoidably embarrassing fact that colleges and universities have played a major role in the industrial devastation wrought on the world roughly in proportion to their national rankings." (Orr, 1996, 13-14) Orr therefore suggests a different catalogue of criteria for judging the overall quality of university education (see box 19).

Box 19: How to identify a sustainable university
The first of these [ranking criteria] has to do with how much of various things the institution consumes or discards per student. Arguably, the best indicator of institutional impacts on the sustainability of the earth is how much carbon dioxide it releases per student per year from electrical generation, heating, and direct fuel purchases. Other ratios of interest would include amounts of paper, water, materials, and electricity consumed per student. (...) A second basis for ranking has to do with the institution's management policies for materials, waste, recycling, purchasing, landscaping, energy use, and building. (...) Third, does the curriculum provide the essential tools for ecological literacy? (...) My fourth criterion has to do with institutional finances. Does the institution use its buying power to help build sustainable regional economies? (...) To what extent are their funds invested in enterprises that move the world toward sustainability? (...) Fifth, institutions might be ranked on the basis of what their graduates do in the world. (...) Are they part of the larger ecological enlightenment that must occur as the basis for any kind of sustainable society, or are they part of the rear guard of a vandal economy? (...) I know of no college that has surveyed its graduates to determine their cumulative environmental impacts. (...) [6.] No (...) shame as yet is attached to graduates who are merely ecologically illiterate and ignorant of how the planet works. There is, I think, only one reasonable course of action, the precedent for which is the practice of recalling defective automobiles at the manufacturer's expense. Likewise, defective minds should be "recalled" and offered an opportunity to return to the institution's tutelage to undergo remedial instruction.

David W. Orr (1994), *Earth in Mind: On Education, Environment, and the Human Prospect* (Washington, DC, Island Press), pp. 90-92.

---

[432] See Philipp Egger, 'Neue Umweltlehrgänge an den Universitäten; gesellschaftliche Notwendigkeiten und inneruniversitäre Möglichkeiten', in *Umweltbildung in Theorie und Praxis*, 1998, 157-165, here 162; this was also the message urged by WPS at the final conference of the *Forum for the Future Higher Education 21 Project* (London, November 1999).
[433] "The results of empirical surveys of students are in agreement that there is not enough teaching provision on environmental topics. Only a minority stated that they have had the option to follow an environmentally relevant course." (Gerd Michelsen, 'Nachhaltigkeit und Hochschulen – Stand und mögliche Perspektiven', in *Umweltschutz und Nachhaltigkeit an Hochschulen. Konzepte – Umsetzung*, 1988, 169-178, here 173)

3. Educating for the Future

> With knowledge now so potent a force for good and for evil, all education becomes moral education. (*Fool's Gold*, 2000, 46)

> The task of education would be, first and foremost, the transmission of ideas of value, of what to do with our lives. (...) More education can help us only if it produces more wisdom. (Schumacher, 1993, 62-63)

It seems obvious to me that any sensible reflection on education for sustainability (EfS) can only start now, after a deliberate attempt to sketch out what the preconditions of a sustainable society are (without making the mistake of defining a blueprint) (chapter 1) and where the limitations of our current, Euro-American way of going about things lie (chapter 2). I believe that it is an illusion to think that we can meaningfully conceptualise an educational approach fit for the future, if we do not at the same time attempt to keep the whole picture constantly in mind. We cannot, as is the case with most literature on the issue, define sustainability in the introduction as the broad concept which it necessarily has to be, encompassing ecology, politics, justice, economy and science/technology, and then proceed by virtually ignoring these areas or their implications for EfS. Such suggestions are always well-meaning, driven by a sincere wish to improve and mostly heart-felt, but good intentions alone are simply not good enough: "Sir, the road to Hell is *paved* with good intentions!" (Samuel Johnson)

I am, of course, not claiming that my analysis above is correct on all accounts. That would be nothing but hubris. But I do think that we have to have a decent stab at understanding the system we live in. We need to revisit and revise the preconceived notions that drive much of our thinking and behaviour; we need to face up to the power structures that govern us. Otherwise, all our attempts to reform education will be futile at best or legitimising and reinforcing unsustainability at worst: "Education cannot do the job of politics."[434] The idealistic notions which drive much of the international educational discourse – namely that we just need to change the way we educate our kids and students, and sustainability will fall into our lap – are both horribly naive *and* utterly unfair on the younger generation.

Firstly, to assume that a bit of nature watching, critical discourse and enlightened teaching will alter the current power structures and that those in power will let that happen without any attempt to defend their privileges only proves, if anything, the political illiteracy of those proposing these things. Illich has said enough on that (and what he says about university reform applies of course to all levels of education): "Any attempt to reform the university without

---

[434] Gertrude Hirsch Hadorn in a personal conversation, Zurich, 23.10.2000; see also Linz, 2000, 5.

attending to the system of which it is an integral part is like trying to do urban renewal in New York City from the twelfth story up." (Illich, 1971, 38) In other words, if we do not, at the same time, do everything we can to change our political, economic and social system into more sustainable structures, we might as well forget about the educational part. Saying this, of course, presupposes one crucial aspect of sustainability, namely that we assume responsibility for and control over our whole lives, from professional activity, to political behaviour, to private consumption, not just one isolated aspect. We therefore do not need any more well-meaning, generalised common learning agendas with wishy-washy goals inscribed in them, more often than not ignoring the economic imperatives determining the shadow curriculum. We need to be precise about any impacts our educational proposals will have on democracy, self-determination, equity and all the rest.

But the second aspect is almost as bad. By investing education with all our hope for change, by stylising EfS as the exclusive saviour from all our unsustainable ills we are in fact delegating the responsibility for this change to the next generation. What a disingenuous thing to say: we, including the generation before us, have pushed the world into the mess it is in, we were evidently capable of doing this, but we now are saying that we are neither able nor willing to pull it back out again. Yet our young people, who are not responsible for the current situation, should be both able and willing to accomplish the task. This position starts from the ever so convenient assumption that "changes in society must be brought about by burdening the young with the responsibility of transforming it" (Illich, 1971, 67). Not much intergenerational justice here, as far as I can see. The strategy, however, is a familiar one. It underpins our current economic structure, which systematically externalises its costs onto other people, future generations and the biosphere. But such an educational strategy is bound to fail, both because the young are very sensitive to such double standards and because our unsustainable lifestyles will be much more effective and influential educators than any education for sustainability programme could ever hope to be.

Still, saying this, and noting the tremendous influence of the hegemony of current power structures, institutions, economic practices and ingrained world views and lifestyles, is *not* exculpating education at all because it influences, in a delayed feedback loop, all these aspects in its limited ways. After all, all the corporate bosses, political leaders, teachers, scientists, economists, journalists have gone through school, most of them through higher education. And this is precisely the reason why we need ecoliteracy for everyone. EfS cannot even be limited to schools and universities; it also needs to influence public awareness, the media discourse and continuing education and training.

In other words, what we need in the educational sector is an approach which is aware of the context within which education operates, but at the same time faces up to its responsibilities, which Schumacher aptly described thirty years ago when nobody was talking about EfS: we need to break through the vicious cycle of education serving the educated only, and insist

> that the educated have taken upon themselves an obligation and have not simply acquired a "passport to privilege". (...) If it is taken for granted that education is a passport to privilege, then the content of education will not primarily be something to serve the people, but something to serve ourselves, the educated. (Schumacher, 1993, 172-173)

Education will therefore need to enable pupils, students *and* educators to invest their energy and imagination into creating sustainable, positive solutions for and within the community rather than using the same life-energy for destruction, cynicism or individualistic self-fulfilment (see McKibben, 1997, 113-114). I will try to develop the necessary parameters for such an education in the following section, but I would like to make a few provisos right at the outset so as not to foster any illusions.

  Any such educational strategy will not be an easy ride because it is an uphill struggle. To name but the most important hurdles:

- *Misconceptions about learning*: since the dominance of TV, learning is supposed to be nothing but fun (see Postman, 1987), which completely ignores that learning is hard work, and work at that which produces satisfaction only after a delay of years, running counter to the cleverly fostered contemporary notion that gratification always has to happen right here and now.[435] Learning is also not about stuffing as much data as possible into students' minds, but about concepts, connections and understanding. It is important to note here that we are *not* in a rush when it comes to education. The current notion that we need to push ever more skills ever faster onto ever younger children is "at odds with much of what research in human biology and psychology reveals about children's intellectual, emotional, social, physical, and spiritual needs" (*Fool's Gold*, 2000, 19): "Educational shortcuts that attempt to bypass the physical and emotional stages of learning defy science." (ibid. 10) And this is equally true for older students. When we get the foundations right, information processing and acquiring new skills and knowledge is not a problem any more.
- *Un-learning what current society teaches us*: Since Euro-American society is so steeped in unsustainable structures, we have the hard task of unlearning what we are constantly taught through media, advertising, business propa-

---

[435] "Most learning isn't fun. Learning takes work. Discipline. Commitment, from both teacher and student. Responsibility – you have to do your homework. There's no shortcut to a quality education. And the payoff isn't an adrenaline rush, but a deep satisfaction arriving weeks, months, or years later. Equating learning with fun says that if you don't enjoy yourself you're not learning." (Stoll, 2000, 12; see also Jane M. Healy (1998), *Failure to Connect: How Computers Affect Our Children's Minds – for Better and Worse* (New York, Simon & Schuster), 54)

ganda, traditional educational institutions (see p. 228ff.) and culturally based language systems (see Bowers, 1995, 75). We need to free our mental and emotional system of most of the values we have been indoctrinated with and which have distanced us from human need and solidarity to facilitate exploitation and greed (see chapter 2.b). Instead we have to start thinking people- and earth-centred and relearning new-old values (see chapter 1.c). In order to achieve this, we need to break out of the limitations industrial society places on our horizons and learn from other cultures and societies what it means to be sustainable. We therefore have to redefine our notion of education, as the following quote from the Ladakh village headman reveals when he reminds us Euro-Americans that "there is education also in farming, nurturing, spinning, running a family" (quoted in Kingsnorth, 2000, 36). We need to seriously reflect on the gap which has opened between real education for life – which includes accepting relevant elder knowledge and community based transgenerational communication (Bowers, 1995, 135-177) – and institutionalised school education as we know it. We need to remind ourselves that any education, just like education through the media, can serve different ends: it can be an ideological tool for those in power, perpetuating abuse, or it can be a tool for liberation and self-determination.

- *Learning complexity*: Real education, as we shall see, would further our innate ability to cope with the messy reality out there in adequate, complex, multi-layered and interrelated ways. Yet much of our culture, and increasingly our educational endeavours are actively limiting, rather than enhancing our capacity to do so. Just one prominent example: compared to real encounters with nature even the most elaborated educational software "presents sanitized or sensationalized versions of nature" (*Fool's Gold*, 2000, 50) which limit, determined by the structure of the programme and background of the programmer, "the range of a child's creative responses" (*Fool's Gold*, 2000, 53). Because it is predetermined, as opposed to real experience, such 'tools' foreclose and limit our understanding of the complexity of the real world, rather than enhancing it. But specialisation, loss of cultural diversity through the imposition of a global monoculture and manipulation of opinion through vested interests are equally counterproductive.

There is no doubt about it: reinventing education will be hard work, and my guess is that the hardest part will be to change our personal habits, rather than institutions. But maybe we can take heart from an observation by the German philosopher Gernot Böhme. He puts forward the suggestion that

> education has to be anticyclical, fostering precisely what is not a manifest trend. So young people do not need to be trained to use computers in school because they learn this faster and in more creative ways in their peer groups. What school

should offer them is the acquisition of traditional skills like reading and writing. In this context we should not just keep in mind that Germany by now boasts an illiteracy rate of near ten per cent, but also that reading and writing – understood in a wider sense – is precisely what first-year students at universities are not competent at any more.[436]

Education for Sustainability (EfS)[437]

> The ultimate subject, of course, is our real world, especially what's most special about our own planet – life itself. (*Fool's Gold*, 2000, 10)

> The crucial task of the educator will be to develop an awareness that sees through the logic of destructive globalization and to combine this with critical skills to resist the rhetoric that now saturates us. (O'Sullivan, 1999, 33)

> In direct opposition to the trend in mainstream culture toward greater specialization, we need to actively promote the generalist – the one who sees connections and makes links across different disciplines. (Norberg-Hodge, 2000, 189)

Since graduates of every discipline (whether as engineers, teachers, politicians, lawyers, architects, biologists, bankers, doctors, managers, or tourist operators, etc.) will have a profound effect on future society, it is imperative that they all have a sound working knowledge of issues of sustainability. For the higher education sector this means that not only disciplines traditionally associated with issues of sustainability and environmental education, but all disciplines, have to contemplate their provision for EfS, including the humanities and social sciences.

This is a tall order. But, as we have seen in the preceding chapters, if we are serious about sustainability, everything has to be judged within a sustainability framework. This really does mean that EfS "must become an integral part of the normal teaching in all the disciplines" (*Essex Report*, 1995, 7) so that *all* graduates acquire a "robust" social/scientific/humanist knowledge: "All students must understand that we are an integral part of nature and that we are co-evolving with all the other species in the biosphere." (Cortese, 1999) The most important reason why this is so is the same which Watzlawick *et al.* long ago stated for communication: "One cannot *not* communicate."[438] If you are in a room with people and don't talk, you are communicating messages through your silence and your body

---

[436] Gernot Böhme, 'Bildung als Widerstand. Was sollen die Schulen und Hochschulen lehren? Ein Versuch über die Zukunft des Wissens', in *Die Zeit*, No. 38, 16.9.1999, 51.

[437] I am focusing in this section on higher education, yet much of what is said applies equally to (and is often drawn from) primary, secondary and adult education.

[438] Paul Watzlawick, Janet Beavin and Don D. Jackson (1967), *Pragmatics of Human Communication: A Study of Interactional Patterns, Pathologies, and Paradoxes* (New York, Norton), 49.

language all the same. Equally there is no such thing as apolitical action: by deciding not to get involved in political action you support inevitably the current political order.[439] And the same goes for EfS:

> All education is environmental education. By what is included or excluded, students are taught that they are part of or apart from the natural world. To teach economics, for example, without reference to the laws of thermodynamics or ecology is to teach a fundamentally important ecological lesson: that physics and ecology have nothing to do with the economy. It just happens to be dead wrong. The same is true throughout the curriculum. (Orr, 1994, 12)

This means that *all* disciplines have to reconceptualise their provision within the responsibility for sustainability and it ought to lead to a situation where EfS is the basis of all teaching. William Scott once made a telling comparison. He noted that it is very strange that there have to be special conferences for EfS where one has to reflect about whether to adopt it or not and how. It should be accepted as common sense that EfS is part of the core of any sensible education, just like maths:

> Everybody knows what maths is; maths is a pillar of the curriculum which no school would be without; maths teachers exist, as do professional associations of mathematicians; the maths curriculum and its aims are relatively uncontentious, and approaches to teaching maths are known and valued; the TTA funds professional development programmes to support new maths initiatives, and maths resources exist in some profusion; the link between maths education and national economic performance is seen as obvious.[440]

We haven't succeeded unless we can say all this for EfS as well, and more: since EfS is not a special subject, as explained above, it has to permeate all subjects and approaches. In other words, we need a new educational philosophy, starting from a sound understanding of sustainability which then can be fused with all the subjects. The remainder of this chapter is an attempt to provide some parameters for such an approach.

Yet on the basis of what we have said about the state of the world in previous chapters, let me stress again that, understood so comprehensively, EfS in the North and the North of the South has one primary responsibility, namely to change the models of overdevelopment and overconsumption in these sectors of

---

[439] See Max Frisch (1985), *Tagebuch 1946-49* (Frankfurt/M., Suhrkamp [=suhrkamp taschenbuch 1148]), 290.

[440] William Scott (University of Bath), 'Presentation at the Council for Environment Education's seminar on "Strategic Approaches to Supporting Education for Sustainable Development – Producing Resources to Support ESD in Schools"', 18.5.2000, quoted in *Schools News. Development Education work in the formal sector*, (July 2000) 6, 4.

global society. This at the same time means that we need a comprehensive over-haul of the power structures, both political and economic, which govern us today.

The paradox of knowledgeable ignorance

The German romantic writer Heinrich von Kleist has provided us with a wonder-ful parable for what we are facing with regard to EfS. In his text *Über das Mario-nettentheater* he visualises human beings before and after the Fall of Man and compares them to puppets and animals. The latter, without self-awareness, con-sciousness and therefore vanity, can move with perfect grace, innocence and beauty. So did human beings before the Fall and still do in their childhood, in 'blissful ignorance' and harmony. Yet after the Fall, when reason, knowledge and self-consciousness take effect, humans lose their immediacy and carefree nature; they start to ponder why and how, cannot reproduce the smoothness of motion of the marionettes and animals any more, but become awkward and graceless. Only through tremendous training, thought and understanding can they finally regain their original grace.[441]

It appears that we have a similar situation with sustainability. We still have a sizeable number of people living on Earth, subsistence farmers and indigenous cultures, who, in their 'blissful ignorance', 'know' what sustainability *is* and *live it*.[442] Yet all the others, having adopted modern society, industrialism, capitalism and all the other blessings of 'civilisation' have actually fallen from grace. The former group has no need for EfS, whereas the latter, in a very difficult process of relearning what previously was given to them, have to reacquire the knowledge of sustainability and the capacity to live accordingly.

As ever, such parables are only correct up to a point. The notion of indige-nous peoples as living in 'blissful ignorance' and a state of grace has a lot more to do with Euro-American arrogance, underestimation and idealisation than the more complex reality. It is undoubtedly true that we can learn a lot from indigenous no-tions of nature as not separate from human beings and much more, yet the stylisa-tion of American Indians as *homo oecologicus* is incorrect.[443] Many of these people, given the destructive spoons of industrial society and the lure of the manu-

---

[441] See Heinrich von Kleist, 'Über das Marionettentheater', in *Sämtliche Werke und Briefe*, ed. by Helmut Sembdner (Munich, Hanser), Vol. II, 338-345.

[442] Sometimes indigenous peoples who have adopted a 'sustainable way of life' out of the necessi-ties of their unforgiving local environments (as the Ladakhis, some North American Indians and others) are accused of having an ecological world view *only* due to these necessities. I would argue that this is not, as normally perceived, a negative, but rather a positive feature, since we all are, globally, in the same situation: biogeochemical limits are a reality. If we were to develop a sus-tainable world view in recognition of these limiting necessities, it surely would be sensible, not lamentable.

[443] See Peter R. Gerber, 'Der Indianer – ein homo oekologicus?', in *Religiöse Wahrnehmung der Welt* (1988), ed. by F. Stolz (Zurich, Theologischer Verlag), 221-244.

factured 'needs' of consumer society, turn into unsustainable Euro-Americans very quickly. And we can't really blame them as long as we do everything to sell our model as the only one to aspire to.

Secondly, Kleist's text itself is more nuanced than that. It makes clear that the second 'state of grace', acquired through understanding and insight, is on a higher level than the first, 'ignorant', one: you are at the same point, but one level up, as in a spiral movement. If we manage the transition to sustainability it will be the same: we will not just live sustainably, but we will know both *that* we are doing it and *why*. And here it becomes clear that many indigenous people who live sustainably now, are on that spiral point already, since they are quite capable of explaining why they are living like that and what connection their lifestyle has to the survival of the biosphere. We, on the most part, have still to accomplish this 'quantum leap' in understanding, the shift from our current "aggressive technology" and society to a precautionary one (see Vester, 1997, 452). At present, most of our tools and technologies resemble Kleist's human beings: they are clumsy, inappropriate, failing attempts to emulate the workings of nature.

It might in fact be that we have the following paradox *within* the unsustainable Euro-American culture. There is no doubt whatever that we have acquired immense new insights into understanding nature, human nature and society, yielded through the various different sciences. Yet on the other hand we also encounter a tremendous, often mind-boggling ignorance among politicians, scientists and economists, acting completely unaware of context, history, culture, impact and real human needs. Usually this illiteracy, backed by economic power, behaves in the old colonial manner: we (fill in the gap: the US, Europe, the OECD ...) are the great saviours of the world, we know everything and we now bring you salvation, and you've got to take it, whether you want to or not. It is characteristic of this approach that the people affected are hardly ever asked or involved in the decision;[444] nor is there ever an attempt to first inquire whether they might have a home grown solution ready at hand, or, indeed, whether locals perceive our 'problem' as a problem.

It becomes even clearer, then, that EfS' first job is to educate the eco-illiterate educated classes of this world, to connect their unsustainable knowledge and behaviour to sustainable learning and action.

---

[444] The recent development of genetically engineered Vitamin E rice must be a prime example: hailed as a true benefit to the poor to counteract a common vitamin deficiency which leads to blindness and, oh mercy, given free to poor farmers, it is only theoretically a nice idea. In practice it ignores both physiological basics (absorption of Vitamin E in the body) and cultural traditions (how rice is eaten, say in India). So, apart from being useless, it also locks poor people into dependency from benevolent seed TNCs which in turn create a market for themselves to safeguard the 'bottom line' (see Vandana Shiva, 'World in a Grain of Rice', in *The Ecologist*, 30 (2000/2001) 9, 51).

Awareness versus action

Unfortunately this need to educate the educated leads us into one of the most serious obstacles facing EfS and into the next paradox. In the so-called highly industrialised countries, primarily responsible for the bulk of environmental destruction on Earth, we consistently find in surveys that there is a high environmental awareness, yet no matching sustainable behaviour. Often on the contrary: the higher the environmental awareness and the higher the income the bigger is the ecological footprint, i.e. the adverse impact on nature.[445]

There is a host of reasons for this mismatch between knowledge and behaviour (and once more we see that [abstract] knowledge alone is not enough).[446]

1. Structural barriers impede the individual's sustainable behaviour
Studies in Switzerland have found that a majority of people don't need to be 'sensitised' to environmental problems and even "deem the general aim of responsible environmental action personally important", yet material and socio-economic structures prevent them from doing so, which leads to a feeling of helplessness as to what they can do to help the environment,[447] and to a tendency to shift the blame onto others.[448] Similar studies in other countries have revealed the same result: knowledge about sustainability issues and awareness of environmental problems does not automatically lead to sustainable behaviour (see, for example, *Education and Learning for Sustainable Consumption,* 1999, 32-33). The ability of individuals to change their behaviour is "limited by the social milieu of a person, in particular by structural and institutional factors".[449] In other words, unsustainable behaviour is not in the first instance linked to a lack of information, but to the fact "that an 'ecological reason' does not match the non-material and material interests of industry, institutions and population."[450]

One important aspect of these structural limitations is the fact that they inhibit a sensible organisation of everyday life according to sustainability principles. In other words, sustainable practice only happens if people perceive this as an individual *and* collective responsibility and, more importantly, if they have

---

[445] See Gerhard de Haan and Udo Kuckartz, 'Umweltbewußtseinsforschung und Umweltbildungsforschung: Stand, Trends, Ideen', in *Umweltbildung und Umweltbewußtsein*, 1998, 13-38, here 22.

[446] See Gerhard de Haan and Udo Kuckartz, 'Zur Einführung', in *Umweltbildung und Umweltbewußtsein*, 1998, 7-12, here 9.

[447] Ruth Kaufmann-Hayoz/Antonietta Di Giulio, 'Bilanz', in *Umweltproblem Mensch*, 1996, 538.

[448] See Hans Spada, 'Mensch und Umwelt: Bewußtsein und Verhalten', in *Oekoskop. Fachzeitschrift der Ärztinnen und Ärzte für Umweltschutz (Schweiz)*, (2000) 2, 19-24, here 20.

[449] Matthias Finger, Silvia Bürgin and Ueli Haldimann, 'Ansätze zur Förderung organisationaler Lernprozesse im Umweltbereich', in *Förderung umweltbezogener Lernprozesse in Schulen, Unternehmen und Branchen*, 1996, 43-70, here 44.

[450] Heiko Breit and Lutz H. Eckensberger, 'Moral, Alltag und Umwelt', in *Umweltbildung und Umweltbewußtsein*, 1998, 69-89, here 70.

adequate scope for sustainable action and potential to influence and structure their everyday lives in such a fashion, i.e. if there is a high degree of empowerment and self-determination.[451] There is a considerable degree of meanness in the current system, which on the one hand allows sustainable moral values, but on the other *prevents* corresponding action: "For example, the structures of production, distribution and settlement do indeed often favour individual mobility by private car."[452]

2. *Ingrained mentality and resistance to change prevent transformation*
Brand/Poferl/Schilling have described convincingly why changes to mentalities are actually so hard to achieve: The transition to sustainable perspectives in mentality and behaviour demands from people "to accomplish, without role models in society, the integration and synthesis of issues". In order to succeed they have to cope with "a (more or less drastic) disruption, questioning and overcoming of ingrained lifestyles, habits and routines":

> These achievements of integration consist equally in reconciling "old" thinking patterns and behavioural habits with "new" ecological criteria and also in bringing different social, cultural and specifically ecological demands and requirements into line. These difficulties are made worse by a fundamental characteristic of everyday life. On the one hand people are integrated into various behavioural constraints, on the other hand there is the tendency to stick to well-established patterns of orientation and behaviour.[453]

In other words, if people define their identity, what they think they are and what society is to see in them, their status and self-esteem via certain "social-moral standards" and "cultural preferences" – for example, certain culturally constructed notions about perfection and cleanliness, or specific, symbolically laden ideas about mobility fixated on private cars, or narrow equations of living standard with material possessions or consumption based forms of social acceptance[454] –, then it is very difficult indeed to change their behavioural patterns, since this would at the same time mean that they would have to create a new identity and, worse, to face up to the fact that their old one was riddled with contradictions, illusions and self-deceptions. We know from historical experience how resistant deeply ingrained identities which defined the entire world view of people are to alterations

---

[451] See Karl-Werner Brand, Angelika Poferl and Karin Schilling, 'Umweltmentalitäten. Wie wir die Umweltthematik in unser Alltagsleben integrieren', in *Umweltbildung und Umweltbewußtsein*, 1998, 39-68, here 54.

[452] Ruth Kaufmann-Hayoz, referred to in Wölfling Kast, 1999, 285.

[453] Karl-Werner Brand, Angelika Poferl and Karin Schilling, 'Umweltmentalitäten. Wie wir die Umweltthematik in unser Alltagsleben integrieren', in *Umweltbildung und Umweltbewußtsein*, 1998, 39-68, here 57.

[454] Ibid. 50.

by fact or moral insights. The inability of the vast majority of Nazi-killers to admit their guilt even many decades after the deed, the difficulty of former communists to accept the dimension and cruelty of Stalinist crimes or, on a smaller scale, the refusal of much of the older generation of Swiss to accept their explicit or tacit complicity in crimes against Jews and Gypsies through support for the relevant government policies are just a few examples. Knowledge and awareness alone are very unlikely indeed to alter such identities. This is true for the average citizen, but even more so for those in power since their retention of power *depends* directly on the continued integrity of their current, unsustainable, identity. To believe otherwise is horribly misguided, as Susan George has pointed out in her annexe to *The Lugano Report* (1999, 181; see above p. 173).

3. Economic incentives foster unsustainable behaviour
We have seen above that in most spheres of life the economic perspective has become the Master Science and that the level playing field is heavily tilted in favour of unsustainable behaviour on all levels, through the tax system, utility pricing structures (which grant you better tariffs the more resources you waste), through open and hidden subsidies, through externalisation of the real costs of goods, services and production processes. This means that sustainable solutions on all levels are, in this skewed system, frequently more expensive, more difficult (or impossible) to obtain and so forth. This acts as a very decisive disincentive to sustainable action.

Consequences for an effective EfS
The above barriers to an effective translation of environmental awareness into sustainable action have to be taken seriously if we want to devise an EfS that is not right from the start rendered ineffective by its illusory character. However, in this process we, yet again, have to be able to cope with a paradox, at least at first glance.

1. Any EfS is obsolete or dispiriting if economic incentives are not altered
Unless we change the economic system so that prices don't lie any more, in other words, abolish all incentives fostering unsustainable behaviour, the only people changing their behaviour in a sustainable direction will be a very small group of idealists. It is sad, but true: "The environment fares best if environmentally friendly behaviour is cheap, convenient and fun."[455] Studies show clearly that eco-

---

[455] Heiko Breit and Lutz H. Eckensberger, 'Moral, Alltag und Umwelt', in *Umweltbildung und Umweltbewußtsein*, 1998, 69-89, here 85-86.

nomic incentives are massively more important for sustainable behaviour than environmental awareness and/or knowledge.[456]

I need to add a small proviso here, since we should avoid a reductionist approach which overemphasises the economic imperative. Today, you will very often find an almost ritual statement in relevant studies that in democratic societies we need to trust the citizens and that bans don't work. Yet it has been shown that bans (for example, banning private traffic from city centres as in Singapore) can in fact be ecologically more effective than economic incentives. This leads to the conclusion that we need a complex mix of moral, political and economic measures.[457]

2. Any EfS is obsolete or dispiriting if institutional structures are not altered
This, of course, does not only apply to economic activity, but to any other structural or institutional features of our unsustainable industrial society. Unless we dismantle those unsustainable structures, the ideologies and propaganda machines propping them up and the institutions perpetuating them, EfS is bound to vanish without trace like a drop in the ocean. If you work in an unsustainable institution, such as the average university or any industrial factory, you are, day in, day out, forced to behave unsustainably; yet on the other hand, if you individually, against all the odds, act sustainably, the institution is not forced to change its ways. It goes even further. The incentives, in terms of careers and earnings, clearly favour unsustainable behaviour rather than the opposite. This is why we need new notions of excellence as discussed above (see p. 250). As long as the structures are stacked against sustainability and people are actually, financially and otherwise, punished for sustainable behaviour a large-scale move towards sustainability is inconceivable. This means that we have to "create basic institutional conditions which allow and make it easier for individuals to act sustainably".[458]

3. Any EfS is obsolete or dispiriting if the dominant ideology is unsustainable
If the dominant values governing economy, politics and society, and the way people view themselves, others and nature are *not* conducive to sustainable behaviour it will not happen and lecturing from any university podium or any amount of knowledge will have no or at best a marginal effect if it is isolated. This demands that these dominant ideologies of the 'market', 'liberal democracy', 'free media' and so forth have to be reworked into sustainable alternatives.

---

[456] See Andreas Diekmann and Axel Franzen, 'Einsicht in ökologische Zusammenhänge und Umweltverhalten', in *Umweltproblem Mensch*, 1996, 135-157, here 156.

[457] See Danielle Bütschi, Hanspeter Kriesi and Daniel Scheiwiller, 'Ökologie zwischen politischer Intervention, Moral und Ökonomie', in *Umweltproblem Mensch*, 1996, 159-180, here 179.

[458] Karl-Werner Brand, Angelika Poferl and Karin Schilling, 'Umweltmentalitäten. Wie wir die Umweltthematik in unser Alltagsleben integrieren', in *Umweltbildung und Umweltbewußtsein*, 1998, 39-68, here 58.

4. Any EfS is obsolete or dispiriting if it assumes that 'one size fits all'

Another problem for EfS, arising out of the above limitations, is that it has to take seriously the fact that different socio-economic realities produce different mentalities, material realities and starting positions relative to sustainability. It is therefore illusory that the same discourse about sustainability will be appealing to all. A relatively poor factory worker whose ecological footprint – not least out of necessity – is relatively small will not find it terribly amusing if he is asked to reduce his relatively meagre consumption in the same fashion as a rich banker's family with a large villa and three cars between them. We also have to remember that lifestyles

> do not just emerge arbitrarily or are free to choose, but are inherently connected with the inner orientation and the exterior position of people. The demand for a change in lifestyles therefore doesn't aim at a change in fashion, but primarily at a change of the social and personal identity.[459]

This means that EfS, in order to be of any value, will have to be closely focused on and shaped by the needs of the actual students.[460] In other words, we have to become sensitive not just to the large gap in the size of ecological footprints between the North and the South, but also to the equally existing difference between the footprints of the North and the South of the North, and the North of the South and the South. Yet this insight shouldn't blur the proportions. The large majority of people in the North and the North of the South live beyond their global allocation of the earth's carrying capacity and therefore have to reduce their footprint. And it is also possible to define limits to certain activities, based on the global average footprint, for example 500-600 kilometres by plane per person and year,[461] and to favour certain activities, such as organic agriculture. So we have to look very much at the details before we use the above arguments to exculpate any social group in 'developed' countries from any obligation to act.

But there is another dimension to this argument that one-size-doesn't-fit-all, and it is a pedagogical one. It has long been acknowledged that there are very different learning types. Some people learn very well via visual means, others better through abstract representation, yet others via tactile and other sensory approaches, and most of us through a complex individual mix of all of these. This

---

[459] Fritz Reusswig, 'Die ökologische Bedeutung der Lebensstilforschung', in *Umweltbildung und Umweltbewußtsein*, 1998, 91-101, here 93.

[460] Ibid. 99. See also Karl-Werner Brand, Angelika Poferl and Karin Schilling, 'Umweltmentalitäten. Wie wir die Umweltthematik in unser Alltagsleben integrieren', in *Umweltbildung und Umweltbewußtsein*, 1998, 39-68, here 62.

[461] Theophil Bucher-König, 'Wird Fliegen zu einer Frage des Gewissens? Die Beurteilung des Flugverkehrs auf Grundlage der Nachhaltigkeit', in *Reisen & Umwelt. Informationsmagazin von SSR Travel*, (September 1999) 7, 16-19, here 17.

means that EfS, if it wants to reach students, has to "allow the unfolding of the different learning types. (...) Everybody has to have the possibility to translate the topics and information offered into the language, the potential of association of their own patterns." (Vester, 1998, 128)

There are other limitations to the possibility of acquiring new knowledge. This is not achievable in every situation, but only within a certain "range of possibilities of assimilation and accommodation". (Wölfling Kast, 1999, 282) In other words, what is too new and unrelated to the knowledge already acquired – which, as we know, might itself be highly dubious– cannot be integrated without tremendous effort. Additionally, how knowledge is 'transported' makes a difference: studies have shown that "information imparted from person to person is more likely to be accepted than if it is provided by media", yet at the same time media have a strong influence over people with "little social contact and a weak social network" (ibid.).

5. Yet without EfS none of the above limitations will go away
As we have seen above when discussing empowerment and real democracy, it would be self-defeating and plain wrong to assume that institutions, ideologies, power structures, mentalities and lifestyles *totally* determine individuals.[462] If that were true any notion of personal freedom would become inconceivable. But on a practical level this notion can be disproved as well: despite all the odds and incentives stacked against them, there are individuals within industrial societies who attempt *and* succeed in living sustainably (see McKibben, 1997 and Schwarz/ Schwarz, 1999).

This has two implications. Firstly, moral values, knowledge and personal convictions are not as unimportant as current common sense in EfS studies would have it. In fact, even environmental awareness is decisively important for public

acceptance of political measures for the environment. Incentives to further ecological behaviour – for example the introduction of eco-taxes – only stand a chance of being implemented in the political process if there is a strong environmental awareness among the population.[463]

Secondly, since it is individuals who make up society, since all the institutions are human-made, we do well to heed Frederick Douglass's advice: "what man can make, man can unmake" (quoted in Zinn, 1996, 176). We cannot unmake unsustainability if we assume we are mere victims of circumstances. And this is where a comprehensive notion of EfS – understood not just as a little exercise at univer-

---

[462] See Heiko Breit and Lutz H. Eckensberger, 'Moral, Alltag und Umwelt', in *Umweltbildung und Umweltbewußtsein*, 1998, 69-89, here 86.
[463] Andreas Diekmann and Axel Franzen, 'Einsicht in ökologische Zusammenhänge und Umweltverhalten', in *Umweltproblem Mensch*, 1996, 135-157, here 156.

sities, but something which ought to pervade all levels of education, including the media and public sphere – comes into its own: we have seen that knowledge alone won't change things, but we also know that ignorance and ecological illiteracy are the preconditions of the current state of affairs. This means that without a comprehensive understanding of our unsustainable present on all levels and in all its dimensions, we will not be able to build a sustainable future. This is a *sine qua non*. Now the crucial point: in order for this knowledge to become embodied in action, we need to look closely at how and in what context it is acquired.

6. *To overcome the gap between awareness and action you just need ... to act*
This is neither a joke nor too easy, but it is the most important and equally most basic aspect of any effective EfS: if we can clearly establish, as many studies have done, that there is a large gap between awareness and sustainable action and if, as psychological studies indicate, "we retain 80 percent of what we do as opposed to 10-20 percent of what we hear and read" (Cortese, 1999) the conclusion seems clear: education has to become a different kind of *action* than it is at present. I will elaborate on this below in the subsection on *experiential learning* (see p. 291), but to begin with let me just quote Smith/Williams from the introduction to their *Ecological Education in Action*:

> Moving away from this [environmental] crisis will require a fundamental transformation of the way we perceive the world and one another as well as the nature of our membership in both the human community and the community of all beings. Such a transformation is not likely to occur as a result of taking courses in environmental studies that are primarily driven by conventional academic concerns. As educators, we need to teach in a manner that aims to transform the way our students interact with the world and one another.[464]

7. *To overcome the gap between awareness and action you just need ... to face the consequences*
One of the major structural devices which enables and perpetuates unsustainability is the divorce of product from real cost, of action from environmental and social consequences, of consumption from production. Therefore a prime strategy for EfS ought to be to devise learning experiences where all these parts of the whole picture are forced together again. Heiko Steffens has developed the following "thought experiment", called "the ecological goods basket", which EfS practitioners should multiply into similar, but real learning experiments: A family wins a 24-hour stay in an ecological model-flat and an additional DM1,000 in a competition. The family is free to use the DM1,000 to purchase any goods they like and perceive as needs. The only difference to the real world is that the amount

---

[464] Gregory A. Smith and Dilafruz R. Williams, 'Introduction: Re-engaging Culture and Ecology', in *Ecological Education in Action*, 1999, 1-18, here 5.

of carbon dioxide and other pollutants generated during the production of the relevant goods is pumped into the family's flat straight after the purchase. In other words, the family is very directly and physically confronted with the consequences and the full costs of their consumption. This leads immediately to a change in behaviour towards environmentally more benign goods and associated production methods.[465]

8. *To overcome the gap between awareness and action you just need ... a*
   working community and lived sustainability
There is another important factor for effective EfS which is closely connected to the last two: the integration of the individual into a working community within which s/he has his/her place and into which s/he is bound by social responsibility.[466] In other words, it is important that an individual feels an obligation to act sustainably.[467] This lends support to the "small is beautiful" theory, because such mechanisms work no longer in large, anonymous industrial centres. However, this social responsibility is seen, particularly in the Europe and the US, after centuries of individualism propaganda, as dangerous and conservative, a step back into the bad old days of social control in the village with all its oppressive and inhibiting side effects. I believe that this is too negative a view. Firstly, within the framework of sustainability, where we have to learn to reaccept limits to our individual freedom (be it when we hit upon the borders of carrying capacity or the borders of our neighbours' freedom), being under social control and forced to accept limits, or else face consequences, doesn't seem such a bad thing after all – provided the legitimacy of the limits and the measures can be established. Secondly, it is conceivable that community control can be liberating, rather than inhibiting, if it is based on open dialogue, transparent structures and rules, and truly democratic mechanisms which allow the rules to be changed if deemed necessary. That is a far cry from the rules of traditions which had to be obeyed not for good reasons, but simply because they existed. Secondly, positive behaviour within a community can have an encouraging and catalysing effect: "The more integrated people are into environmentally friendly social networks and the better integrated the

---

[465] See Heiko Steffens, 'Voraussetzungen für ökologischen Konsum', in *Ökointelligentes Produzieren und Konsumieren*, 1997, 56-69, here 60-63.

[466] See Huib Ernste, 'Kommunikative Rationalität und umweltverantwortliches Handeln', in *Umweltproblem Mensch*, 1996, 197-216; Urs Fuhrer and Sybille Wölfling, 'Von der sozialen Repräsentation zum Umweltbewusstsein und die Schwierigkeiten seiner Umsetzung ins ökologische Handeln', in *Umweltproblem Mensch*, 1996, 219-235, here 232f.; Hans-Joachim Mosler, Heinz Gutscher and Jürg Artho, 'Kollektive Veränderungen zu umweltverantwortlichem Handeln', in *Umweltproblem Mensch*, 1996, 237-260, here 257.

[467] See Christian Jaeggi, Carmen Tanner, Klaus Foppa and Stephan Arnold, 'Was uns vom umweltverantwortlichen Handeln abhält', in *Umweltproblem Mensch*, 1996, 181-195, here 194, and Hans Spada, 'Mensch und Umwelt: Bewußtsein und Verhalten', in *Oekoskop. Fachzeitschrift der Ärztinnen und Ärzte für Umweltschutz (Schweiz)*, (2000) 2, 19-24, here 23.

neighbourhoods of the surveyed are, the larger is the extent of ecological behaviour."[468]

Going for the big picture

We have seen in chapter 1 that sustainability cannot be properly understood, let alone achieved unless we grasp it in its full complexity and multidimensional nature. For EfS this has the consequence that we have to judge *all* teaching and research against the elaborated principles of future-proofness, complexity, diversity, acceptance of limits, slowness, impact and prudence, and assess it within the dimensions of ecology, empowerment, equity, economy and equipment.

This has dramatic implications, particularly for university teaching. As can be seen from the discussions above, it is clear that no discipline, no single human being can declare themselves unaffected and not implicated. There is no scientist who can evade answering the ethical questions arising from his research or the technical applications flowing from it; there is no architect who shouldn't be brought to task about the energy efficiency of and building materials used in the houses s/he builds; there is no economist who can run away from the questions of social distribution of power, exploitation and inequity; there is no teacher who doesn't bear some responsibility for the ecoliteracy of her/his pupils. This has to lead to a reorientation both of the content and the practice of university education across the board, as Anthony Cortese from *Second Nature* has clearly spelt out:

> The education of all professionals will reflect a new approach to learning and practice. The university will operate as a fully integrated community that models social and biological sustainability itself and in its interdependence with the local, regional and global community. The content of learning must embrace interdisciplinary, systems thinking to address environmentally sustainable action on local, regional and global scales over short, medium and intergenerational time periods. Education must have the same "lateral rigor" across the disciplines as the "vertical rigor" within the disciplines. The context of learning must change to make the human/environment interdependence and values and ethics a central part of teaching in all the disciplines, rather than isolated as a special course or module in programs for environmental specialists. (Cortese, 1999)

---

[468] Andreas Diekmann and Axel Franzen, 'Einsicht in ökologische Zusammenhänge und Umweltverhalten', in *Umweltproblem Mensch,* 1996, 135-157, here 150.

Transdisciplinary, not specialist knowledge

> Western culture depends on experts whose focus of attention
> grows more and more specialized and immediate at the ex-
> pense of a broader, long-term perspective. (Norberg-Hodge,
> 2000, 5)

If there is one single reason why we are so incapable of finding sustainable solutions, then it has to be our infatuation with simplistic solutions, our conviction that
a good specialist idea is a sensible solution in the real world. Most of our problems stem from our inability to see the bigger picture, the context, the implications. The problem of TV and other mainstream media is that they cut up the
world into unconnected bits of information, thereby making it impossible to understand how these bits fit together to form the puzzle world. The problem of
nuclear power, quite apart from safety issues, is that neither the resource nor the
waste question was seriously considered *before* we launched headlong into it. The
problem with technological fixes for 'development' issues is that they ignore the
cultural, social and political dimensions, vital for implementation. The problem
with the private car is that nobody ever paused to think about resource, space and
pollution implications when such a technology is used by masses of people. The
list could be continued endlessly. What is common to all these examples is that a
'solution', which in a very narrow sense might have been a very clever one, turns
out to be utterly irrational and counterproductive when generalised and confronted
with the side effects and unforeseen consequences it produces in the complex web
of real life. Or in Vester's words: "If scientific solutions want to be real solutions,
a new insight into the political, social and ecological consequences of scientific
findings has to grow in equal measure beside scientific education." (1997, 489)

This is why Fritjof Capra states that: "to become ecologically literate, we
must learn to think systemically – in terms of connectedness, context and processes".[469] Yet education in the last half a century has failed to take account of this:
"Fragmented subject areas taught in schools engender segmented, disconnected
knowledge without an organic understanding of our connection to nature and to
one another."[470] The failure to notice this is, in a way, rather surprising since we
know from research in many disciplines that children only grow up healthy when
fed a 'sustainable' educational diet:

> a balance of freedom, secure limits, and generous nurturing of the whole child –
> heart, body, and soul, as well as head. The child grows as an organic whole. Her

---

[469] Fritjof Capra, 'The Challenge of our Time', in *Resurgence*, No. 203, November/December
2000, 18-20, here 18.
[470] Joseph Kiefer and Martin Kemple, 'Stories from Our Common Roots: Strategies for Building an
Ecologically Sustainable Way of Learning', in *Ecological Education in Action*, 1999, 21-45, here
28; see also Orr, 1994, 20.

emotional, physical and cognitive development are inseparable and interdepend-
ent. Brainimaging studies are instructive on this point. They indicate that experiences of every kind – emotional, social, sensory, physical, and cognitive – all
shape the brain, and are shaped by the brain and by each other. Healthy human
growth, in other words, is profoundly integrated. (...) Complex intellectual tasks
and social behaviors proceed from a successful integration of a wide range of
human skills, not just a narrow set of computational and logical operations.
(*Fool's Gold*, 2000, 6)

As the latter part of the quote indicates and as we also know from biology (see
Rose, 1998; Vester, 1997, 115), this is not only true for children but for human
beings of all ages. In fact, it is simple common sense, since we know very well
that we wouldn't be capable of surviving one single day were we not adept at integrating complex and diverging information, judging it by our moral standards in
order to be able to act.

In medicine, autism is not surprisingly classified as an illness, because it
very often focuses all the capabilities of the individual on a very narrow set of
abilities, at the expense of all the others. Yet in academia, education and business
we still seem to think and we certainly act as if autism was the norm whereas a
fully integrated human being, capable of complex intellectual and social behaviour, seems so dreadful that it ought to be avoided at all costs. How come we seriously expect genetic engineers – who might be the world's best in their field but
who know nothing about the social causes of hunger and the political, cultural and
historical context of poor societies – to solve the problem of hunger (as happens
with GE food)?[471] How come we expect economists who haven't even heard of
the second law of thermodynamics and the fact that our biosphere is a closed,
materially non-growing system to develop sensible strategies for our economic
behaviour?[472] How come we expect humanities students who can't even tell you
what WTO stands for to understand the implications of the PR slogans they are
creating in their first job at an advertising agency, accelerating the wheel of excessive and unsustainable consumption? How come we trust people who, in Orr's

---

[471] Are these people aware, for example, of the direct link between trade liberalisation and increased food insecurity for the poor? A survey of 27 studies on this link has clearly shown that
"liberalised trade, including WTO trade agreements, benefits only the rich while the majority of
the poor do not benefit but are instead made more vulnerable to food insecurity." (John Madeley
(2000), *Trade and hunger – an overview of case studies on the impact of trade liberalisation on
food security* (Stockholm, Forum Syd [=Globala Studier No. 4]), 7 [http://www.forumsyd.se/
globala.htm])

[472] And precisely these illiterate economists dominate the highest spheres of power. Just read this:
"These results are consistent with the hypothesis that high levels of environmental protection are
compatible with, or possibly even encourage, high levels of economic growth, though they do not
prove it." (World Economic Forum (2000), *Pilot Environmental Sustainability Index. An Initiative
of the Global Leaders for Tomorrow Environment Task Force* (Davos, WEF), 4)

words, have been educated in "disciplines and subdisciplines that have become hermetically sealed from life itself" (1994, 11)?

It is plain that sustainable solutions, in whatever area of life, can only come from a collaboration of all disciplines. This means in particular cooperation between the natural and social sciences and the humanities.[473] Hilary Rose has made clear both why many people are still fascinated by the simplistic 'single-track' solutions and why it is irresponsible to be so:

> Some, exhausted by the difficulties of confronting our uncertain and sometimes ugly world, turn with an almost audible sigh of relief to such fundamentalist diagnostics and pick-your-own politics [such as Evolutionary Psychology]. Meanwhile attempting to explain genocidic conflict, globalisation, the ecological crisis, mass rape as a weapon of war, famine and disaster, new infectious diseases or the growing gap between rich and poor requires an array of analytic tools from many disciplines. Confronting such horrors and finding a political route beyond them requires both social courage and imagination. In this situation giving up responsibility for grappling with cultural and social complexity and embracing facile evolutionary universalism is a moral and intellectual cop-out.[474]

This, of course, applies to any other simplistic 'solution' applied anywhere and everywhere in uniform fashion. We simply have to learn to transcend our artificial carving up of problems in order to make finding a 'solution' easier. We have to face up to the fact that this reductionist strategy, applied since the beginning of modern Euro-American science, is a dead-end road:

> Cartesian science rejects passion and personality but ironically can escape neither. Passion and personality are embedded in all knowledge, including the most ascetic scientific knowledge driven by the passion for objectivity. Descartes and his heirs simply had it wrong. There is no way to separate feeling from knowledge. There is no way to separate object from subject. There is no good way and no good reason to separate mind or body from its ecological and emotional context. (Orr, 1994, 31)

---

[473] See Ruth Kaufmann-Hayoz, 'Der Mensch und die Umweltprobleme', in *Umweltproblem Mensch*, 1996, 7-19, here 10; Andreas Diekmann and Axel Franzen, 'Einsicht in ökologische Zusammenhänge und Umweltverhalten', in *Umweltproblem Mensch*, 1996, 135-157, here 135ff.; and Paul Burger, 'Ein engeres Band zwischen Natur- und Sozialwissenschaften knüpfen', in *GAIA*, 7 (1998) 3, 234-237, here 236-237; UNESCO, 1997, 29, §90; *Die Zukunft denken – die Gegenwart gestalten*, 1997, 12.

[474] Hilary Rose, 'Colonising the Social Sciences?', in *Alas Poor Darwin*, 2000, 106-128, here 125.

Towards transdisciplinarity in all disciplines: focus on function, not method

Yet if we take this seriously and also Cortese's call, quoted above, that EfS "must have the same 'lateral rigor' across the disciplines as the 'vertical rigor' within the disciplines" (1999), then this implies a radical rethink of what we understand under "scientific". It means that transdisciplinary knowledge has to become the norm and specialist knowledge only the servant. It is undoubtedly true that we will continue to need sound disciplinary knowledge, but we have to learn to accept that such knowledge is *only* turned into useful knowledge if it can prove to be so in a sustainability context. However technically excellent and scientifically brilliant an idea or invention might be, if it fails the sustainability test it has to be consigned to the scrapheap of worthless creations of the human mind (such as nuclear energy, large dams, the private car as a mode of mass transport etc.). But that has repercussions. We then need to have the guts to abandon these technologies and also the relevant teaching which sells them as solutions. It does not, however, mean abandoning all teaching about them, since an important aspect of EfS has to be the history and social study of science, and grotesque failures such as the above mentioned can serve as very good examples to elaborate how we should *not* proceed. Transdisciplinary EfS must include a thorough critique of the scientific-technological, economic, political and social models which led to an unsustainable present in the first place. This, in fact, feeds back into the provision of specialist knowledge. It is routinely asserted that first we need sound specialist knowledge, and then, in a second step only, we can proceed to transdisciplinary interrogations. Yet, unless the individual scientific disciplines are remodelled into 'sustainable sciences' themselves, and engage, as called for above, in a thorough discourse of self-reflection, this won't work, since students will be immersed in specialist notions, which we know today are unsustainable, and that will then make it far more difficult for them to unlearn these notions when proceeding to the transdisciplinary stage. It is therefore vitally important that any disciplinary education is, from the very start, only conducted within the larger framework of EfS, i.e. all disciplines would need to become transdisciplinary from the start.

Here again, we can learn from nature. While our division of learning into disciplines is very much focused on different methods of inquiry into problems, this is *not* the case in biological organisms. It is interesting to note that even the smallest cells never specialise in one method or discipline, but only ever in a theme or task:

> No cell specialises only in chemistry, but always also engages in physics and electronics, information processing and mechanics. It writes and reads ("works mentally"), influences other cells and is influenced by them ("deals in politics"), turns over enormous amounts of material and energy ("trades") and produces and dismantles material itself ("works physically"). (Vester, 1997, 232)

In fact, it goes further. Our single-discipline approach is utterly unable to explain and cope with complex systems such as organisms, the biosphere or life. No amount of specialism, however correct in itself, will ever be adequate to explain "the complex behaviour [of say a cell] in its interaction with its environment" (Vester, 1997, 29). Only systems approaches which try to understand the "complex activity *between* things" (Vester, 1997, 37) can help us here. We also know from experience that this is true. The most perfect individual activity can turn into absurdity, if applied *en masse* or if confronted with its unforeseen consequences (see Vester, 1997, 77).

What we urgently need, then, is a *redirection* of research and teaching. We do not need ever more specialised isolated knowledge that blindly follows its supposedly aimless expansion. We need to start from the concrete, existing problems and then bring all the implicated disciplines together in an attempt to solve them. We do not need unflinching adherence to one particular disciplinary method which almost by necessity must lead to limited and therefore, in application, limiting knowledge. Rather, we should start from real problems and then assess what each individual method can, in its limitation, contribute to the solution. Any such transdisciplinary, problem related approach shows up "the limitations of specific disciplinary perspectives more clearly" (Stieß/Wehling, 1997, 123). And such a transdisciplinary approach does, according to Dieter Steiner's formula, *"transdisciplinarity = interdisciplinarity + participation"* (Steiner, 1998, 308), always include the integration of disciplinary knowledge, context, consequences and democratic dialogue with affected people (see "socially robust knowledge", above p. 74).

Towards new career plans and a new academic excellence

Fostering transdisciplinarity, however, means that we need to come back to our call for new notions of excellence (see p. 250; see also Vester, 1997, 465). A review in Switzerland has made clear – and I wouldn't know of any data which would contradict these findings from other countries – that the way university education works, the way institutions are organised, the way careers are planned, facilitated and enabled does the following: it ensures that academic success is mostly possible within *one* discipline only;[475] customs of publication still demand a purely *intra*disciplinary list if you want to apply for jobs; research, let alone teaching funding is difficult to obtain, even if most funding bodies now carry a declaration in their first paragraph that they specifically want to fund interdisci-

---

[475] A professor in philosophy recently told me in private that, in terms of career planning, he would actively discourage anybody from undertaking inter- or transdisciplinary research, since the candidate would not stand a chance of getting a chair afterwards. Numerous other people in Germany and Switzerland, confronted with this piece of advice, confirmed that this would still be the normal advice given today.

plinary work (see Yetergil *et al.*, 2000, 155-158). This stacks the odds massively against transdisciplinary projects and teaching, or in the words of a recent UNESCO study: "the frontiers between academic disciplines remain stoutly defended by professional bodies, career structures and criteria for promotion and advancement." (UNESCO, 1997, 29, §89) The consequence is that the vast majority of current research projects and the overwhelming majority of currently taught courses at universities are still firmly discipline driven in a narrow sense, i.e. not problem, society or function orientated. This in turn leads to the absurd consequence that the only people who can 'afford' to pursue sustainability as a main theme in their research and teaching are senior, often semi-retired, academics who only came to be such through success in their discipline. In other words, in much of academia the career of the lecturer is the primary motive, not the service of science to society, not the solutions to pressing problems.

The motivation for transdisciplinary projects, on the other hand, is very often solution orientated: "As primary motivation for inter- and transdisciplinary projects *social responsibility* and the idealistic motive *'To make the world a better place'* is mentioned. Society should benefit from its expensive science." (Yetergil *et al.*, 2000, 158) If we want to reorientate teaching and research towards sustainability, we need to restructure all these elements. Awarding of jobs, of degrees, of Ph.D.s, of research projects and of promotion should primarily be based on the contribution of the relevant individual's work towards transforming our society into a sustainable one. Yet this means commitment, also in financial terms. Many institutions report that "significant progress" in curriculum greening was only possible due to the "monetary support available to attract faculty time and interest through supplemental salary payments" to develop the necessary multi- and transdisciplinary curricula indispensable for ecological literacy. This is the case because

> it is considerably easier to multiply environmental courses within disciplines, departments, and schools within a university than it is to devise new courses and curricula across them. And yet no single discipline has hegemonic purchase on analysis and resolution of environmental problems.[476]

---

[476] Steve Breyman, 'Sustainability Through Incremental Steps? The Case of Campus Greening at Rensselaer', in *Sustainability and University Life*, 1999, 79-87, here 81.

Teaching complexity is easy

> Integrated thinking is exorcised vehemently in school already
> which is also responsible for the fact that we have, for exam-
> ple, difficulties in understanding the human-environment rela-
> tionships. (Vester, 1997, 56)

All of the above arguments seem to point into the same direction. One of the prime reasons for the unsustainable mess we are in is our incapacity to adequately deal with complexity. School, university education, the media, politics, our inherent drive to black and white solutions (which has driven social exclusion, pogroms and genocide for centuries); all of these constantly hammer into our brains the message of simplistic reductionism.

While reductionism as a scientific method has undoubtedly yielded staggering results (both negative and positive) and has provided us with the ever bigger spoons to dig into the planetary chocolate cake, it has thoroughly disabled us from dealing with the real, complex issues that confront us.

Yet whenever somebody calls for "education for complexity" (as EfS could be called as well), two objections are raised: a) complexity is complicated and b) students and people in general cannot cope with complexity (because of a)). Both assumptions are wrong.

Vester and other systems theorists have explained that to understand complex systems like ecosystems or the biosphere, you do not need full and accurate data for every little detail of the system. To generate sufficiently robust knowledge you need a model that is correct in principle, not in every detail:

> A series of studies of big and small ecosystems as well as economic processes
> prove that a model which is correct in principle and which allows the detection of
> stabilising tendencies, risks and weaknesses can be developed even with missing
> data, inaccurate data and wrong estimates. (Vester, 1997, 45)

Looking at the development of 'open systems' such as an organism we can see that "disturbances, mistakes and feedbacks occur to which the system has to react". This makes it vitally important that the development path is not quantitatively determined from the start since "changing *one* figure would distort all other figures as well so that the programme would abort": "For the functioning of complex systems it goes without saying that it is far more important and safe to guarantee the essential functions and not that certain measures are matched exactly. This, of course, is true for all complex dynamic systems." (Vester, 1997, 148-149) For example, as long as the human ear can hear, it doesn't matter whether the shape and/or size of it is the same in all individuals.

This insight is certainly borne out when dealing with most environmental problems and it is also encapsulated in the precautionary principle. If, based on a complex, systemic assessment it is clear that nuclear waste cannot be stored safely forever, it becomes rather irrelevant whether experts agree that certain containers hold for 345 or 416 years. Also, if the overall pressure on an ecosystem leads to deterioration and collapse, it again doesn't make much difference how many species are lost. Whatever the number is, it is too many unless there are very, *very* good reasons to justify the loss. In terms of carrying capacity, the usual assumption is that the globalisation of the average American lifestyle would necessitate another 3 to 6 planet Earth as resource bases. Yet whether it is 2 or 8 doesn't make much difference, since we only have one anyway.

I believe that, rather than being too complicated, complex thinking is simply too strange. If we were trained in systems thinking right from early school days, nobody would claim it was complicated. We are so used to our ways of reductionist thinking that *any* different thought patterns would strike us as difficult.

This, in fact, is an experience frequently made in reality. Once you trust pupils or people to be capable of grasping issues in their complexity, it turns out they can. Paul Krapfel, involved in environmental education of different sorts, has described his own learning process:

> The important point is that kids respond better to the real thing with all of its complexity than to a simplification. When I began this work, I was trying to make the world more understandable by simplifying it for students. I now believe this was a mistake. I have great faith in children's ability to grapple with the complex as long as they are given opportunities to talk about it and do follow-up investigations. In fact, I have come to believe that complexity attracts and invigorates their minds. Participating in these discussions makes me believe deeply that the human mind evolved and is right at home within nature's incredible complexity.[477]

More often than not, this suggests, it is our own fear of complexity which makes us engage in simplification rather than the student's inability. Or even worse, we might have unlearnt this innate ability, Krapfel suggests, by having gone through the system, whereas students are still very much capable of it.

Yet, as an UNESCO study has noted, it is irresponsible to engage in such reductionism since it actively prevents sustainable learning and solutions:

> Problems related to sustainable development are characterized, *inter alia*, by their complexity. This complexity must be communicated and understood. (...) The simplification of complex issues – so often observed today – is not only fraudulent in that it misrepresents reality, but also irresponsible on the part of those who understand these issues. It is here that the scientific and intellectual communities bear a particular moral responsibility, to ensure that decision-makers as well as

---

[477] Paul Krapfel, 'Deepening Children's Participation through Local Ecological Investigations', in *Ecological Education in Action*, 1999, 47-64, here 53.

the public are fully cognizant of the multiple dimensions of the problems they face. (UNESCO, 1997, 33, §107)

The report might overestimate the degree of understanding of the scientific community, but the basic assumption is correct: only if we face up to, and teach, the complexity of sustainability problems will we find adequate solutions. Or to look at it from the consequences: "The most important policy message from the scientists among ecological economists is: We cannot afford to accept simplistic answers to complex questions; the stakes are too high."[478] (see box 20)

Box 20: Reductionism: The scientific heart of darkness

The German author Christa Wolf has written one of the most intriguing and insightful texts about what the reductionism of science and technology does to us, both in its powerful positive and destructive negative applications. Set on the day of the Chernobyl nuclear disaster, *Accident: A Day's News* follows the narrator through her daily activities. She is caught between the deep hope and belief that the wonders of modern technology deliver on their promises – since her brother is on that day undergoing brain surgery to remove a tumour – and the realisation that the same modern technology can wreak untold havoc. While realising that talking about a cloud and trusting in one's own vegetables will never be quite the same again due to the nuclear fallout, the narrator starts to contemplate the reasons why the modern scientific method can, at the same time, have such benign and malign effects. A profound analysis of contemporary society as well as the structure of scientific research in it leads the text, in analogy and dialogue with Joseph Conrad's *Heart of Darkness*, to propose, metaphorically, the existence of 'blind spots' in our brains. Rather than facing up to the full complexities of life, to our roots as animals, to our embeddedness in nature and our desire for love, the blind spots cover up these dimensions and the tensions within ourselves which we would have to come to terms with. Instead, they redirect all this suppressed energy into a surrogate life, such as scientific research, far removed from real life and real needs, and into surrogate feelings, such as belief in progress or technology. So the fear of liberation, of an acceptance of our human nature and our deepest feelings, leads us to 'liberate' the atoms in nuclear chain reactions and so forth. Yet the text, as can be seen from the underlying conflict, is never a simple accusation against scientific-technological progress, since the reflection on our problems is turned back onto the narrator-writer in unabated form, in the radical questioning of literary writing as a means of understanding the world. Thus, it is a call for all of us to detect our blind spots and come to terms with them.

Christa Wolf (1989), *Accident: A Day's News* (New York, Farrar Straus & Giroux).

I have mentioned already that we know from another recent development that grasping complexity is not so difficult for 'naive' and 'ignorant' lay people. The examples of lay juries to decide on the acceptance or rejection of genetically modified food in India and Britain teach us that contrary to the presumption that such complicated issues should be dealt with by experts, the lay members of the jury very quickly got to grips with the issues in question (see p. 76). This, again,

---

[478] Faye Duchin, 'Ecological Economics: The Second Stage', in *Getting Down to Earth*, 1996, 285-299, here 297.

corroborates the view that we need a much more participatory approach to learning and education than we are used to.

There is one last thing to be said on the complexity-simplicity issue. Maybe we should strive for simple complexity. Precisely on the basis of the democracy issue involved in "expert knowledge" and "knowledge is power" we have to be constantly aware that there are people using 'complexity' – not in the systemic sense, but in the sense of utterly complicated and difficult to understand – as a shield to protect their privileged positions in the knowledge industry. The effect is to make issues appear to be more complicated than they really are, thereby necessitating the involvement of experts to solve them (see Bourdieu, 1992). Lerner, the governor of Curitiba and architect of the comparatively sustainable transport system there, explains that if you are not sure of yourself and don't know what you want "you'll listen to the complexity-sellers, and the city is not as complex as they would like you to believe" (McKibben, 1997, 74). Let's heed the advice attributed to Albert Einstein: "Let's make everything as simple as possible. But not simpler."

Overcoming information centred education

We have elaborated above on the crucial difference between information and wisdom (see p. 205), but I need to add two aspects to the discussion. One has to do with what has been called deep learning, the other with moral values.

Neil Postman has very forcefully made a point about education which is worth pondering: He refers to an observation

> expressed in an essay George Orwell wrote about George Bernard Shaw's remark that we are more gullible and superstitious today than people were in the Middle Ages. Shaw offered as an example of modern credulity the widespread belief that the Earth is round. The average man, Shaw said, cannot advance a single reason for believing this. (This, of course, was before we were able to take pictures of the Earth from space.) Orwell took Shaw's remark to heart and examined carefully his own reasons for believing the world to be round. He concluded that Shaw was right: that most of his scientific beliefs rested solely on the authority of scientists. In other words, most students have no idea why Copernicus is to be preferred over Ptolemy. If they know of Ptolemy at all, they know that he was "wrong" and Copernicus was "right," but only because their teacher or textbook says so. This way of believing is what scientists regard as dogmatic and authoritarian. It is the exact opposite of scientific belief. Real science education would ask students to consider with an open mind the Ptolemaic and Copernican worldviews, array the arguments for and against each, and then explain why they think one is to be preferred over the other. (Postman, 1999, 169)

We might pour as much information into our students' (and, via the media, citizens') brains, but as long as we haven't equipped them with the mental tools to evaluate, contextualise and critically analyse this information, we haven't taught them anything. As Postman goes on to say, we equally haven't taught them anything if we teach them to use computers. This is completely unnecessary: "Forty-five million Americans have already figured out how to use computers without any help whatsoever from the schools. If the schools do nothing about this in the next ten years, everyone will know how to use computers." What education should provide students with, on the other hand, and what we never learnt "about everything from automobiles to movies to television, is what are the psychological, social, and political effects of new technologies." (Postman, 1999, 170) Only a comprehensive contextualisation of new technology within the "economic and social alterations that technology inevitably imposes" (ibid. 171) can enable students to live sensibly in today's society. Otherwise, as in the example above, they are totally at the mercy of experts, usually connected to vested interests, who tell them what to believe: hardly a knowledgeable society, more a society steeped in quasi-religious belief and dogma. That is why Schumacher claimed that education's primary task is "the transmission of ideas of value, of what to do with our lives". The transmission of "know-how", on the other hand, is only secondary, "for it is obviously somewhat foolhardy to put great powers into the hands of people without making sure that they have a reasonable idea of what to do with them." (Schumacher, 1993, 62)

The second aspect is that we all are always value-driven, yet unfortunately this goes mostly unnoticed. Schumacher has shown this for two important examples, the natural sciences and economics:

> The sciences are being taught without any awareness of the presuppositions of science, of the meaning and significance of scientific laws, and of the place occupied by the natural sciences within the whole cosmos of human thought. The result is that the presuppositions of science are normally mistaken for its findings. Economics is being taught without any awareness of the view of human nature that underlies present-day economic theory. In fact, many economists are themselves unaware of the fact that such a view is implicit in their teaching and nearly all their theories would have to change if that view changed. (Schumacher, 1993, 73)

The conclusion from this has to be twofold. On the one hand, we have to enhance self-reflection and transparency so that the underlying values of thought systems and their consequences become clear. On the other hand, there is a need to foster value systems which are conducive to sustainability. That the currently dominant values of highest profit – which privileges those "who manage to exploit their fellow human beings and natural resources faster, more sweepingly, more cunningly and more unscrupulously than anybody else" (Dürr, 2000, 188) – and excessive consumption are clearly not compatible with living lightly on Earth we have shown above.[479] But in order to change this situation a mere transfer of information or knowledge cannot be enough. We need to foster feelings and connectedness to the earth, if we really care for it:

> Here's my take on how ethics works. The thing that builds an ethical human being, an ethical child, is not reasons, but *feelings*. The energetic base of morality is compassion which requires being in touch with things, feeling empathy. That's not just environmental ethics, it's any ethics. And you cannot feel empathy toward something until you're intimate with it somehow. Intimacy is the key word here.[480]

Promoting truth over opinion: against individualistic 'intelligence' and 'creativity'

This leads us immediately into deep water. As soon as you state that sustainability is necessarily connected to certain values, certain principles, certain ideas (as elaborated in chapter 1.c), people immediately accuse you of propaganda, indoctrination, of an attempt to establish an eco-dictatorship and so forth. These are such grave accusations that they have to be taken seriously.

Some critics – such as Bob Jickling (1992, 1999), Jickling/Spork (1998), John Elliott (1995), *Umweltbildung im 20. Jahrhundert* (2001) – see EfS as indoctrination and manipulation precisely because EfS has a purpose, a direction, content and is a framework that can be filled with specifics. In response to that, they talk of the need to free environmental education from "dogmatic teachers" (Elliott, 1995, 137), from "apostles of a scientific and cultural change of thinking, of a radical paradigm shift" (*Umweltbildung im 20. Jahrhundert*, 2001, 241) and advocate that we should "abstain from nurturing an ecological world view" (Elliott, 1995, 139).

Beyond the rather harsh tone, I believe that they have a point. Jickling, for one, reminds us that "we must enable students to debate, evaluate, and judge for themselves the relative merits of contesting positions" (Jickling, 1992, 8). As just elaborated above, EfS does not have the job of implanting a predefined 'sustainable knowledge' package into the heads of students, forcing them at gunpoint to believe it without asking questions. The teachers should constantly re-evaluate what they are teaching, and one of the prime goals of the process should be the

---

[479] Heiko Breit and Lutz H. Eckensberger, 'Moral, Alltag und Umwelt', in *Umweltbildung und Umweltbewußtsein*, 1998, 69-89, here 86.

[480] Ed Good, teacher, Washington Village Elementary School, quoted in Joseph Kiefer and Martin Kemple, 'Stories from Our Common Roots: Strategies for Building an Ecologically Sustainable Way of Learning', in *Ecological Education in Action*, 1999, 21-45, here 42.

kind of enabling of students Jickling is talking about. Knowledge that cannot transparently justify itself is also illegitimate knowledge.

Yet, unfortunately, and this is mostly forgotten by the above critics, this is not even half the story.

1. There is no value-free education
The first problem with the above line of argument is that it is based on the truly postmodern conviction that there is such a thing as value-free education and that conversely all values are equally valid. This starts from two basic assumptions which are both plainly wrong. The first is the claim that all positions are equally interesting and, since one cannot distinguish between right and wrong, deserve equal teaching time allocated to them and so forth. The second assumes that students are already fully formed and informed ideal citizens who, if presented with any array of different information and values, are perfectly capable of making up their own minds and can, somehow out of the black hole of their inner self, produce their own values, uninfluenced by anything else (see, for example, Elliott, 1995, 141). This is a very nice, cosy, ideal world. It sounds liberal and democratic. We are ever so tolerant and we don't care in the least which values our students adopt, as long as they adopt them independently and we as teachers don't impose the values on them.

I will deal with the first claim that all positions are equally valid below under 2. Let me here consider the other aspect, the assumption that education has to be a value-free zone. Firstly, this is plainly wrong. All education is value-led and we can very clearly see this with the most recent shift in emphasis in the university sector in the UK. ICT (Information and Communication Technology) and employability are being forced upon all institutions as core transferable skills. Learning and teaching strategies all over the country had to be rewritten to accommodate the predominance of these perspectives. Yet as we have seen above, this is clearly *not* happening in a value-free void, on the contrary: ICT and employability embody a very specific set of values, cultural preferences and assumptions (that information technology is by definition beneficial to education, that the commodification of education is the dominant educational priority and so forth). Shirley Ali Khan has noted the irony in these charges against EfS:

> Ironically, those who are quickest to rail against value-laden, purposeful education have not hesitated to implement a whole range of national and institutional action agendas such as increased use of information technology through improved IT skills; increased participation in Europe through improved language skills; and increased competitiveness through the development of enterprise

skills. The competitive enterprise culture is particularly highly charged with a specific set of values. (Ali Khan, 1995, 7-8)[481]

The only problem here is that those other approaches never reflect on the underlying values that drive and determine them and therefore assume that there aren't any. And this is precisely the trouble with the assumption that all teachers need to do is to heap piles of information from different perspectives in front of the students, strictly unbiased, without the use of their value-judgement; and students will then be perfectly capable of judging this data on their own. We have seen above how strongly we are influenced and indoctrinated by a whole array of values and assumptions (about 'free markets', 'liberal democracy', 'development', 'science as progress' etc.), through the media, through our upbringing, school and the "shadow curricula" of our daily lives. It is therefore simply a naive assumption that students would be able to cope out of the blue with the whole range of positions advanced. Bowers has forcefully shown that this misconception stems from the Euro-American idea of an entirely individual-centred creativity and intelligence, but "this image of the creative individual does not take account of the influence of culture on thought, identity, and expression. It also represents the individual as separate from the environment." (1995, 10) To put it very bluntly: Just as human tinkering with ecosystems is unlikely to improve them, after evolution has tested and perfected them for billions of years (see Carter, 1999, 21-22), it is highly improbable that the average pupil or student, out of their inner self, their personal interests and current needs, will make a lasting, original and crucial contribution to a sustainable society. Rather, the accumulated body of knowledge on how to live sustainably first has to become, through learning, the framework within which individual intelligence and creativity then *might* (but, again, in most cases won't in any significant way) contribute to the common good and improve facets of understanding. Or, to quote Bowers again:

> One of the great ironies surrounding the modern educator's ideal of creativity is that the type of individualism they wish to foster has not stood out as an exemplar of ecological sensibility or, for that matter, as a critic of the culture's assault on the environment. (1995, 11)

It is plain to see that the position of those fearing indoctrination by EfS is driven both by an over-emphasis on individualism and a reliance on the postmodern belief that any opinion is as valid as any other. This has led to the educational ortho-

---

[481] John Huckle reinforces this point: "Those who accuse advocates of critical ESD of seeking to indoctrinate students with a particular meaning or outcome, ignore the democratic safeguards within critical pedagogy. In reality, it is more often light green reformist ESD that indoctrinates by offering support for the greening of capitalism or the status quo." (in *ESDebate*, 2000, 15) See also McKeown, 1999, 1-2.

doxy – which of course is ideologically convenient since it perpetuates the status quo – that all educators need to teach are "mental processes, techniques, and procedures" while "the content of the curriculum should be determined by the personal interests and preferred learning style (and mood) of the student" (Bowers, 1995, 106). This assumes both that students *know already* what is important and relevant, even before they have learnt it, and that individual creativity and intelligence – akin to the liberal notion of freedom (see p. 53) – is an end in itself and "need not contribute to the well-being of society" (Bowers, 1995, 55). On the other hand, the approach denies that there is a "privileged body of knowledge or values essential to the educated person"; it denigrates the accumulated wisdom of their own and other cultures, of elders and "mentors who have made outstanding contributions to raising the quality of life" (Bowers, 1995, 106) and it presupposes arrogantly that whatever students create out of their own limited experience, perspective and knowledge is superior to the cultural heritage of humankind. This leads to such absurd results that "students' thoughts and behaviors are judged as manifesting 'intelligence', even if they contribute to degrading the environment" (Bowers, 1995, 15) and is part of a broader Euro-American assumption that "new and experimental knowledge should have more authority in people's lives than forms of knowledge that have evolved over generations" (Bowers, 1995, 12). In other words, we need to redefine our notion of intelligence: being intelligent in an ecological view would be judged against "long-term sustainability of the Earth's ecosystems as the primary criterion", rather than individual autonomy, and would therefore "not [be] dependent upon the whim or insight of individuals who often express their autonomy in ways that reflect current cultural fads" (Bowers, 1995, 132).

This leads directly to an even more important argument. Value-free education is neither intellectually legitimate nor desirable. There are certain clearly identifiable values which are conducive to sustainability. And it is not a matter of indoctrinating students with these values. It can be *shown* that and why they are conducive to sustainability, *as opposed* to other values and positions. It is an open, transparent discourse which has to convince students of this, but it is simply incorrect to posit that it doesn't matter which values we adhere to. Or in other words, not to teach values would, in 'developed' countries, just teach the students the underlying values of unsustainable capitalist society.[482]

To show this, there is a neat intellectual tool which I call the flip test. If all values are equally *okay* in order to achieve sustainability the following person must be as well suited as anybody else: The person owns a large house with air conditioning, two cars, travels around the world about twice per year, is biophobic

---

[482] Bowers has shown that the same is true for so-called academic freedom. Not setting sustainability limits to academic freedom will just lead to a perpetuation of academic research which is guided by the, largely unconscious, moral values of the current unsustainable system (1997, 210-218).

and perfectly happy to kill any creature near his house and hates all living things, holds a fascistic conviction that starts from the assumption that anybody not of North American citizenship should be exterminated, believes in unlimited growth and aggressively enacts this belief through the investment portfolio of his wealth of several million US dollars. On top of that the person believes that it is his natural right to exploit mineral resources around the world for his benefit, irrespective of the damage done to people living near the mines....

According to the position of Elliott, Jickling and others, such a person is as correct as anybody else. Yet would you want such a person to be in charge of sustainability education for your children? Such an extreme counter position makes it instantly and palpably obvious that there cannot be any EfS nor a survival of the human race on Earth without a *specific* set of values. As Donna Haraway clearly states: "The point is to make a difference in the world, to cast our lot for some ways of life and not others." (1997, 36, see also 37)

As an aside, such a position also assumes that learning does not work via models. It assumes that teachers who refuse to take a position and stand up for values will be a good role model for students who then should assume such values and be prepared to stand up for them.

2. Truth, not eternal truth

The reason why a lot of people have difficulties as soon as you start to argue that there are certain facts which are true, certain values which are correct in comparison to others, stems from the absolute relativism advocated by postmodernism. Driven by the, certainly correct, insight that "the grand narratives" (such as communism and fascism; capitalism, as the last grand narrative still holding sway over us, is rarely mentioned) always claimed to be true and advanced blueprints of how people had to behave, with devastating consequences, postmodernism jumps to the opposite conclusion that there is no truth at all. And this, of course, is demonstrably false.

In chapter 1.c I have tried to show just that. There are certain facts which are non-negotiable, such as the fact that the biosphere is a materially non-growing, closed system or the second law of thermodynamics. If people then claim that there are no environmental problems in reality, except those which are socially constructed, you just want to ask them whether the hole in the ozone layer or the destruction of topsoil or the deforestation of the Amazon are social constructions as well, which can be made to disappear if the media report them differently and people believe them (you could, for instance, use Orwell's "Newspeak" to achieve this). Or listen to this: "There are as many environments as there are systems, and this alone makes clear that an understanding of *the* environment is wrong." (*Umweltbildung im 20. Jahrhundert*, 2001, 236) In other words: depending on your point of view the laws of gravity do apply or do not!? It just seems to me that, while accusing others of being value-laden, these advocates of a value-free EfS have not quite grasped how deeply their positions are in fact informed by specific Euro-American values of individualism, value-free science, limitless freedom etc. This discussion is lacking exactly what I have tried to show is imperative for any sensible discussion about sustainability, namely a comprehensive integration of all the 5 Es (ecology, empowerment, equity, economy, equipment); instead it takes place with an almost complete disregard of current power structures and the state of our societies. It doesn't once question the *homo economicus* view of humankind (see for example in *Umweltbildung im 20. Jahrhundert*, 2001, 7-11). And because there is no notion whatsoever of how dominant social paradigms are established, Kyburz-Graber *et al.* can claim in *Umweltbildung im 20. Jahrhundert* that "control pedagogy" has failed (2001, 11) without even mentioning, let alone being capable of explaining, the fact that control and conditioning of the majority in democratic societies, through mass media, conventional education and institutional practice, do work smoothly to perpetuate the current, unsustainable power structures.

It is simply impossible to envisage how any talk about sustainability should even be possible if you "abstain from nurturing an ecological world view" (Elliott, 1995, 139). Equally it seems an almost unbelievable disregard of basic facts (from the latest IPCC report to any other attempt to map the way to a sustainable society) when somebody claims that the transition can happen *without* a radical paradigm shift in values *and* behaviour.

The positions above also reveal a very strange, static understanding of truth. They seem to assume that if you claim something as true you have to carry on claiming this for the indefinite future. While we can assume that certain basic facts such as the material limits of the biosphere are 'true forever' and therefore have to necessarily inform and frame our thinking about sustainability, we should not assume so for other statements. The point is that we need a processual, historical understanding of truth. In other words, we have to continually re-evaluate whether the facts and information which led us to assume a certain position are still true. As long as this is the case, and as long as we are capable of showing this in a transparent process of reasoning we can, with all legitimacy, claim that a position is true. If we can show that contesting positions are false, we have not proven that our position is eternally true, but it is true until it is proven wrong (akin to Popper's principle of falsification). It is, for example, true that cultivation of cotton, while occupying "around 5 per cent of the world's cultivable land area, some 34 million hectares" uses "about 25 per cent of all pesticide applications", which leads to "considerable health problems" for field workers and people living close by (Madeley, 1999, 109). The water and pesticide demand for conventional cotton, the rising demand from increased consumption is not sustainable, whereas organic cotton production is. This can be established through calculation, observation, and comparison of different growing methods. Yet the statement is only true as long as most cotton is produced conventionally; if consumption pressure were to ease and most cotton were to be produced organically, we would face a different situation, so that the above assertion would need rewriting. The extreme relativist position would say, since we can't know forever, we don't really know, whereas the pragmatic relativist would insist that we have to act on the comparatively best knowledge we have. There is no totalitarian danger in this whatsoever, as long as the discussion process is kept open and level. Positions can and ought to be readjusted whenever it is warranted. Within the limits of nature, sustainability is not mapped out as a blueprint. It is completely open how it will develop, but it will have to be compatible with the basic parameters.

3. Have the cake and eat it
But this is precisely the problem for many people. Accepting the basic ecological truths about life on Earth and the principles of impact limitation that flow from them, are inconvenient for a perspective that would in theory like to promote sustainability, but in practice doesn't want to do anything for it. If you look closely enough, as I have tried to do above, you have to acknowledge that certain values, certain behaviours, certain positions, certain systems are *incompatible* with sustainability. The inevitable logical conclusion is that you will have to change that system if you want to achieve sustainability.

Yet often people – even though they might claim to be "critical of society" – are unwilling, as Bowers has shown with such clarity (1995), to address precisely those underlying assumptions and dominant values of Euro-American society which turn it into an unsustainable one. An indication of this unwillingness to face up both to natural and social facts can be seen in the demand for the "dissociation of sustainable development from a natural science-ecology approach" (Kyburz-Graber *et al.*, 2000, 291). Even granted that in the past twenty years the natural science focus on environmental education has been damaging in reducing the complexity of the issues involved, this clearly is going too far in the other direction. It means: let's forget about the facts, let's just talk about the social construction of environments.

The danger of such a position is probably most neatly encapsulated in a sentence by Jickling, which lays bare the utter eco-illiteracy underpinning it: "The real challenge for growth in the next century is to go where sustainability cannot – to go beyond sustainability" (Jickling, 1999, 6). In a limited system there is nowhere to go for growth, there simply is no "beyond sustainability", unless Jickling can somehow magically produce another planet Earth.

Enabling self-determination and critical citizenship

> The future is not some place we are going to, but one we are
> creating; The paths to it are not found but made; The making of
> those pathways changes both the maker and the destination.
> (Australian Commission for the Future, quoted in: Ali Khan,
> 1995, 9)

This much we can conclude from the above: EfS has to walk a tightrope between the most serious and intellectually honest attempt to provide students with the most accurate analysis of our current situation, its problems and possible remedies *and*, at the same time, the attempt to enable them to produce such an analysis, which might differ from the teacher's, on their own. So on the one hand we shouldn't shy away from the rather difficult task of penetrating our contemporary society and on the other hand we should do so with tools and with a transparent process which puts students into a position to do the same for themselves. They should grow into independent, critically minded citizens, fully aware of the traditions, histories and ideologies which formed them and respectful of their responsibilities and obligations towards their communities and the biosphere. If, as we have seen above, such self-determination and empowerment are crucial pillars of any sustainable society, we need to facilitate both through the learning process, i.e. the learning process has to be an exercise in empowerment and self-determination. Yet this doesn't mean that we as teachers are absolved from showing our true colours. We cannot, as it were, shift the whole burden onto the students and expect that they will be able to do what we refuse to do if we just pile information up before them and refuse to interpret the world.

This process of self-determination which we are supposed to encourage is, it seems from studying children, in fact a crucial element for a healthy development: "The child only discovers and learns to deal with her potential and her capabilities, but also her limits through self-motivated action. Initiative and control over her activities should therefore, as far as possible, rest with her." (Largo, 1999, 246) This self-motivation seems to be naturally strong: Children have an innate desire to learn about the world, are curious, interested, and they keep at things until they get them right (see Holt, 1991, 37). And with that curiosity comes a sense of empowerment, of being in control, of being able to change and influence things. The trouble is that school seems to kill this off by disempowering these urges through the I-tell-you-and-you-don't-find-out-yourself approach (see Holt, 1991, 34). Modern child development theory seems to support very much the idea that self-determination, self-pacing and independence are very important when learning such important skills as walking, talking, reading and writing (see Largo, 1999):

> When she [Lisa, the observed child] learned to read, it was going to be by her
> own choosing, at her own time, and in her own way. This spirit of independence

in learning is one of the most valuable assets a learner can have, and we who want to help children's learning, at home or in school, must learn to respect and encourage it. (Holt, 1991, 132)

If that is true it means that learning is about trusting the individual (see Holt, 1991, 297), but only within a community with responsibilities towards each other: learning, in other words, ought to be a very democratic process where learner and tutor are in the same boat; they complete a process together. We have, though, to be clear about the fact that during this process the tutor has to impart knowledge to the learner and cannot just be a guide by their side (see p. 283). Yet it is crucial to "recognise that students are key agents in their own learning". EfS has to develop the responsibility of students for their own learning; otherwise they will never be able to take on responsibilities towards others and/or nature. Nurturing this thoughtfulness sets EfS apart from traditional teaching

> in which the ownership of the knowledge is firmly in the hands of the course designers, and in which students are allowed little opportunity to participate in the construction and transformation of the study materials in ways that are meaningful in the particular socio-political contexts in which they live and work. (Plant, 1998, 110)

Malcolm Plant describes his experience in running the MA in Environmental Education at Nottingham Trent University (UK) as corroborating this view, stressing the insight that learning is as much about making the community work as about individual 'success':

> If students are to be encouraged to be reflective and adopt a socially critical approach to environmental education they must be part of the learning environment that provides them with a sense of social belonging both within the course processes and in their own communities. (Plant, 1998, 113) [483]

It is important to stress this. It points to the fact that education, just like the media, can have two faces. It can be a massive feat of social engineering, or it can provide the learners with tools and instruments for self-enlightenment and self-determination. If we stick to the latter notion of education it actually is an important means to achieve freedom. Quite opposed to the customary lament in EfS studies nowadays that sustainable behaviour is not possible since the institutional structures prevent us from its attainment, education as empowerment gives learners the tools to take control of their lives in relation to an overwhelmingly

---

[483] See also Kyburz-Graber, 1999, 430 and Regula Kyburz-Graber, Lisa Rigendinger and Gertrude Hirsch, 'Problemorientierte Umweltbildung als partizipativer Prozess', in *Umweltproblem Mensch*, 1996, 313ff.

unsustainable society. It is important to remember that – a rather existentialist thought – we all are free to act, even against barriers and odds. Otherwise we sacrifice any notion of freedom and we would also not be able to explain why people like Nelson Mandela, human rights activists in oppressive regimes, Gandhi or the widows of the disappeared in Latin America did and do act even though the institutional structures and dominant value systems are even more decisively stacked against them than the unsustainable processes are in 'developed' countries. Lummis has described this using a story by Vaclav Havel in which a greengrocer suddenly stops going along with state propaganda, no longer puts up the required signs, stops voting in the fake elections, starts speaking his mind and showing solidarity with likeminded people:

> It [the decision of the greengrocer in Havel's story to start living "within the truth"] is an act of the mind: a decision. Of course "a decision" means a decision to act; it comes not before but just at the moment the greengrocer stops putting up the sign. At that moment a free space has been created, without any organizational or institutional change whatever. After the greengrocer begins to act on his decision – and especially if he finds fellow actors – new organizations may emerge. (Lummis, 1996, 34)

We are all at liberty – as the example from such an oppressive regime shows – to open up such "free spaces" which also shows up a lot of the complaint about the lack of such spaces as, often, a feeble excuse for lack of action. EfS, I believe, should facilitate such free spaces, not dictate behaviour; it should point up alternatives to customary behaviour, particularly where official political, economic or media discourses claim there are none. It should equip learners with the tools to problematise the basic values of our society, for instance our understanding of freedom which is connected to property which, in a sustainability context, is highly questionable. This understanding, then, could be juxtaposed with the understanding of property by indigenous peoples where not individuals, but only the entire community can own land, for example.

What this amounts to is that learners, with the help of tutors, should start to accept responsibility for their own lives, their own learning, in other words, attempt self-determination, but within the community of other people and nature. This, of course, has profound implications, as elaborated above in chapter 1.c and 2.a, with regard to our political system. It is a *sine qua non* for constructive and active citizenship. And since empowerment is such an important plank in a sustainable society, "a curriculum reoriented towards sustainability would place the notion of citizenship among its primary objectives" (UNESCO, 1997, 23, §68), more specifically:

> Students need to learn how to reflect critically on their place in the world and to consider what sustainability means to them and their communities. They need to

practice envisioning alternative ways of development and living, evaluating alternative visions, learning how to negotiate and justify choices between visions, and making plans for achieving desired ones, and participating in community life to bring such visions into effect. (UNESCO, 1997, 24, §71)[484]

Otherwise, if we stick to the old model of handing knowledge down to mere recipients "the goal of qualifying the coming generation for a democratic society has a rather hollow sound" (Finn Mogensen in *ESDebate*, 2000, 30).

Experiential Learning: Reconnecting to reality

> Experiential learning is based in messy reality, with all its paradox and untidiness, its ever-changing pattern, its refusal to conform to our expectations. As such, it inevitably leads to humility. (Norberg-Hodge, 2000, 190)

Based on the above insights that EfS has to teach complexity and should enable learner empowerment and self-determined learning in a transdisciplinary setting, the question is how this can be achieved. We are, after all, faced with an increased dependency on media, electronic and other tools, with specialism in disciplines and subdisciplines and with the fact that most people in industrialised countries "over the course of their lives (...) spend only four to five percent of their time out-of-doors".[485] This alienates and distances us ever more from nature, from a sense of place, from a feeling of connectedness with and being part of the life-support system Earth. Against this background it has been argued that the best available strategy is *experiential (or active) learning* which is to say learning by experiencing, by doing things (see *Education and Learning for Sustainable Consumption,* 1999, 20; Allan, 1998, 10).

There are powerful arguments from the biology of learning and from psychology for this claim. Cortese relates one such fact which shows that learning by doing is much more effective than learning by listening or seeing.

> Educational psychologists tell us that we retain 80 percent of what we do as opposed to 10-20 percent of what we hear and read. Therefore the process of edu-

---

[484] A good example where this empowerment still has to happen is technology literacy which has to be part of eco-literacy: "In a democracy, the point of technology literacy is to prepare students to be morally responsible citizens, actively participating in shaping the nation's technological future, rather than merely reacting to it as passive consumers. (...) Considering the importance of preparing young people for the moral responsibilities of making decisions about technology, it seems scandalous how little space this issue gets in public discussions of education." (*Fool's Gold*, 2000, 68)

[485] Gregory A. Smith and Dilafruz R. Williams, 'Introduction: Re-engaging Culture and Ecology', in *Ecological Education in Action*, 1999, 1-18, here 7.

cation must emphasize active, experiential learning and real-world problem solving on the campus and in the larger community. For example, the learning experience for students should include:
- working on actual, real-world problems of communities, government and industry as part of the curriculum;
- working in groups so that they will be able to effectively collaborate as future managers and leaders. (Cortese, 1999)

It is also the case, according to the biology of learning, that we understand much better, much more comprehensively and can retain the resulting knowledge much easier if as many different senses or channels of perception as possible are involved in the learning process. We learn better if we are offered more ways of explanation, more perspectives on the issue, more associations and connections to integrate it with already existing knowledge. Brain imaging technologies have shown that knowledge acquired in such a way is 'saved' in more areas of the brain and the neural connections between these areas are strengthened (see Vester, 1998, 51; 83-84). This is done most easily when we actually experience something. Whereas it is usually the case that we have to repeat information various times before we can retain it, a real-life experience 'burns' itself into our memory often in an instant and can be recalled throughout our life (Vester, 1998, 83): "The more impressions we retain, the more thought connections will be produced and the better the chances that their interplay will produce new ideas. The one who experiences more will have more ideas." (Vester, 1998, 109) Incidentally, this is a very clear verdict on the sedentary 'learning' styles in front of electronic media of whatever sort.[486]

We learn better if we are exposed to reality, rather than mediated abstractions and reductionist images of it, because we learn, at once, in a complex way:

> Instead of working with notions of things we should work with the things themselves, their interactions, their relationship to the environment. And immediately the notions would engrave themselves not scarcely, but in multiple ways in our brains. They would use the visual, tactile, emotional and auditory channels in equal measure and thereby provide much stronger possibilities for associations than swotting out of touch with reality. (Vester, 1998, 126)

In fact, learning for life (such as walking or talking) does not happen according to the 'software download' model, but is integrated into the process of living:

> Humans, like other animals, get to know the world directly by moving about in the environment and discovering what it affords, rather than by representing it in

---

[486] "Lack of exercise is bad for learning. Child development experts emphasize that moving in three-dimensional space stimulates both sensory and intellectual development." (*Fool's Gold*, 2000, 25)

the mind. Thus meaning, far from being added by the mind to the flux of raw sensory data, is continually being generated within the relational contexts of people's practical engagement with the world around them. (...)
The novice engaged in such trials [learning to walking, for example] is not "acquiring culture", as though it could be simply downloaded into his or her head from a superior source in society, but is rather embarked upon the process that anthropologist Jean Lave has called "understanding in practice".[487]

There is *no other way* to learn such advanced skills than by doing them. No amount of textbook studying will do. So the scientific case for integrated, experiential learning which through "the development of physical skills can help foster an intense emotional commitment to learning" is supported by "a wide range of research and case studies": It seems that what's most beneficial is "the dynamic synergy released by the 'fusion' of movement, thought, and feeling" (*Fool's Gold*, 2000, 13, FN 8). This leads Frank R. Wilson to conclude: "The clear message from biology to educators is this: The most effective techniques for cultivating intelligence aim at uniting (not divorcing) mind and body."[488] That should also make us very cautious when virtual reality is sold to us as a 'sensible' learning environment. All this research seems to indicate (and experiences from virtual, distance learning institutions such as the Open University certainly bear this out) that education with real people is much more effective education, yet our culture generally relies more and more on virtual, mediated contact.[489]

Since the case for it seems so compelling it is rather surprising that integrated teaching is so infrequently practised:

In order to find solutions to complex, multifaceted problems related to living sustainably, today's students must be exposed to techniques of teaching that directly involve them in diverse interdisciplinary collaboration, the creative process, advanced technology, cross-cultural communication, economic development, and environmental ethics and values, to name a few. (Flint *et al.*, 2000, 192)

---

[487] Tim Ingold, 'Evolving Skills', in *Alas Poor Darwin*, 2000, 225-246, here 238 and 241.

[488] Frank R. Wilson (1998), *The Hand: How Its Use Shapes the Brain, Language, and Human Culture* (New York, Pantheon Books), 289.

[489] "'The most important gift that parents can give a child to spur their mental development', Greenspan adds, 'is not a good education, elaborate educational toys, or summer camp, but time – regular, substantial chunks of it spent together doing things that are naturally appealing to the child.' (...) But by 1997, parents [in the US] were already spending about 40 percent less time with their children than they had 30 years before. (...) A 1999 study by the Kaiser Family Foundation concluded that children ages 2 to 18 spend on average about 4 hours and 45 minutes a day outside of school plugged into electronic media of all kinds. About 65 percent of the older children, ages 8 to 18, had televisions in their bedrooms, and 21 percent had personal computers. Another recent study estimated that children between the ages of 10 and 17 today will experience nearly one-third fewer face-to-face encounters with other people throughout their lifetimes as a result of their increasingly electronic culture, at home and school." (*Fool's Gold*, 2000, 28)

The more varied, and the more complex the teaching environment the more we enable our students to come to terms with complexity. And what better teaching environment than reality itself? If we want to come to terms with the complexity of real-life problems why not research them in real-life situations? As advocated above for research, teaching complexity has to start from real problems within real social contexts, rather than isolated environmental or other issues.[490]

Educating for a sense of place

> Knowledge of place is cumulative and thus intergenerational – enriched by elder knowledge and communal experiences renewed through stories, ceremonies, and mentoring. (Bowers, 2000, 68-69)

In order to ground education in reality and to give students the chance to connect to a place, develop emotions, a sense of belonging and thereby some sense of responsibility and care for this place, we have to reconnect education to locality, i.e. the place where education takes place.

David Orr even argues that it is crucial to give students this sense *before* we equip them with the big spoons of abstract knowledge: only then will they be able to adequately use them:

> I believe that we should introduce students to the mysteries of specific places and things before giving them access to the power inherent in abstract knowledge. I am proposing that we aim to fit the values and loyalties of students to specific places before we equip them to change the world. (...) I am proposing that we make them accountable in small things before giving them the keys to the creation. (Orr, 1994, 96-97)

This means that we have to resist the "great homogenizing force undermining local knowledge, indigenous languages, and the self-confidence of placed people" which conventional education, based on "abstractions, generalized knowledge, and technology", has become: "The fact is that modern education has contributed greatly to the destruction of local cultures virtually everywhere. Locality has no standing in the modern curriculum." (Orr, 1994, 129) We have to revitalise the appreciation of local places, since it only makes sense, as Lummis has argued in connection with democracy, to make the here and now, the place where we live, sustainable (obviously with a firm eye on the fact that by doing so we do not make life unsustainable for others elsewhere). Another consequence is that the globalisation of education and (mono)culture is not an option any more, as Paul Krapfel has shown:

---

[490] See Regula Kyburz-Graber, Lisa Rigendinger and Gertrude Hirsch, 'Problemorientierte Umweltbildung als partizipativer Prozess', in *Umweltproblem Mensch*, 1996, 310, 321.

If "teaching about place" means learning about what is right around students, then this concept creates reverberations throughout the educational infrastructure. For example, local knowledge, by definition, could never be incorporated into nationally distributed standardized examinations. Therefore, assessment of students and evaluation of a school's program could not rely simply on tests of standardized content. Evaluation would have to look more closely at the local and nonstandardized. (Krapfel, 1999, 60)

In other words, EfS focusing, in the first instance, on the real place around students would nurture diversity[491] and would make a mockery of much of the quantifiable, but qualitatively useless knowledge that count as education nowadays. Indeed, "one of the most destructive characteristics of high-status knowledge", according to Bowers, "is the emphasis on diagnosing problems and framing solutions as models that can be replicated in various cultural contexts" (2000, 75).

But if this strategy of localised education is to succeed, we need to start early (and this argument is in line with what was said about the biology of learning above) and we need to confront the fact "that young people and adults are increasingly being isolated from direct contact with nature" (Plant, 1998, 17). And this is true not only for virtual reality and the media, but also for education:

There is a tendency in much of environmental education to separate nature and human perception into two realms, to view the environment as if it is "out there" rather than as something which is inextricably linked with our lifestyles and emotions, and that inevitably shapes our human identity. In conclusion, I have argued that favourable childhood experiences spent *in* nature are a basis for cherishing this identity with nature in later life. (Plant, 1998, 19)

This is why Smith and Williams, in their list of principles for EfS, stress that the actual experience of interconnectedness – and not only between human beings and nature, but also between human beings themselves – is crucial:

- Development of personal affinity with the earth through practical experiences out-of-doors and through the practice of an ethic of care

---

[491] Bakko makes the point that "colleges and universities that own or have access to rural land holdings have opportunities and options to provide experiential learning for their students. (...) Work in sustainable agriculture offers unique possibilities to gain knowledge and understanding of a critically important aspect of our society that would be difficult to learn or appreciate in the more traditional classroom setting. Restoration of natural habitats offers our students opportunity for practical field experience, better understanding of the complexities of ecosystems and instils in them an appreciation for a land ethic." (Eugene B. Bakko, 'Sustainable Agriculture and land management in the liberal arts: A case study', in *Sustainability and University Life*, 1999, 243-251, here 249-250)

- Grounding learning in a sense of place through the study of knowledge possessed by local elders and the investigation of surrounding natural and human communities
- Induction of students into an experience of community that counters the press toward individualism that is dominant in contemporary social and economic experiences.[492]

The last point in particularly seems important since 'normal' education forms us into solitary individuals (see Bowers, 1995, 92-134), whereas it is clear that human beings can only survive in groups (see Vester, 1998, 170). We therefore need to re-emphasise cultural forms of intelligence, rather than individual ones (see Bowers, 2000, 152-158).

In order to add a bit of flesh to these ideas and to show what they can look like in practice I would like to present two successful and innovative examples of an integrated approach which starts from real-life problems and then teaches all secondary school subjects through that practical experience (see boxes 21 and 22). It is to be hoped that more institutions, particularly in the university sector, follow suit by adopting and adapting this idea.

Box 21: Building canoes – building an integrated EfS curriculum
The Northampton Environment Legacy Program links studies of the historic culture of the [US] Eastern Shore life with an awareness and understanding for the importance of environmental quality in this region. Legacy students are given the team task of building their own boats (kayaks and canoes), and then are provided the opportunity to use the boats they have built to explore and scientifically measure the coastal marine environment that surrounds them, as well as to learn about the historical culture that depended upon this environment. (...) The principle instructional methods used in the Northampton Legacy Programme are the "experiential learning" concept advocated by John Dewey (...) and the Socratic techniques used in the "Paidea Seminar" method (Dewey, 1963). (...) The instructors use the boat and its construction to teach team work, math, grammar, writing, biology, and history. (...) Typically, the boat explorations center on the salt marsh, tidal flat, and barrier island ecosystems. Following a winding gut through the marsh, these young explorers might find themselves on a tidal flat in a broad lagoonal bay. The instructors can use this opportunity to give a lesson in the biology of a clam. After having instruction in one particular "clam sign" (...), students prowl the mud and sand with bent back on hands and knees. When repeated attempts fail to expose the elusive clam, they return for more detailed instruction until one or two acquire the knack and then they become instructors. And soon a seemingly barren expanse of mud and sand, something to be religiously avoided by most recreational boaters, has become a gold mine of learning and life forms. Experiences such as these lead to a natural connection between multiple academic disciplines. It becomes the fodder for reading and writing, the touchstone for historical understanding, and the genesis for scientific curiosity. (...) The answers to the questions are real, and they are connected by the boat to the clam, to a parent, to a job, to the future. They relate to the necessities of daily life and the difficulty in making trade offs in order to preserve and protect. But it does not stop there. The students visit a school board meeting and make

---

[492] Gregory A. Smith and Dilafruz R. Williams, 'Introduction: Re-engaging Culture and Ecology', in *Ecological Education in Action*, 1999, 1-18, here 7.

an unrehearsed presentation of their program of study. They are invited to educational and environmental conferences to assemble boats and explain the connection between boat building and the larger issues of education, environment, and culture. (...) And finally, the students' realization through class reports and essays that the changing regional landscape has significantly changed the culture of their ancestors was found to be an important feedback loop in the students' connecting the importance of their heritage to their present lives.

R. Warren Flint, William McCarter and Thomas Bonniwell (2000), 'Interdisciplinary Education in Sustainability: Links in Secondary and Higher Education. The Northampton Legacy Program', in *International Journal of Sustainability in Higher Education*, 1 (2000) 2, pp. 196-200.

Box 22: Dry toilets yield rich educational fertilizer

At the school, *Secundario "Instituto Patria"*, in Xico-Chalco, different groups of teachers and students have started several projects for sustainability and ecological literacy: daily composting all organic matter; recycling nonorganic school waste; growing and cooking organically grown macrobiotic foods, etc. Complementing these projects, we initiated an interdisciplinary curriculum and pedagogy for ecological literacy: to study and research the production, distribution, and improvement of *sanitario ecologicos/latrinas secas* (dry latrines). This project develops the knowledge, skills, attitudes, dispositions, and habits required for regenerating communal soil – physical and cultural. (...) The dry latrine project simultaneously develops theoretical and practical skills. History, English, Spanish, social studies, natural sciences, journalism, community development, recycling, waste "management", and other disciplines are taught in the integrated manner essential for engaging in moral education and ecological literacy. (...) The dry latrine project involves "learning by doing". The school day is divided into two parts. During the afternoon, teams constituted of students, supervising teachers, parents, and neighbors install at least one dry latrine a week for the entire academic year in the homes of community members. This is key to communal learning. The mornings focus on theoretical explorations of this postmodern technology. The theory and practice of producing, promoting, constructing, and installing dry latrines involves several elements: (a) curing themselves of the ignorance and apprehensions regarding dry latrines – including fears of odor, disease, plagues, dysfunctionality, underdevelopment, or backwardness; (b) liberating themselves from their blind faith that the flush toilet of the developed world is superior or more desirable; (c) emancipating themselves through the discovery that even if the flush toilet were made available to the population of Xico-Chalco, it would not be desirable; that, in fact, it would prove to be a catastrophe for a whole set of ecological, economic, political, health, moral, and educational reasons; and (d) developing leaders in the community for the production, promotion, installation, and construction of alternative postmodern ecologically friendly technologies.

Madhu Suri Prakash and Hedy Richardson, 'From Human Waste to Gift of Soil', in *Ecological Education in Action*, 1999, pp. 65-78.

Keeping an eye on the Shadow curriculum: effective EfS

Based on the insight that EfS can only work as a comprehensive, transdisciplinary endeavour and in the knowledge that only actions, not awareness, count, sustainability education cannot just concentrate on the curriculum, its content and pedagogical approaches. Again, we have to bear the bigger picture in mind. Since we have identified the reductionism and dysfunctional disjointedness of much of current education as one of the main problems, EfS has to reflect this and attempt to integrate all areas which have a bearing on it. This attempt produces five criteria which EfS will have to satisfy, if it is to succeed.

1. Change the dominant shadow curriculum of society

In view of the overwhelming dominance which the unsustainable paradigms of Euro-American society have over much of the globe, educators and students will have to work not just on isolated change within individual academic courses, but in the wider context of their lives. The shadow curricula of the media, the economic imperative, the political structure and many of the national and international institutions are preaching everyday unsustainable messages of 'growth', 'liberal democracy', 'scientific progress', 'development' and the like. These curricula have to be fought and counteracted through active sustainable citizenship by both educators and students. This, for example, means that we have to be very careful in selecting material for any courses. Today it is still predominantly the case that tutors use any old text, say in language studies, if they want to teach a certain grammatical point. But using propagandistic texts, uncritically perpetuating unsustainable mentalities, just reinforces the shadow curriculum. Unless we start to seriously take this other curriculum into account, EfS at a school or a college is a mere drop in the ocean and rendered either ineffective or worse turned into a fig leaf for justifying the unsustainable status quo.

2. Institutional greening

It is quite obvious that one of the key requirements for EfS is to "promote a whole institution approach to environmental practice." (Howard *et al.*, 2000, 84) Anthony Cortese from *Second Nature* has summarised what such an approach entails and why we need it:

> Higher education must "practice what it preaches" and make sustainability an integral part of operations, purchasing and investments, and tie these efforts to the formal curriculum. The university is a microcosm of the larger community. Therefore, the manner in which it carries out its daily activities is an important demonstration of ways to achieve environmentally responsible living and to reinforce desired values and behaviors in the whole community. By focusing on itself, the university can engage students in understanding the "institutional metabolism" and ecological footprint of materials and activities. Students can be made aware of their "ecological address and footprint" and they can and should be actively engaged in the practice of environmentally sustainable living. (Cortese, 1999)

If we don't attempt to do this, we run the danger that "the pedagogical influence of institutional structures and operations", i.e. the "shadow curriculum", undermines or even renders useless whatever is taught about sustainability (see Ali Khan,

1995, 3): it is essential for educational institutions to "link [their] core educational mission to the daily life of [their] institutions".[493] Orr speaks of "the power of examples over words" (1994, 13) – and we have seen above that this is borne out by the biology of learning. That is why it is so important that educational institutions adopt environmental or sustainability policies – which cover all aspects of their operations, from energy, waste and water management to teaching, research, funding, investment, procurement[494] and local biodiversity – *and* implement them. The universities need to be embodied role models of sustainability – or at least visibly on the way towards such a state – if they are to be places where sustainability can be meaningfully taught. Otherwise, we unwittingly teach our students lessons not at all conducive to sustainability:

> Students hear about global responsibility while being educated in institutions that often spend their budgets and invest their endowments in the most irresponsible things. The lessons being taught are those of hypocrisy and ultimately despair. Students learn, without anyone ever telling them, that they are helpless to overcome the frightening gap between ideals and reality. (Orr, 1994, 13-14)

If we want active citizens fighting for sustainability the shadow curriculum has to teach them lessons more favourable to empowerment.

3. Appropriate architecture, learning atmosphere and culture

> I propose that the way in which learning occurs is as important as the content of particular courses. (Orr, 1994, 14)

However, it is not enough to introduce an energy saving regime and recycle wastepaper, however important that may be.[495] We have to consider the entire environment within which learning takes place, in other words we have to check how much our lecture halls, seminar rooms, libraries, malls, student buildings, cafés and computer rooms fulfil the criteria of what Orr calls "ecological design" (1994, 104; for a comprehensive elaboration see box 23).

Box 23: Ecological design: bringing experience back into 'liberal' arts
Ecological design is the careful meshing of human purposes with the larger patterns and flows of the natural world and the study of those patterns and flows to inform human purposes. Ecological design competence means maximizing resource and energy efficiency, taking advantage of the free

---

[493] ULSF (University Leaders for a Sustainable Future), 'Learning from the "Shadow Curriculum"', in *The Declaration* 1 (1996) 2 [http://www.ulsf.org/pub_declaration_othvol12.html].

[494] Big educational institutions such as universities have considerable purchasing power for which they bear responsibility (see UNESCO, 1997, 32, §100).

[495] Universities in the UK have been relatively keen to embark on institutional, rather than curriculum greening since it is easier to implement, especially in the current financial climate, when you can save thousands of pounds with effective energy management and the like.

services of nature, recycling wastes, making ecologically smarter things, and educating ecologically smarter people. It means incorporating intelligence about how nature works, what David Wann (1990) called "biologic", into the way we think, design, build, and live. (...) Good design everywhere has certain common characteristics including the following:
• right scale,
• simplicity,
• efficient use of resources,
• a close fit between means and ends,
• durability,
• redundance, and
• resilience.
(...) Good design promotes
• human competence instead of addiction and dependence,
• efficient and frugal use of resources,
• sound regional economies, and
• social resilience.
(...) Third, poor design results from poorly equipped minds. Good design can only be done by people who understand harmony, patterns, and systems. Good design requires a breadth of view that leads people to ask how human artefacts and purposes "fit" within the immediate locality and with the region. Industrial cleverness, however, is mostly evident in the minutiae of things, not in their totality or in their overall harmony. Moreover, good design uses nature as a standard and so requires ecological intelligence, by which I mean a broad and intimate familiarity with how nature works. (...) Design competence requires the integration of firsthand experience and practical competence with theoretical knowledge, but the liberal arts have become more abstract, fragmented, and remote from lived reality. Design competence requires us to be students of the natural world, but the study of nature is being displaced by the effort to engineer nature to fit the economy instead of the other way around. Finally, design competence requires the ability to inquire deeply into the purposes and consequences of things to know what is worth doing and what should not be done at all. But the ethical foundations of education have been diluted by the belief that values are relative. (...) The liberal arts have come to mean an education largely divorced from practical competence. Inclusion of the ecological design arts in the liberal arts means bringing practical experience back into the curriculum in carefully conceived ways.

David W. Orr (1994), *Earth in Mind: On Education, Environment, and the Human Prospect* (Washington, DC, Island Press). pp. 104-109.

We can again see the importance of the overall learning atmosphere from the way learning occurs. The quality *and* content of learning is influenced not just by the information provided by the tutor, but also by the atmosphere in the classroom, the way students sit and interact, the relationship between students and tutors, the emotional state students are in while learning (see Vester, 1998, 128): "Unfortunately it is generally not properly recognised – and therefore not taken into account in teaching – that information absorbed while learning does not only consist of the topics but also of all the other concurrently occurring perceptions." (Vester, 1998, 135)

David Orr has succinctly summed it up:

The problem is not just that many academic buildings are unsightly, do not work very well, or do not fit their place or region. The deeper problem is that academic buildings are not neutral, aseptic factors in the learning process. We have assumed, wrongly I think, that learning takes place in buildings, but that none occurs as a result of how they are designed or by whom, how they are constructed and from what materials, how they fit their location, and how – and how well – they operate. My point is that academic architecture is a kind of crystallized pedagogy and that buildings have their own hidden curriculum that teaches as effectively as any course taught in them. (Orr, 1994, 112-113)

4. Change tutor and student lifestyle
A logical consequence both from what we have said about effective, experiential learning and about greening the "shadow curriculum" is that we have to focus teaching and institutional practice on a change of lifestyle. First and foremost that would mean that tutors would have to adopt sustainable lifestyles: What we desperately need are tutors and "administrators who provide role models of integrity, care and thoughtfulness" (Orr, 1994, 14). If tutors lecture students on the ecological necessity of cutting down on private car use and the students see the same tutor drive into the university every day, even though s/he lives only two miles away, on a direct bus route, they will not learn to use public transport, but the lesson will be: don't trust your tutors (which is good) and: if s/he doesn't, why should I? (which is rather counterproductive).

Unfortunately, according to Gerd Michelsen, this area is the one most thoroughly neglected: "The fewest activities [in the context of sustainability and universities] are to be found with regard to making lifestyles more sustainable." He goes on to say that unfortunately, based on evidence from behavioural studies, there is no reason to assume that university tutors are any more environmentally friendly in their *actions* than comparable social groups, which means that "changes in personal environmental behaviour are found, if anywhere, almost exclusively on the student side."[496]

To facilitate such a change in behaviour you need to structure the incentives correctly and offer an environment that is conducive to such a change. This would include sustainable transport concepts for educational institutions, tracking the life-cycles of resources consumed, evaluating innovative energy and waste solutions, and screening the purchasing of academic departments, computer centres, services, restaurants, cafés and shops on campus. But this is not enough in itself; you also need to communicate the reasons for restructuring, and best of all involve both staff and students in the process.

---

[496] Gerd Michelsen, 'Nachhaltigkeit und Hochschulen – Stand und mögliche Perspektiven', in *Umweltschutz und Nachhaltigkeit an Hochschulen*, 1998, 169-178, here 173.

5. Demonstrate changes in attitudes and behaviour

Success in EfS cannot be demonstrated by glossy end-of-year environmental reports, a wide variety of different courses on sustainability topics in most disciplines and transdisciplinary activities, even though these would be a good start. The only yardstick for successful EfS is real change in the ways in which the institution is operated, the ways tutors and students live their lives and interact with each other, and in the ways they take on the responsibilities for their lives and the impact thereof. This, as Orr has observed, is a very good basis for a definition of wisdom leading straight into the heart of sustainability since

> the consequences of one's actions are a measure of intelligence and the plea of ignorance is no good defence. Because some consequences cannot be predicted, the exercise of intelligence requires forbearance and a sense of limits. In other words it does not presume to act beyond a certain scale on which effects can be determined and unpredictable consequences would not be catastrophic. (Orr, 1994, 50)

Being serious about this has implications for the way we measure success. Rather than believing, as is the current craze in UK universities in the context of so-called Quality Assurance, in shelves and shelves of documents, in proclamations of what institutions are doing, in degree programs and written declarations full of aims and objectives, or in students' declarations of intent in their coursework, we should measure real behaviour both of the institutions and students against the framework of a sustainable society developed in chapter 1.c.[497] This has to include a long-term perspective on what the institution and what the students are doing. We must ask whether the graduates of a university are

> part of the larger ecological enlightenment that must occur as the basis for any kind of sustainable society, or are they part of the rear guard of a vandal economy? (...) I know of no college that has surveyed its graduates to determine their cumulative environmental impacts. (Orr, 1994, 92; see box 19 on p. 252)

---

[497] Gerhard de Haan and Udo Kuckartz, 'Umweltbewußtseinsforschung und Umweltbildungsforschung: Stand, Trends, Ideen', in *Umweltbildung und Umweltbewußtsein*, 1998, 13-38, here 29.

Why do we need humanities in EfS?

> No student or academic institution can be up-to-date without a working knowledge of issues like global climate change, tropical deforestation, ozone depletion, and toxic waste. Furthermore, these issues pose a unique opportunity to tie together the values and concerns of the humanities and sciences. (*Alvaro Umana, Professor, Central American Management Institute, Costa Rica*)

> Even in the humanities we may be bogged down in a mass of specialised scholarship furnishing our minds with lots of small ideas just as unsuitable as the ideas which we might pick up from the natural sciences. (Schumacher, 1993, 70)

Like most other EfS proponents, I have claimed that EfS *must* be a multi- and transdisciplinary endeavour. There is simply no way that we can come to terms with the complexity of real life (and therefore real-life sustainability) with a single discipline approach.

It is also relatively uncontroversial that we need the natural sciences for EfS. Without a good understanding of how ecosystems, nutrient cycles or the carbon cycle work, we have little hope of making clear to unsustainable farmers, industrialists, managers and the general public why exactly they should change their practices. But we have also seen that we do need the social and political sciences, since there is no sustainability without empowerment and equity. But what on earth do we need the humanities and liberal arts for? Interestingly enough this question is not so much asked from outside, but from inside the humanities. The outside perspective very often actively approaches the humanities, as in the case of concerned scientists who want help, for example, with ethical questions because they see that an exclusively natural science approach is inadequate. Inside the humanities, however, a very strong view is that these questions of sustainability have nothing to do with them; either they are claimed to be completely irrelevant, or it's back to the 'us-against-them' game: it is not us, the humanities, who are destroying the planet, it's those bad scientists with their destructive technologies. Maybe this has a lot to do with Orr's "ecological illiteracy", in other words, that people do not really understand what is meant by EfS. For when I posed the question "Do you agree that ensuring the integrity of the biosphere on which humanity depends qualifies as a legitimate educational purpose in the humanities?" in my questionnaire to humanities faculties in four European countries (for more details see next section), all respondents answered with "yes". Phrased like this it seems entirely uncontroversial. When asked "why", respondents either replied "because it is self-explanatory" or stated that we need to foster "responsibility for our natural resource base", related ethical norms and necessary "responsible action".[498] One respondent stated that only humanities could deal adequately with "horizons of meaning". Despite the craze for employability and ICT skills at universities everywhere a majority of respondents to the questionnaire argued that EfS in the humanities was "more important in the long run" than such transferable skills. When asked why one simply stated: "because EfS safeguards the elementary conditions of life".

So it does seem clear that the various academic disciplines need each other, or more precisely that all specialist sciences are in urgent need of a sound working knowledge of the other sciences:

> A sober account has to acknowledge that many natural scientists have only a rudimentary command of social sciences knowledge. The reverse, of course, is equally true. It is often frightening to see how little fundamental natural science knowledge social and cultural scientists possess. (Burger, 1998, 237)[499]

Of course, natural scientists, "since they influence the development of societies more and more, [need] social and cultural competences"[500] which many of them lack to a worrying degree. Yet it is a complete overestimation of the influence of natural sciences and of the workings of contemporary society if we assume that economists, teachers, journalists, bank managers, insurance brokers and so forth have no influence on the path of society and can therefore be safely excluded from the list of those responsible for the unsustainable present. We have shown above that everybody is implicated, particularly in the 'developed' countries with their rampant overconsumption, and it is hard to see why a teacher who hammers home the growth and progress myths (see Bowers, 1997, 84-93) or a broker shifting investment around the globe to achieve the highest return rate should be any less to blame than a scientist who might, through the economic imperative, have been driven to prioritise research which otherwise would not have been touched. Indeed, it can be argued, as Bowers has convincingly done, that all humanities disciplines lack a coherent sustainability framework as a consequence of their dominant anthropocentrism which keeps them rooted in the myths of the Industrial Revolution (see 1997, 62-77).

This is precisely why we need compulsory EfS courses for all students, providing a sound working knowledge of all relevant aspects of sustainability from a perspective of all disciplines.

---

[498] All unacknowledged quotations in this section and the next are taken from responses to my survey.

[499] See also: George Monbiot, 'Beware the appliance of science', in *The Guardian*, 24.2.2000, 22.

[500] Hartmut Böhme, 'Wer sagt, was Leben ist? Die Provokation der Biowissenschaften und die Aufgaben der Kulturwissenschaften', in *Die Zeit*, No. 49, 30.11.2000, 41-42, here 42.

Yet if it is the case that the humanities ought to *contribute* to EfS, then the question arises what they have to offer in principle.[501] There are four main reasons why we cannot do without an integration of humanities perspectives into EfS.

1. Training our imagination for alternatives

We have seen with regard to many issues above that what makes transition to a more sustainable way of life so difficult is not that we don't know any solutions or that proposed solutions are mere theoretical ideas. More often than not our dominant world view supposes with a vengeance that the way it is going about things is the only possible way. This "monoculture of the mind" (Vandana Shiva), which can only ever perceive reality through one specific set of ideas ('growth', 'progress', 'liberal democracy', 'development' etc.), is in fact a combination of arrogance and blindness. It is either oblivion or refusal to change one's point of view, refusal to search in other cultures for practices which might be sustainable, but also require a change in attitude by 'developed' countries: "Often enough we do not want to know, even if the necessary knowledge would be available." (*Zukunftsfähiges Deutschland*, 1996, 125) It is illiteracy with regard to the cultural diversity of our planet, to the myriad forms of daily practices which have evolved, some of them much more useful than others when viewed in a sustainability framework. The Euro-American cultural hegemony does everything in its power to destroy and obliterate this diversity, but if we want to find practical, tried and tested sustainable solutions we need to look carefully to other cultures.

Yet in order to do that, to be able to transcend the limitations of one's own world view, we need to train specific abilities: "If we are trying to change behaviour, to loosen up fixed ideas in order to avoid catastrophes (...) training in dealing with the unexpected and in questioning the well-known acquires a really important role." (Vester, 1997, 460-461)

If the study of literature, of fiction, of poetry, and of dramatic writing is good for anything, then it is precisely for this. As the Swiss writer Max Frisch has stated: "Art remains in opposition to power. (...) Literature preserves what in politics is time and again necessarily lost, namely utopia."[502] There are two aspects to this. On the one hand fiction, not bound by the laws of physics, can train what in German is called *Möglichkeitssinn*, i.e. the ability to see and imagine what is possible, beyond the confines of actually existing society and the limitations imposed by the accompanying ideology: "Literary texts can themselves be imaginary states of nature, imaginary ideal ecosystems, and by reading them, by inhabiting them, we can start to imagine what it might be like to live differently."[503] As Nussbaum

---

[501] That most existing humanities courses are as deficient in terms of EfS as any others we shall see below (see also box 23 on p. 299).

[502] Max Frisch in *Die TAT*, 6.3.78, 22, quoted in Heinz Lippuner and Heinrich Mettler, 'Literatur: der Ort des Nirgendwo?', in *Utopien - Die Möglichkeit des Unmöglichen*, 1987, 205.

[503] Jonathan Bate in *The Times Higher Education Supplement*, 4.7.1997.

rightly says, "this knowledge of possibilities is an especially valuable resource in the political life" (1998, 86), thus making it an important skill for responsible citizenship. Of course, factual reports of other ways of life and cultures can be just as helpful here; and the study of as wide-ranging a variety of them as possible should be encouraged on all accounts (and that is where we need journalism, sociology, anthropology etc.).

Yet there is the other aspect of fictional literature which distinguishes it from factual accounts: Arundhati Roy explains in an interview:

> "A writer" (...) – meaning a novelist, a creator of fiction, rather than a journalist –
> "has licence to write things differently... As a writer, I have the licence, and the
> ability I guess, to move between feelings and numbers and technical stuff and,
> you know, to tell the whole story in a way which an expert doesn't seem to have
> the right to do. And in this case [of the Narmada dam] I think that's crucial." Roy
> sees the connections between the economics, the politics, the ecology and the
> human story of the Narmada as the key to the problem. "When I went to the val-
> ley", she says, "I realised that what has happened is that all these experts had
> come in and hijacked various aspects of it, and taken it off to their lairs. They
> didn't want people to understand." Roy, on the other hand, wanted to tell the
> whole story. (Roy, 2000, 31)

Fiction, as opposed to scientific or historical accounts, involves us as whole persons, not just speaking to our intellect, to our knowledge, but also to our feelings, our empathy: "Literature presents 'whole' human models as benchmarks to which individual readers can compare – and evaluate – their own attitudes and behaviours."[504] It involves us in the life of the characters, so that we can see, feel and understand an issue through their perspective, *as if* we were standing in their shoes, allowing "involvement and sympathetic understanding" (Nussbaum, 1998, 88). Nussbaum goes on to explain why this is important for the cultivation of values, such as compassion (which, we might add, are conducive to sustainability):

> In these various ways, narrative imagination is an essential preparation for moral
> interaction. Habits of empathy and conjecture conduce to a certain type of citi-
> zenship and a certain form of community: one that cultivates a sympathetic re-
> sponsiveness to another's needs, and understands the way circumstances shape
> those needs. (...) Compassion, so understood, promotes an accurate awareness of
> our common vulnerability. It is true that human beings are needy, incomplete
> creatures who are in many ways dependent on circumstances beyond their control
> for the possibility of well-being. As Rousseau argues in *Emile*, people do not
> fully grasp that fact until they can imagine suffering vividly to themselves, and
> feel pain at the imagining. In a compassionate response to the suffering of an-

---

[504] Vernon Owen Grumbling, 'Literature', in *Greening the College Curriculum*, 1996, 151-173, here 152.

other, one comprehends that being prosperous or powerful does not remove one from the ranks of needy humanity. (Nussbaum, 1998, 91)

Nussbaum – I have noted the deficiencies above (p. 246f.) – forgets to include nature and other species into her definition of compassion, but if we allow ourselves to include them, the point is a very important one, crucial both for empowerment and equity of a sustainable society: "A society that wants to foster the just treatment of all its members has strong reasons to foster an exercise of the compassionate imagination that crosses social boundaries, or tries to. And this means caring about literature." (Nussbaum, 1998, 92)

We can now specify "the political promise of literature" more succinctly: "It can transport us, while remaining ourselves, into the life of another, revealing similarities but also profound differences between the life and thought of that other and myself and making them comprehensible, or at least more nearly comprehensible." (Nussbaum, 1998, 111) It allows us to become aware of 'the other', to transcend "our frequently obtuse and blunted imaginations" and acknowledge the existence "of those who are other than ourselves, both in concrete circumstances and even in thought and emotion" (Nussbaum, 1998, 112).

In fiction we can, then, look into the characters' brains and hearts, as it were, something which is almost impossible to do, if one is limited to available historical accounts or scientifically verifiable data. We have seen above, when discussing the barriers to sustainable behaviour (see p. 261), that factual knowledge about issues is not enough to motivate people to change their behaviour. We need to convince the whole person, so to speak, not just their mind, but also their heart and feelings, that the transition to sustainability is both necessary and possible. Only if we in the 'developed' countries are able to feel what happens to the whole life of a person poisoned by mercury from gold mining will we understand what is meant by talking about impacts of our actions. Mere abstract figures will not do that. This is precisely why *Zukunftsfähiges Deutschland*, one of the most influential studies on sustainability in Germany, contains qualitative models of a sustainable society in order to make it palpable what sustainability would mean in practice, rather than theory (see 1996, chapter 4, 149-285).

While reading literature, in other words, the reader "configures a *possible course of events or a possible state of affairs*, (...) s/he risks an assumption about the structure of worlds." (Eco, 1987, 143) Characters are living, thinking, feeling in a microcosm, an environment. Reflecting about literature can therefore encourage readers "to think seriously about the relationship of humans to nature, about the ethical and aesthetic dilemmas posed by the environmental crisis, and about how language and literature transmit values with profound ecological implications."[505] It encourages the reader, as Roy has said above, to view the whole story, the emotional, intellectual, economic, social and political dimensions, something which scientific accounts, if at all, tend to do in separate studies only.

There are also various literary devices which we often unconsciously use and which allow us to understand issues in a more comprehensive way, since they enable contextualisation with other knowledge: we can use "analogy, for example, 'spreading like a blanket shaken over a bed' to help us grasp the experience, to awaken within us the sense of vitality of the setting which an objective scientific account would be hard pressed to achieve." (Plant, 1998, 14)

Yet let's not get carried away. What I have presented above are *potentials* embodied in the "narrative imagination", as Nussbaum calls it. In EfS we need to set these potentials free. We won't get any nearer to a sustainable mind set just by reading literature without enabling and encouraging the reflection on values, interactions and behaviour. After all, most tyrants and dictators, as well as many CEOs of TNCs have been or are keen readers of world literature.

Nevertheless, this enabling of (social, political and economic) imagination seems a crucial plank in EfS, but the current climate at universities worldwide is not conducive to such things. For this we are paying a high price, Nussbaum argues:

> It now seems to many administrators (and parents and students) too costly to indulge in the apparently useless business of learning for the enrichment of life. Many institutions that call themselves liberal arts colleges have turned increasingly to vocational studies, curtailing humanities requirements and cutting back on humanities faculty – in effect giving up on the idea of extending the benefits of a liberal education to their varied students. (...) People who have never learned to use reason and imagination to enter a broader world of cultures, groups, and ideas are impoverished personally and politically, however successful their vocational preparation. (Nussbaum, 1998, 297)

When the primary aim of education is 'employability' – in today's unsustainable economy, that is – and therefore all aspects which are not seen as quantifiably contributing to this aim are cut back, study times are shortened and spaces for reflection are cut out, it seems that we are depriving ourselves of the chances to build an educational system which can support and enhance sustainability.

2. Humanities as the reflective sciences needed to deal with complexity
The humanities are often seen as the reflective sciences because in them humans reflect on themselves, their thoughts, actions and belief systems – philosophy

---

[505] Cheryll Glotfelty, 'Introduction', in *The Ecocriticism Reader*, 1996, xv-xxxvii, here xxv; see also Vernon Owen Grumbling, 'Literature', in *Greening the College Curriculum*, 1996, 151-173, here 152.

being the *primus inter pares* amongst them. In the context of EfS, self-reflection on what we are doing, what consequences (including delayed feedback) and (long-term) impact our actions have, particularly on other people and nature, what our responsibilities are in respect of what we are doing, is absolutely vital. We have noted above (p. 185) that the natural sciences by and large lack this self-reflection. It is therefore important that this element, via EfS, becomes central for all science education as well. Without self-reflection we will not come to terms with the complexity of real life. We need it to re-evaluate our theories, terms and concepts with which we view our world on a constant basis. The humanities can help us here.

3. Developing empathy, awe and love for nature and humans
Many scientists have attested, and the biology of learning seems to suggest, that you will only be prepared to act in order to preserve something you know intimately and for which you develop respect and awe. This, of course, is best achieved through experiential learning out in nature. Yet non-scientific approaches to nature can help to nurture these feelings and values. As Williams and Taylor state: "Developing a sense of awe and wonder about nature – that is, appreciating the *poetry* of nature – is just as important as comprehending its *science.*"[506] Quite literally, literature, *if* a supplement to real experiences of nature, can help us to appreciate the poetry of the biosphere.

4. Human made problems require humanities (and social sciences) approaches
One of the groundbreaking aspects of the concept of sustainability over a 'merely' ecological perspective is the insight that we do not just need to know how nature, how ecosystems work, but also how humans interact and interfere with these systems, altering, even destroying them. This is a prime reason why a merely scientific EfS, however important it is, will not do. Further: "Environmental problems include, over and above anthropogenic changes in nature, also the anthropogenic *reasons* for these changes. This requires an inclusion of perspectives (and disciplines) from social sciences (and humanities)."[507]

We need to recognise "that current environmental problems are largely of our own making, are, in other words, a by-product of culture."[508] This means we need to understand the workings of culture, of society, of economic processes, of mentality and ideology, of the media, of connections between mentality and behaviour, of political and other forms of power structures and so on, all things the

---

[506] Dilafruz R. Williams and Sarah Taylor, 'From Margin to Center: Initiation and Development of an Environmental School from the Ground Up', in *Ecological Education in Action*, 1999, 79-102, here 83.
[507] Regula Kyburz-Graber, Lisa Rigendinger and Gertrude Hirsch, 'Problemorientierte Umweltbildung als partizipativer Prozess', in *Umweltproblem Mensch*, 1996, 303-323, here 309.
[508] Cheryll Glotfelty, 'Introduction', in *The Ecocriticism Reader*, 1996, xv-xxxvii, here xxi.

so-called hard sciences can tell us little about.[509] And we need to understand the ethical dimensions of our behaviour and lifestyles. Sachs clearly points this out, with a view to the Master Science, and calls for contributions from the humanities:

> A full appreciation of resource productivity in the substantive sense calls for a debate on civilization, rather than a debate on technology. It requires us to wonder how productive society's economic output is in terms of welfare, use value, beauty and meaning. What is all this effort worth? And what do we want? (...) Satisfaction of people in society, however, is part of a larger picture – it depends on society's aspirations, shared narratives and institutional values. It is at this point that – next to sciences and engineering – the humanities must join in the research on resource productivity. (Sachs, 1999, 181)

Steiner has pinpointed the crucial question even more sharply: "What we need is not improved environmental protection in an unchanged society, but a society changed in such a way that environmental protection is superfluous." (Steiner, 1998, 309) Because, Steiner and others argue, we know relatively well what the tolerance limits of ecosystems are, we do not primarily need more scientific studies to establish this; what we urgently need is a discussion, with the help of the humanities, about the type of society which can live within these limits (see Steiner, 1998, 309-310 and above, chapter 1.c). Humanities are important here, as one respondent to my EfS questionnaire noted, "because they are essentially about man and his place in the world" with all the associated fundamental problems. Another expanded:

> In my view issues of sustainability cannot be resolved by technical or political solutions alone. They can only be solved by a cultural change and a willingness to make that change. Humanities are uniquely placed to encourage debate on this topic and, maybe, to enable some people to develop as opinion formers.

Yet it seems that the funding structures are focus neither on the humanities nor on sustainability. To give just one example: of the ten special foci of publicly funded research in Switzerland from 2001 onwards none are from the social sciences or humanities; one is focusing on climate change, one on plant protection in ecosystems. There is not a single large research project on sustainability or EfS in the wider sense, but four out of ten projects are in the euphemistically named 'life sciences' with a more than questionable sustainability record (at least so far).[510]

---

[509] And why, incidentally, the claim by evolutionary psychologists and sociobiologists that their scientific model of the world can explain everything, is so ludicrous (see Hilary Rose, 'Colonising the Social Sciences?', in *Alas Poor Darwin*, 2000, 106-128).

[510] See Claudia Wirz, 'Festere Strukturen der Forschungslandschaft', in *Neue Zürcher Zeitung*, No. 296, 19.12.2000, 11.

The current state of EfS (with a particular focus on humanities) in Europe

Judged against the "planetary emergency" (Orr, 1994, 27) and the responsibility of universities to educate ecoliterate leaders for tomorrow, it has to be said that curriculum and institutional greening, let alone the greening of the shadow curriculum, have hardly started at all. A comprehensive review of the efforts in the UK in 1996 found that "the vast majority of FHE [further and higher education] institutions have not yet developed environmental policies. (...) Hardly any progress has been made in respect of curriculum 'greening'" (Ali Khan, 1996, 2). Similar conclusions can be reached for the US (see Bowers, 1997, 11-16), while an overview in Germany concluded that EfS "still tends to take place at specific events or in passing; it does not yet have any substantial influence on everyday life in the educational sector."[511] And the more you move from primary into higher education, the less integrated and transdisciplinary EfS becomes.[512]

The experience in the US that literally billions of US dollars are being spent "bringing computer technology into the nation's classrooms" whereas "the efforts of environmentally orientated professors and classroom teachers are generally undertaken as part of their work overload"[513] seems to be reflected elsewhere: As noted in chapter 1 there is an impressive array of international declarations of intent, agreements and documents which all emphasize the crucial importance of EfS particularly in tertiary education, yet the practical actions and the financial commitment of governments, educational institutions and practitioners in no way match these 'good intentions'.

We are nowhere near where we should be, namely with EfS being naturally counted and practised as a core component of any kind of education (such as Maths or English).[514] Yet what can we learn from institutions which have started the journey or those which would like to but can't, or those which refuse to? This was the purpose of a research project I undertook in the winter of 2000/2001, supported by a Study Abroad Fellowship of the Leverhulme Trust (UK).

Since I wanted to focus specifically on the humanities – largely because the humanities have been the slowest to embark on EfS, yet, as we have argued above, are crucial for a successful EfS – I surveyed all the humanities faculties in

---

[511] Gerhard de Haan, 'Von der Umweltbildung zur Bildung für nachhaltige Entwicklung?', in *Umweltbildung in Theorie und Praxis*, 1998, 11-28, here 17. The same is true for research into EfS which in German speaking countries only happens to "a very restricted extent" (ibid. 26).

[512] Ibid. 15.

[513] C.A. Bowers, 'Changing the Dominant Cultural Perspective in Education', in *Ecological Education in Action*, 1999, 161-178, here 170.

[514] See William Scott (University of Bath), 'Presentation at the Council for Environment Education's seminar on 'Strategic Approaches to Supporting Education for Sustainable Development – Producing Resources to Support ESD in Schools'. 18.5.2000', quoted in: *Schools News. Development Education work in the formal sector*, Issue 6, July 2000, 4.

Germany, the Netherlands, Switzerland and the UK with the help of an internet based questionnaire.

The response to the questionnaire has been disappointing. The prime reason for this, so my informed speculation, is the view quoted above that sustainability does not have much to do with the humanities. But there might be some other aspects to it. Since it was an internet based questionnaire and the invitation to fill it in (partly by letter, partly by email) did not itself contain the questions, there might be an argument for looking at the ratio 'accessed/filled' in rather than 'total sent out/filled in' (3.8 per cent of total; 18.5 per cent of those who accessed, i.e. read it). It is clear that other pressures (time, teaching/ research/ administrative commitments), lack of interest, top-down selection (I sent the questionnaire to the dean of the humanities faculty with a request both to fill it in and to pass the internet address of the questionnaire on to other relevant members of staff) might have adversely affected the response rate on top of the added barrier that the respondents had to start up their internet browser and call up the questionnaire.

The number of humanities faculties surveyed were as follows: UK: 107, Germany: 132, The Netherlands: 10, Switzerland: 12 (total: 261). 53 faculties or individual staff accessed the questionnaire (some individuals up to three times and up to four different staff from the same institution) while 10 filled it in (and took between 7 minutes and 2 hours and 13 minutes to finish it; average time spent: 50 minutes), resulting in the following returns: UK: 9/3 (accessed/filled in), Germany: 43/7, The Netherlands: 0/0, Switzerland: 1/0 (total: 53/10).

Because the number of responses is so small one certainly cannot draw any statistically meaningful conclusions from them. Yet I believe that we can see some tendencies and certainly use them qualitatively, not least because the respondents came from a varied background of disciplines (education, geography, social science, philosophy, theology, history, cultural studies, language and literature studies, comparative literature and peace/conflict studies).

Here are the most interesting results:

1. There is no university wide strategy to EfS

Question	Yes	No	Total responses
Does your institution have an EfS policy commitment?	20% (2)	80%(8)	10

Only two institutions (Gerhard-Mercator-Universität, Duisburg, Germany, and University of Bath, UK) claim to have an institutional policy attempting to integrate EfS into the curriculum of the entire institution (but in neither case was I given any details about the policy, even when following it up). The UK (but not the German) institution has an academic member of staff who is in charge of implementing this EfS policy.

Question	Yes	No	Total re-sponses
Has your institution defined a common learning agenda for sustainability relevant to all students, regardless of which course they are on?	0%(0)	100%(10)	10
Are there core modules on sustainability/ responsible global citizenship that every student has to take, regardless of their scheme of study?	0%(0)	100%(10)	10

None of the institutions has defined a common learning agenda nor are there any core modules delivered across the disciplines. This seems to indicate that the process has just been started and is still at the level of policy formulation rather than concrete planning of how such a policy can be put into practice. The German example seems to bear this out. At present there is just a recommendation from October 1999 by the vice-chancellor and senate of this university to introduce sustainability modules into all schemes of study.[515] In September 2000 the commission for teaching, study and study reform recommended introducing a module called "sustainable development". This module, which has been temporarily shelved due to a university reform, is intended as a compulsory one, akin to a *studium fundamentale*. All faculties are asked therefore to integrate sustainability into their curriculum and make it part of their examination requirements.[516]

The majority of other institutions rely on individual members of staff to pursue "some aspect of environmental questions as a specific interest". Often, this is in the context of individual modules, in masters programmes, or the institutions have specialised units, such as the interdisciplinary, centrally funded International Centre for the Environment (ICE) at the University of Bath (UK), the International Centre for Development Studies at the Edge Hill College (UK) or similar. Another tool often used is interdisciplinary lectures, such as the series in Duisburg on "Sustainable development and sustainability". In terms of the educational value for students, it is highly doubtful that this is an effective way of teaching EfS (see sections on experiential learning and the shadow curriculum); but such events at least might help to raise the profile of the issue and get a larger public interested in it. It has to be said, though, that these "add-on" approaches are often little more than "tokenism", as Bowers calls it (1997, 15), with the purpose of avoiding a fundamental rethink.

---

[515] Note that other universities, such as Edinburgh, made such a recommendation much earlier: "All undergraduates, at some time in their course, should be exposed to teaching about the wider and more fundamental issues of society's relationship to the environment, including complex social, economic and ethical questions." (The University of Edinburgh Education Policy Committee, 1991, quoted in HE21 (1998), *Trail Blazer Stories: University of Edinburgh* (London, Forum for the Future), [2].

[516] Additional information by Cäcilia Tiemann, Gerhard-Mercator-Universität, Duisburg (email 27.12.2000).

Nevertheless, these fragmented activities can add up. The University of Oxford (UK) claims, for example, that "of the 5,000 or so teaching and research staff in the University, over 500 regard environmental questions of one kind or another as a main focus of their work."

Question	Yes	No	Total responses
Has your institution an Environmental Policy Statement (or similar) in place?	20% (2)	80%(8)	10

It is interesting to note that the same institutions which have an EfS policy tend to have an overall environmental policy for the institution, indicating that there are synergies to be gained if institutions try to further the issues both in teaching and institutional practice. Further indication for such an assumption can be seen in the fact that the only institution with both an environmental *and* an EfS policy (University of Bath) is also the only one which has offered staff development courses on sustainability. Teaching, research and institutional commitment to sustainability correlate strongly. Yet half of the institutions with either policy type have no implementation strategy. If you do not commit the institution to a tight implementation plan, with specific targets, deadlines and indicators, then the policies tend not to be worth the paper (or web space) they are written on. This, at least, is the experience of many institutions with a policy but no perceptible change (see the progress report for the UK by Ali Khan, 1996).

There is also another dimension to institution-wide EfS. If we ought to make sure that all students not only hear about sustainability, but are also enabled to adopt sustainable lifestyles it is evident that higher education institutions should care about students' career paths after graduation. My questionnaire therefore contained the following question: "How, if at all, does your institution encourage students to consider sustainability issues when choosing career paths?" None of the respondents could report anything positive on this. Ideas like the "Graduation Pledge" (see below p. 319) seem not to have taken hold in Europe at all.

2.   *EfS is not seen as significant for all disciplines*

A number of respondents obviously felt that EfS was of concern only to some disciplines and that others should not be asked to introduce EfS teaching requirements into their "syllabuses, irrespective of their significance for the particular discipline". This clearly implies that there are some disciplines where sustainability is a relevant topic (presumably Biology and Geography), whereas for others (such as the humanities) it has no importance. Such a position bears witness to the widespread ecological illiteracy amongst some academics who quite clearly have never seriously contemplated the impact of their discipline, their own lifestyles and their institutions.

3. Institutional environmental impact has nothing to do with EfS
As we have seen, many institutions don't even have an environmental policy which aims at turning the institution into a sustainable role model for others (see also Ali Khan, 1996, 9). "You would look in vain for a written policy commitment", was one response. Most respondents don't even understand the connection between what you do and what you teach. This works both ways. Either respondents are completely oblivious to the shadow curriculum – typical response was to claim that an institution's "own environmental impact" was a "largely unrelated matter" with regard to EfS – or respondents felt that all you had to do was "green the institution", but that you shouldn't teach sustainability. Again, a typical response was that we should "concentrate on getting institutions to act in a sustainable manner NOT on getting them to try and include teaching on this topic."

4. Barriers
Even those sympathetic to the idea acknowledge that EfS isn't the easiest task in the world. The questionnaire aimed to establish the main barriers to EfS implementation and tried to elicit ideas about how to overcome them.

Question (multiple response)[ordered by relevance]	Selected	Percentage
What are the major barriers to integrating sustainability themes into Humanities subjects?		
Lack of staff expertise	9	30%
Perceived irrelevance by staff	4	13.33%
Perceived irrelevance by students	3	10%
Insufficient time to update courses	3	10%
Perceived lack of academic rigour	3	10%
Financial restrictions	3	10%
Institutional structure	2	6.67%
Others	2	6.67%
Confusion about what needs to be taught	1	3.33%

Four of the six reasons selected most often would support Orr's analysis of ecological illiteracy. There is simply a vast knowledge deficit both on the part of staff and students which impedes any transition. If nearly all the respondents (9 out of 10) select "lack of staff expertise" and only one "confusion about what needs to be taught", this seems to indicate that EfS itself is very broadly acceptable, yet we need a major effort to integrate it into every curriculum *and* into every tutor's professional development in order to be able to advance. Some of the other reasons frequently selected (insufficient time, financial restrictions, institutional structure) point very forcefully to the shadow curriculum in the broadest sense, which clearly needs to be addressed should encouragement be more than half-hearted.

The following suggestions were thought to be the most promising when it comes to encouraging EfS implementation:

- Use restructuring initiatives (such as change of degree schemes into international bachelor or masters programmes) which are currently under way; this is much easier than changing long established degree schemes; but once done it can feed through to older structures.
- Make concepts and materials which can be easily reused or adapted in other institutions readily available.
- Introduce a compulsory module for all.
- Establish a dedicated chair for EfS. The history of gender studies or applied ethics, which like EfS can be seen as relevant to all disciplines, seems to indicate that establishing a chair facilitates, within the current HE institutional structures, the acceptance and proliferation of the respective ideas.[517]
- Encourage inter- and transdisciplinarity wherever possible.
- Establish an interdisciplinary working group or committee which draws together expertise and committed individuals from the whole institution.

---

[517] This, at least, was the conclusion from evidence presented at the workshop 'Interdisziplinäre Themen in Fachstudiengängen: Umweltwissenschaften, Ethik und Gender Studies' (see *Interdisziplinäre Themen in Fachstudiengängen* (2000).

Conclusions

> There are two types of master bricklayers: a master to make it happen, or a master to tell you it can't be done. (McKibben, 1997, 79)

Rather than summarising what has been said – if it hasn't become clear what the book is about by now, even such a summary conclusion will be a waste of space – I have decided to heed the advice of various practitioners and end with a list of practical solutions, inspirational examples and ideas which – at various levels and with varying degrees of ease of implementation – take up the spirit of McKibben's motto. He clarifies this call for practical solutions and positive approaches – we don't need more overdeveloped, resource hungry cynics – with a quote from Lerner, the governor and 'architect' of Curitiba, who turned this Brazilian city into one of the most sustainable cities on Earth:

> The hardest habit to break, in fact, may be what Lerner calls the "syndrome of tragedy, of feeling like we're terminal patients." Many cities have "a lot of people who are specialists in proving change is not possible. What I try to explain to them when I go to visit is that it takes the same energy to say why something can't be done as to figure out how to do it." (McKibben, 1997, 113-114)

If you keep your eyes open, if you attempt to transcend the limiting perspective of your own culture, ideology and education, you will find that the solutions are there, somewhere on Earth, lived out, ready to be picked up, adapted, put into practice. We *do* know very well that the problem is not that we don't know enough, don't know the reasons for our unsustainable present, wouldn't know the direction in which to go or wouldn't know the solutions. The problem, as ever, is political will and the complacency of the overdeveloped elites of this world.

But that is not all. As we have seen when discussing institutional barriers to more sustainable pathways, a lot of us are hemmed in, almost immobilised from doing sensible things by institutional structures which obstruct change. In the unsympathetic context of a hostile dominant educational paradigm and sandwiched between too many demands on a tutor's time – exploding administrative and bureaucratic burdens on teachers, implementation of new parodies of panaceas into teaching such as ICT and employability,[518] pressure to "publish or perish" etc. – it is very hard to mobilise time and energy to re-conceptualise one's pedagogical approach, one's teaching practice, rewrite courses and so on. Practical examples and the notion that people elsewhere have managed to achieve change might therefore both motivate and make it easier to engage in such positive

---

[518] Stephen Sterling talks of "the market view of education" in this context which reinforces "dominant values of egocentricism and instrumental rationality" ('Education in Change', in *Education for Sustainability*, 1996, 18-39, here 27).

practice ourselves.[519] All these examples have in common the fact that they don't rely on new inventions, new discoveries or groundbreaking new scientific insights. All they need for implementation is people with the will to act.

28 practical strategies to foster EfS

1. The polluter pays (or at least doesn't get any money)
At governmental, regional, local and institutional levels, teaching and research funding, promotion, the development of career paths and appointments (most importantly the highest posts) should be linked to sustainability criteria (Ali Khan, 1996, 4). For sets of such criteria see box 19 above on p. 252 and the sustainability indicators developed by the HE21 project.[520]

2. Invest in people, not technology
The most promising start-up for curricular change towards EfS has been the provision of financial incentives and staff time to reorganise courses and degree programmes, with an emphasis on transdisciplinary work.[521] This could easily be financed through a freeze on unnecessary IT spending, brought about by a serious rethink on educational priorities and a new focus on tried and tested, human solutions, not unproven information technology which is draining resources (both financial and natural) (see *Fool's Gold*, 2000, 77).

3. Train the trainers
Provide staff development courses to green the curriculum, by using practitioners from institutions which have started implementation of EfS. We have seen above that universities are not just victims of an unsustainable society, but active perpetrators. The questionnaires have also shown a rather worrying degree of eco-illiteracy amongst academics. Therefore, the task of educating the educators, of educating the illiterate elite of today in ecoliteracy, might in fact be the much more difficult *and* much more important task than developing EfS for students (see p. 241ff.).

4. Redefine notions of excellence
Both in the public mind as well as in the way academic disciplines, professions and services see themselves, we need to redefine what we mean by excellence in a sustainability context (see p. 250). Unfortunately, it seems, the higher you go up

---

[519] See Regula Kyburz-Graber, Lisa Rigendinger, Gertrude Hirsch, 'Problemorientierte Umweltbildung als partizipativer Prozess', in *Umweltproblem Mensch*, 1996, 311f.

[520] See HE21 [Higher Education for the 21st Century] (1999), *Sustainability Indicators for HE* (London, Forum for the Future).

[521] Steve Breyman, 'Sustainability Through Incremental Steps? The Case of Campus Greening at Rensselaer', in *Sustainability and University Life*, 1999, 79-87, here 81.

the hierarchy ladder in our society, the bigger the snobbery and resistance to change: "It is no accident that environmental education and, more recently, education for sustainable development, has progressed more rapidly at the secondary and primary levels than within the realm of higher education." (UNESCO, 1997, 29, §89; see also Bourdieu, 1992) Success and intelligence of individuals, institutions and nations will have to be judged by the level of (long-term) impact they have on the survival of the planet and all its inhabiting species (see Bowers, 1995, 132), not by their monetary wealth, false notions of status or utterly misleading measures such as GNP.

5. Expose the eco-illiterate lecturers and courses

Bowers has made a very useful suggestion which can encourage the transition to more sustainable university education. Part of the problem is that lecturers and professors can still carry on teaching courses which foster unsustainable attitudes, thought patterns and behaviour and this fact is largely hidden from public view. To counteract this, Bowers advocates a strategy which in the past has been quite successful with regard to sexist and racist courses, namely "public exposure of course content that reinforces the mythic underpinnings of modern culture" (1997, 238), and he encourages students and faculty to foster discussion about the following questions:

> Which classes are still based on the cultural assumptions that have contributed to ecologically and culturally destructive practices in the past? (...) Who benefits from the forms of knowledge promoted in classrooms and research settings? What will be the impact on cultural groups that have developed more in the areas of community and environmental responsibility? Do the forms of knowledge promoted in universities have self-limiting guidelines that recognize the rights of other forms of life as well as the rights of the seventh unborn generation? (Bowers, 1997, 225-226)

6. Commit your students: Graduation Pledge

A very good example with the aim to encourage such a redefinition of excellence in higher education is the Graduation Pledge idea, quite a successful and popular scheme originating from the US. At Manchester College, where the national efforts of the Graduation Pledge Alliance are coordinated, typically 50-60 per cent of all graduates sign the pledge. Other universities run the scheme as well, including Harvard University, Humboldt State University, Indiana University, Massachusetts Institute of Technology (MIT), Tufts University, University of Chicago, University of Kansas, and University of Vermont. Students sign a pledge which reads: "I [name of student] pledge to explore and take into account the social and environmental consequences of any job I consider and will try to improve these aspects of any organization for which I work." Taking the Pledge is voluntary and allows students to determine for themselves what they consider to be socially and environmentally responsible. There are different ways of integrating the pledge into graduation ceremonies, but the most important aspect is visibility, i.e. to communicate to the parents of the students, the tutors, the fellow graduates and the outside world that the signatories are prepared to make a stance for sustainability, not just in theory but also in their professional careers. At Manchester College, supporters and supportive faculty wear green ribbons during the graduation ceremonies and the Pledge also appears in the printed graduation ceremonies program.[522]

7. *Enable self-determination in learning*
If we are serious about sustainability – and that means the empowerment of people to take control of their lives within functioning communities, in full acceptance of the biogeochemical limits of the earth and the responsibility for the impacts of their lifestyles – we need to take seriously the demand to turn education into a practical example of such empowerment. Yet empowerment, as we have seen (p. 53), is about freedom in the context of a community and within the limits of the biosphere. This has consequences: On the one hand education should not just be an isolated transmission of specialist information, but a process whereby students are empowered to take their learning into their own hands. On the other hand, this learning comprises not just the development of the learner's unique potential, but also the acceptance of responsibilities towards others, the earth and future generations as well as respect and recognition of relevant and legitimate elder knowledge. Or, as Sterling phrases it, the educational process "should itself be socially sustainable in the sense that it is based on meaningful rather than token empowerment, participation and ownership"[523] (see p. 288ff.).

8. *Do, and you will learn*
Integration of education into real life, and more specifically, turning the educational process into an experience adequate to the complexity and multidisciplinary nature of reality, is a necessary basis for any effective learning. EfS should therefore aim to be experiential learning, starting from real problems and grappling with the multidimensional nature of these in an attempt to come to real, rather than reductionist solutions (see p. 291ff., esp. boxes 21 and 22).

9. *Practise what you preach*
Sustainability only makes sense as an integrated, complex concept, whereas isolated notions and solutions contribute, through long-term impacts, feedback loops and so on, to a worsening of the planetary situation. We therefore have to bury the

---

[522] For more information, practical guidelines for introduction of the scheme at your institution and joining the Graduation Pledge Alliance see http://www.manchester.edu/academic/programs/departments/Peace_Studies/files/gpa.htm.

[523] Stephen Sterling, 'Education in Change', in *Education for Sustainability*, 1996, 18-39, here 35.

notion that we can teach in one way and behave in another. If we want to progress towards sustainability, we have to practise "education *as* sustainability" (*ESDebate*, 2000, 41). In education as in real life we have to go for the big picture and integrate teaching, research, personal behaviour and professional activity within the framework of sustainability. In *ESDebate*, the international debate on education for sustainable development, "most participants agree that ESD can only be successful if those who are initiating ESD projects and processes are also changing their behaviour and practices." (*ESDebate*, 2000, 32) We have to be well aware that activities in one area might cancel out endeavours in others if we do not try to keep an eye on the overall situation. In other words, as IUCN has stated, "learning must take place at various levels, individual, organisational and institutional, as the path to sustainability requires changes in our systems as well as in people." (*ESDebate*, 2000, vii)

10. Shadow curriculum: be clear about its likely impact
We have seen that any potential educational impact is held in check by what is called the shadow curriculum (see p. 297ff.). These checks are formidable forces and educators should harbour no illusions about the possibilities of learning to counteract them. The most important ones to keep an eye on are:

- the economic structure which is currently based on the exploitation of people and nature and aims for growth within a closed system; it finds its expression in the economic imperative which increasingly determines peoples' lives around the globe (maximising profit, minimising expenditure);
- institutional structures, tax systems, cultural values ('growth', 'progress', 'development') and ideological beliefs encouraging and entrenching unsustainable behaviour; and
- all forms of media and other propaganda denying access to real knowledge about the real world (see chapter 2.b).

Realising this should trigger a dialectical reaction. On the one hand educators should be rather cautious about exaggerated beliefs in the power of change through education. Any official claims of that nature should be carefully screened: are they not in fact attempts to turn the focus away from social areas where change would be far more necessary *and* effective? Yet at the same time, knowing that systems are made of people, educators should counteract the shadow curriculum wherever and whenever they can: by changing the institutional structures of their educational institutions, by exercising their responsibilities as citizens. But there is also action called for on the microlevel of everyday teaching, if you like: we have to learn to be careful with the selection of course material and screen it so that it doesn't reinforce the shadow curriculum. This is not a call for censorship but one to raise tutors' and learners' awareness of how this process of reinforcing unsustainable values works. To do this we must be aware of it ourselves – we need to be ecoliterate ourselves – and make this transparent in the way we deal with material in teaching, i.e. showing how and why these processes work.

11. Building the teaching building that teaches you
An important part of the shadow curriculum is constituted by the architecture, setting, atmosphere and day-to-day running of the educational institutions (see p. 297ff.). David Orr has tried to fuse these insights with the recognition of the importance of experiential and self-empowering learning. Guided by Lyle's question "is it possible to design buildings and entire campuses in ways that promote ecological competence and mindfulness" (quoted in Orr, 1999, 230) Orr and 25 students set about planning an environmental studies centre for Oberlin College. First, they clarified whether there was a need at all for a new building, bearing in mind the following facts:

> By some estimates, humankind is preparing to build more in the next half-century than it has built throughout all of recorded history. If we do this inefficiently and carelessly, we will cast a long ecological shadow on the human future. If we fail to pay the full environmental costs of development, the resulting ecological and human damage will be irreparable. (Orr, 1999, 234)

In a two-semester class the students and a dozen architects established the criteria for the new building – amongst them: no wastewater discharge, generating more electricity than is used, use of sustainable building materials in an efficient way, and promoting a building which "in its design and operations" is "genuinely pedagogical" (Orr, 1999, 231). On this basis the building was designed with the help of two former graduates, an architect, various sustainability experts (such as Amory Lovins) and an open planning process involving "some 250 students, faculty and community members" in order to design a building which reflected the needs and knowledge of its anticipated users. Apart from the empowerment aspect, this was done for a simple reason: "No architect alone, however talented, could design the building that we proposed." (Orr, 1999, 231) What resulted in the end was a genuine example of real learning: a concrete project, arising out of need, is picked up and, within a framework of sustainability parameters, is brought to fruition in a democratic and transdisciplinary process, always oriented towards the goal, rather than any specific vested interests of any discipline involved. The initial building process was completed in 2000, though the building is designed to evolve over time.[524]

There are more general lessons to be learnt from this which can be applied everywhere. Whenever there is a need for a new building for an educational in-

---

[524] For a full account see Orr, 1999 and the website of the building which contains information on building facts, the innovative technologies used, the design philosophy and the Living Machine (water purification plant) [http://www.oberlin.edu/newserv/esc/Default.html].

stitution we should remember the importance of the shadow curriculum and, rather than engage in lofty pedagogical goodwill, make sure we provide the most appropriate environment for our "leaders of tomorrow", be they seven year old pupils or twenty year old students. It also means that any building maintenance tasks, any planning of alterations should be done from this perspective. Often, particularly at higher education institutions, there is a wealth of untapped local knowledge available which can be put to good use. Students can perform waste, energy and water audits, develop green transport plans, plan refurbishments, help with landscape gardening etc.; there is plenty of scope for real learning which can feed back into the running of institutions. This was precisely the approach taken by the University of Sonora, Mexico, when it developed the "sustainability cell" where "students should put into practice what they learned by initiating a program of Sustainable Development for the University itself." (Velazquez *et al.*, 1999, 362) Between 1996 and 1999 1,150 students participated and engaged in such projects as social communication, efficient use of water and energy, ecological toilets, composting, paper and inorganic materials reuse, recycling of plastic, industrial development and quality of life (Velazquez *et al.*, 1999, 365-368).

12. We are the problem, not the solution

There is one basic insight which we have to try to foreground in all of the discussions about EfS and sustainability in general. We have to accept, since it is demonstrable beyond doubt (see chapters 1.c and 2), that the only reason why we have a sustainability problem in the first place is the economic, political and cultural system of Euro-American origin, which through globalisation has now spread across our planet. I do not think that it makes much sense to shy away from this basic insight, since if we do we will not be able to come to sensible solutions. They would be based on principles which led to unsustainability in the first place and would therefore be unable to contribute to a solution to the problem. Plant has summarised most reasons why the Euro-American ideology is the problem:

> *Poverty* is not caused by the absence of the American way of life but by its wealth; *overpopulation* is caused, not cured, by modernisation destroying the traditional balance between people and their environment; [an] *open international economic system* (...) will extinguish cultural and ecological diversity; the problem of *externalisation of pollution* is not solvable by pricing the environment but by reversing the closure of the commons so there is nowhere to "externalise" to; the imperative to transfer *Western technological expertise* across the world smacks of scientific imperialist arrogance. (Plant, 1998, 61; emphasis in the original)

Let's say it as clearly as we can, and let's not hide behind "the system is bad and forces us to...": the cornerstones of *un*sustainability can be defined: the Euro-

American materialistic world view,[525] perpetual overconsumption in the North and the North of the South,[526] specialism and fragmentation of society, alienation from nature, from a sense of place, from community and other humans.

If we are not prepared to confront this "home perspective", as Sachs calls it (1999, 86) we would do better not to talk about EfS at all.

13. Close the feedback loops!
As we have seen above (p. 267) a good strategy to motivate people to face up to their responsibilities, see the consequences of their actions and change their behaviour accordingly, is to make them experience these impacts *directly and immediately*. There are many interesting ideas about how to achieve this. I am quoting some examples as related in *Factor Four*:

For example, many factories sited on riverbanks traditionally drew their water supplies into one pipe and discharged their wastes out another. Where this arrangement still exists, why not simply put the intake downstream of the outfall, so that whatever pollution the plant emits, it gets right back again? (…) If wastes are OK to put into the river whose water others use, why aren't they OK in the water that the plant itself uses? (...) If a laboratory building is discharging hazardous solvent fumes from its fume hoods up a stack so that they are diluted and rebreathed by everyone else, why not press the operators to explain why they are willing to make everyone outside breathe what they are unwilling to breathe themselves? This logic rapidly leads to zero discharges. (…) If people are concerned that an oil company is not safely operating a refinery or a petrochemical plant, why not ask it, as a token of its good faith, to require that all the plant's senior managers, and their families, live at the downwind site boundary, so that they are the first to be exposed to whatever releases occur? If a maker of nuclear power stations claims its new plant designs will be perfectly and "inherently" safe, why not ask it to waive its legal exemption from unlimited liability for accidents caused by that plant? After all, if it's really safe, the company has nothing to worry about. Or are they unwilling to bear them-

---

[525] Karl-Werner Brand, Angelika Poferl and Karin Schilling, 'Umweltmentalitäten. Wie wir die Umweltthematik in unser Alltagsleben integrieren', in *Umweltbildung und Umweltbewußtsein*, 1998, 39-68, here 49-50.
[526] Possibly contributed to by boredom, as Bassey's interesting theory has it: "But there is also a second figure casting a shadow across the world – the spectre of boredom. I believe this too contributes to rampant consumption. I suspect that at the level of consumers it is boredom rather than greed that operates. People in affluent circumstances, like most of the inhabitants of the western countries, are culturally conditioned to become bored with their existing possessions – clothes, furniture, appliances, cars, etc. – and to buy new ones. Boredom seems to be something that is learned in school and then pervades society." (Michael Bassey, 'Greed, Boredom, Love and Joy and the Ecological Predicament: a Call for Educational Research into the Learning of Green Moral Values', in *Environmental Education for Sustainability: Good Environment, Good Life*, 1998, 149-162, here 160)

selves the supposedly nonexistent risks to which they wish to expose the public? (1997, 183-184)

14. Open your eyes: Kerala, Curitiba, Ladakh, Nigeria, Papua New Guinea...
I have stressed above that one of the most important things we have to re-learn and develop is our imagination, our ability to visualise and implement sustainable solutions. The best way to train this imagination is to look beyond what is presented to us in the mainstream media, beyond the limits of the imagination of our culture, our history, our ideological perspective. We need to start seeing sustainable ways of living, wherever they are practised, and to see whether we can adopt and adapt them. As soon as you start to look, you will find examples. In terms of sustainable agriculture you will find that the people from Ladakh in the Himalayas have developed a carefully balanced system which gives them self-determination over their livelihood even in very hostile environmental conditions (see Norberg-Hodge, 2000). Or you will come across Nigerian market women who have realised that the capacity of their families to provide the means for survival steadily decreases with every step towards greater trade liberalisation: "These women knew and felt that there can be no global market, or any kind of market, that takes global care of subsistence, but only ever a local marketplace and local market-trading" (Bennholdt-Thomsen/Mies, 1999, 120). In order to regain control of their lives they took their fate into their own hands and fought off economic and ecological devastation by international oil TNCs. Or, in a bit more detail, see the following three examples:
- *Papua New Guinea:*

> While 85% of the population live in rural areas and have access to the benefits of this [communally owned] land usage system directly in day to day life, most of the 5% of the population that live in towns and the 10% that live in rapidly growing urban shanty settlements can return at any time to their ancestral areas and use the land. Because of this system, hunger, homelessness and unemployment are unknown, an achievement that should make Papua New Guinea a much more convincing case and model for true developmental success than other countries which, in the name of development, have reduced their populations to landless, homeless, hungry paupers, desperate to sell their bodies and their work at any price. (Nicholas Faraclas, quoted in Bennholdt-Thomsen/ Mies, 1999, 145; see 145-149)

- *Kerala, India:*
Kerala in the South of India is a particularly inspiring example. In spite of tiny footprints – in the order of 0.2 to 0.3 hectares per capita; remember there are roughly 2 hectares available and the average American occupies more than 9 hectares – the people of Kerala enjoy demographic stability, low child mortality, a long lifespan, and literacy rates as high as in highly industrialized countries. Also,

they can pride themselves on a lively democracy with the active participation of women. What they have achieved is really the challenge for tomorrow: *how to squeeze lots of quality of life out of small footprints*. McKibben summarises the lessons we can learn:

> Kerala, unfortunately, offers no answer to the question of our will. But it sharpens the debate, cutting off one line of escape. It gives the lie to the idea that *only* endless growth can produce decent human lives. It's as if someone had shown in a lab that flame didn't require oxygen or that water could freeze at sixty degrees. Suddenly, in the light of Kerala, whole new chemistries of people and society and money and happiness seem at least conceivable. (McKibben, 1997, 127)
> Kerala demonstrates that a low-level economy can create a decent life, abundant in the things – health, education, community – that are most necessary for us all. Remember, Kerala's per capita GNP in 1986 was $330, while the U.S. GNP was $17,480. And GNP is an eloquent shorthand for gallons of gasoline burned, stacks of garbage tossed out, quantities of timber sawn into boards. Every dollar spent, remember, means burning half a liter of oil. (ibid. 163)[527]

- *Curitiba, Brazil:*
How do you limit, for example, growth in private car use, in practice? One of the most interesting examples to show how intelligent ecological solutions can be combined with social ones, if the necessary political will is there, is the Brazilian city of Curitiba:

> Jaime Lerner, the architect who has transformed Curitiba into one of the world's greenest cities, has been twice re-elected to run the city and is now state governor. Lerner's creed is revolutionary: "The poorer you are, the better the services you should have." When he first became mayor, Curitiba was mushrooming as the rural exodus of the seventies sent people into the cities and the transport system was heading for chaos: 50 bus companies competed in the city centre, the jams worsening every day. Something drastic had to be done. A subway system cost too much, and would take too long to build. So Lerner's planners identified what made an underground system fast and applied it to the bus service. Huge red articulated buses purr speedily up special lanes stopping at tubular steel and glass stations where passengers buy tickets before boarding. As the buses stop, ramps descend from their doors and boarding time is minimal. Neat little lifts in the pavement raise handicapped passengers to the platform. Lerner has produced an

---

[527] For more details see McKibben, 1997, 117-169; *Kerala: The Development Experience*, 2000, Richard W. Franke and Barbara H. Chasin, 1994 and William M. Alexander's site with data on Kerala [http://www.jadski.com/kerala/]. Recently, the village Pattuvam in Northern Kerala has "created history by declaring its absolute ownership over all genetic materials currently growing within its jurisdiction" (Shiva in Frederick Noronha, 'Is Globalization Killing the Environment? An Interview with Vandana Shiva', in *MNC Masala*, Corporate Watch Feature, (February 1998) 8 [http://www.corpwatch.org/trac/feature/india/interviews/shiva.html]).

efficient, passenger-friendly service. Bus jams never happen, vandalism is un-
known. "People don't vandalise it because they like it. They feel respected, they
show respect," says Carlos Ceneviva, president of Urbs, the municipal company
which collects fares and regulates 10 private companies. No subsidies are paid:
80 per cent of people go to work by bus; 28 per cent of car owners take the bus
instead, which has led to a 20 per cent drop in fuel consumption. This had three
effects: since most people take the bus the fares are so cheap that even poor people can afford it. Additionally, the drop in private car use has made Curitiba the
Brazilian city with the cleanest air and the highest percentage of parks and green
areas, since the demand for parking spaces is lower. Thirdly, for low-income
housing a clever scheme has allowed poorer families to live along the bus tracks
in the city centre. Lerner says: "The less importance you give to cars the better it
is for people. When you widen streets for cars you throw away identity and
memory."

Curitiba has also revolutionised the concept of waste: it can mean food,
books or even Shakespeare. At one point in 1996, for example, 700 schoolchil-
dren each paid four kilos of recyclable rubbish to watch King Lear, performed by
one of Brazil's best theatre companies. They came from the city's poorest areas
and it was the first time they had been to the theatre. And 35,000 low-income
families exchange recyclable waste for food once a fortnight. Each four kilos
mean one kilo of fruit and vegetables. In one month, the 54 exchange points collect around 282 tonnes of waste at a cost of $110,000 – lower than before; the
fruit and vegetables are bought at market prices from small farmers. Benefits include a better diet for citizens and less risk of flooding from rubbish in streams
and canals. Nearly 70 per cent of waste is now recycled and sold.[528]

15. Don't forget the others: Greening the professions

EfS cannot just be about making the educational sector more sustainable. Even if
the entire formal educational school and university sector were to go it alone, it
wouldn't be enough. We need to turn all the professions, all the companies – from
small and middle-size to TNCs – into sustainable units. And there, the formal education sector can, I believe, learn a lot from the voluntary sector and NGOs which
are often far more advanced in EfS. A good example I have encountered is the
training centre of the WWF Switzerland in Berne. The overall aim is to make sectors of the economy, of the market more sustainable. To achieve this the training
centre works on the political level with professional bodies of branches of industry to rewrite the mission statements for individual qualifications and also to
rewrite the curricula of vocational schools to include EfS. These schools are compulsory for everybody in Switzerland doing an apprenticeship. But the attempt is
also made to reach training institutions which retrain or further qualify people already in employment or unemployed people trying to get back in. The second

---

[528] Jan Rocha, 'Urban Renewal: Let them eat cake', in *The Guardian*, 5.6.1996, G2, 18; see also
McKibben, 1997, 57-115, *Factor Four*, 1997, 126-128 and *Natural Capitalism*, 2000, 288-308.

thrust is to target specific branches, such as the transport sector, tourism, electricity utilities, estate agents or the retail sector, to work in co-operation with them to get sustainability criteria accepted and built into the normal operation of these branches.

But it is not just the overall aims that are interesting, but also the pedagogic approach behind them. At the training centre they are working with a concept of EfS which prioritises empowerment and capacity building; for 40 per cent of the time they concentrate on subject specific competence and for the remaining 60 per cent they focus on social competence, which means that they place a lot of importance on the actual transfer of knowledge: in workshops, fora and consultations. Important is the process of communication, the relevance for the daily job and life, and participation of the learners. And they practise what they preach by getting ISO 14001 certification, using 100 per cent renewable energy and 100 per cent recycling or FSC paper; they also have a mobility concept for employees of the centre which doesn't allow air travel in the EU etc. They feel that they could not legitimise themselves as an EfS training centre, if their content was not matched by the method of delivery and the institutional day-to-day running – something most universities still have to latch onto.[529]

16. Getting used to complexity
One of the more interesting attempts at integrating EfS into a university has been undertaken in Basle, Switzerland. In 1991 a private foundation was established, called Mensch – Gesellschaft – Umwelt (MGU) (Humankind – Society – Environment), with the aim of providing transdisciplinary teaching and research for students of all faculties of the university of Basle. There was a very conscious attempt to bridge the gap between the sciences, not to deteriorate into the "science wars", but to offer a view of sustainability integrating social, legal, human and natural sciences. Teaching comprises five main themes, namely nature, perception, ethics, development and technology, and tutors are also consciously drawn from different disciplines and faculties.

All students of the university of Basle can take the scheme as their first subsidiary subject. It is specifically intended to complement a sound disciplinary degree with the contextualisation necessary to approach sustainability issues. The scheme itself falls into three parts:
• A compulsory first part where the students, irrespective of their home department, have to take courses on the scientific fundamentals of sustainability both in the social and human as well as the natural sciences; on scientific methodologies – where students are confronted with different scientific approaches

---

[529] Personal conversation with Ueli Bernhard, director, training centre, 26.10.2000 in Berne (Bollwerk 35, CH – 3011 Bern, Switzerland, tel: ++41-(0)31-3121262, fax: ++41-(0)31-3105050, email: wwfbildung@bluewin.ch).

and their relative merits to approach specific questions; and on "images of nature" – where historically and culturally different conceptions of nature are analysed to sensitise students to our specific perception/ construction of nature.

- A second part where students can choose transdisciplinary courses on any of the five main themes, yet they have to select courses outside their home discipline so that they learn to understand connections between various disciplines.
- The final third part is, in my view, the most interesting one. Students with a different disciplinary background form small groups to work on a transdisciplinary project which has to be a 'real' project from outside the ivory tower and it has to involve participation with outside groups. Past projects include: living in Basle, experiencing the greenhouse effect, ecotourism in Alghero, an eco-audit of the university of Basle, an eco-fridge, and resource management in the kitchen.[530]

Since the aim is not to produce specialisation but the capacity to work together with other specialists and laypeople in order to address complex real problems, the scheme focuses very consciously on social competences. It is not good enough to tell people to work together in a transdisciplinary project; you have to equip them with the necessary skills to co-ordinate groups, to work together with outside interests, to communicate, to organise team processes. These skills are taught throughout the scheme in so-called competence modules. Another indication that these skills – which one needs to successfully complete transdisciplinary projects – are taken seriously is the fact that of the 22 points you can score in the exam, 6 are awarded for these social and organisational skills.

Yet there is another aspect to the MGU scheme. Starting from the insight that learning doesn't work very well in traditional lecture delivered courses, the provision focuses very much on interactive, exemplary learning in groups which tries to foster in students a sense of responsibility for their own learning.[531]

If we bear in mind the importance of EfS for *all* students a scheme like MGU's seems like the ideal candidate for a *compulsory* part of any degree scheme. Yet in terms of the number of students MGU reaches in Basle the current situation is nowhere near this ideal. In 1999/2000 the university of Basle had 7783 students in all faculties. MGU teaches about 250 to 280 regular students per year and about 100 to 150 who just follow individual parts. The distribution between faculties is not as even as it would be in an ideal world either. While humanities and social science students make up about 32 per cent of the whole student population, they account for about 60 per cent of MGU students. Natural science stu-

---

[530] See http://www.unibas.ch/mgu/publikationen/Publ3MGU.html.

[531] Personal conversation with Paul Burger, MGU, 15.12.2000, Basle; Pascale Vonmont, 'Aufbau und Integration eines neuen Umweltlehrganges an der Universität Basel: Programm Mensch – Gesellschaft – Umwelt (MGU)', in *Umweltbildung in Theorie und Praxis*, 1998, 167-175; Helmut Brandl *et al.*, 1999; and the MGU website [http://www.unibas.ch/mgu/].

dents, accounting for 19.5 per cent of all students, make up about 30 per cent of MGU students. But economics and law, arguably faculties which would need a good dose of redirection towards sustainability, see much lower ratios. Economics students account only for about 3 per cent of MGU students (12.3 per cent overall), while for law students the ratio is around 1 per cent (14.7 per cent overall) (Yetergil *et al.*, 2000, 151).[532]

17. Get your institution a policy

A very good way to move educational institutions towards more sustainable behaviour is to get an environmental or sustainability policy accepted *and* implemented. The best way to guarantee implementation is the accreditation to a nationally or internationally recognised environmental management scheme (such as the EU's Eco Management and Audit scheme or ISO 14001; see Ali Khan, 1996, 3, 11-12). It is important to get the support of as many student groups, faculty and administrators as possible, and to institutionalise this support in an environment or sustainability committee, with representation from all these groups. Experience in most institutions also seems to indicate that the best progress can be made if there is a dedicated post of environmental/sustainability co-ordinator. Otherwise the voluntary members of the committee rapidly reach their limits in terms of time and capacity. You might find it interesting to view other environmental/ sustainability policies of universities; and then get in touch with the people who wrote them to find out about the implementation process.[533] *Sustainable Development on Campus* also provides a handy "Policy Evaluation Instrument".[534]

18. Make EfS a core of all teaching

A more radical approach than MGU, but akin to the attempts referred to above which start from concrete problems and then organise teaching around the complexity of the issues at hand (e.g. canoes, dry latrines, see boxes 21 and 22 on p. 296f.), is tried in a handful of institutions around the world. They organise the whole body of teaching *within* the framework of sustainability, focusing every discipline on the environment or at least forcing every student to take compulsory courses in that area. I would like to present four examples:

---

[532] Figures for the University of Basel taken from the annual report, section "Studieren" [http://www.zuv.unibas.ch/jb/1999/029_studieren_1.pdf].

[533] See Environmental Policies of Universities [http://www.swan.ac.uk/german/modulepages/green.htm#EnvPoliciesUniversities]; Sustainable Campus Policy Bank [http://iisd1.iisd.ca/educate/policybank.asp]; Campus Greening and Institutional Transformation [http://www.second nature.org/resource_center/resource_center_profiles.html].

[534] See http://iisd1.iisd.ca/educate/evaluate.asp.

- *Green Mountain College*, Vermont [http://www.greenmtn.edu/]

    All Green Mountain College students are required to complete the 37
    credit Environmental Liberal Arts Program (ELAP). ELAP combines the
    skills and content of a strong liberal arts course of study with a focus on
    the environment. Green Mountain College believes that a thorough un-
    derstanding of natural and social environments, and our relationships
    with them, coupled with the skills, knowledge, and courage necessary to
    act as responsible citizens in a globally interdependent world, are central
    to the development of a person's intellect and character. The core theme
    of the environment, in both a local and global sense, is thus embedded in
    and defines the philosophy of a liberal arts education at the College. (...)
    The four interdisciplinary core courses provide a common learning ex-
    perience and body of knowledge that fosters a sense of community. (...)
    In recognition of the complexity of the linkages between humans and the
    natural world, each of these courses taps expertise and skills from a vari-
    ety of disciplines.[535] The program consists of four core courses [Year
    One: Images of Nature (ELA 1000); Voices of Community: First Year
    Writing Seminar (ELA 1500); Year Two: Dimensions of Nature (ELA
    2000); Year Four: A Delicate Balance (ELA 4000)] *that all students*
    complete and eight additional courses chosen from four distribution cate-
    gories [Scientific Endeavor (e.g. Environmental Science), Social Per-
    spectives (e.g. Simplicity and Sustainability), Humanities (e.g. Envi-
    sioning the Good Society), Health and Well Being (e.g. Health and the
    Natural Environment)]. [http://www.greenmtn.edu/learning/ela.asp](my
    emphasis)[536]

- *International Independent University of Environmental & Political*
  Sciences (IIUEPS), Moscow:
  IIUEPS is the first university in Russia and the CIS countries to focus all
  disciplines taught on sustainability. The pedagogical foundation is an
  "ecological-humanitarian" concept which starts from the assumption that
  you have to provide sound scientific *and* social-humanitarian education in
  order to enable ecological literacy and to contribute "to the formation of a
  new ethics necessary for the reconsideration of Man and Nature interrela-
  tions". The most interesting aspect of IIUEPS is the integrated nature of
  teaching, starting from the assumption that every discipline has to face up

---

[535] "In our 'Images of Nature' course, for example – which uses literature to introduce students to
several important ecological concepts – I taught with a philosopher, two biologists, two artists, a
music teacher, an economics teacher, a sociologist, and two from the education department."
(Laird Christensen, Assistant Professor of English Literature, Green Mountain College, email conversation, 5.3.2001)

[536] According to Laird Christensen, other colleges that have an ELA core include Prescott College,
Northland College, Unity College, and College of the Atlantic (email conversation, 5.3.2001).

to their impact on the environment and work towards solutions. The other dimension, from which the more rationally orientated Euro-Americans might do well to take heed, is that due to this comprehensive approach "environmental education is oriented towards *spiritual and moral values, a responsible attitude to nature, and a combination of careful attitudes with aesthetic perception."*[537]

- *EMS (Environmental Middle School)* (Portland, Oregon):
  EMS follows a similar approach, but at secondary school level. The school is entirely organised around the theme environment, and thereby also engages in a sustainable pedagogy, including a strong focus on experiential learning, as outlined above; all other curricular elements (maths, literacy, poetry, history, ecology, etc.) are developed out of concrete garden work, restoration work in the community, explorations of rivers etc. The report in *Ecological Education in Action* gives a detailed account of how the school was established and how such education works in practice. Many of the examples could be easily adapted to other contexts (such as teaching maths through an assessment of the waste thrown away at home: at university level this could easily be turned into a more comprehensive waste impact assessment). The following conclusion allows a more detailed appreciation of the approach at EMS:

> By moving environmental science from the margin to the center of education, and by grounding students' experiences in their own place, students at EMS are provided with a framework that enables them to create context-based knowledge and to "feel" their place. (...) Comments by parents and students indicate that this awareness is being translated into action. At EMS, practical wisdom is obtained through hands-on community service and interaction with elders. Instead of becoming "information processors" about the environment and about environmental degradation, the approach taken by EMS enables the school community to form and to develop civic sense, efficacy, and responsibility.[538]

---

[537] Irina Arefieva (IIEUPS), 'The Experience and the Praxis of Integration (Endogenisation) of Environmental Education and Environmental Health Sciences into the Programs of IIUEPS as a Model for the Environmental Education at Universities', in *Interdisziplinäre Themen in Fachstudiengängen* (2000), 44-61, here 44, 49; emphasis in the original. Contact address: International Independent University of Environmental & Political Sciences (IIUEPS), Krasnokazarmennaya-str., 14, box 20, 111250 Moscow, Tel: (095) 362-7373, Fax: (095) 361-1313, email: info@iiuep.mpei.ac.ru, website: http://iiueps.ru/english.htm.

[538] Dilafruz R. Williams and Sarah Taylor, 'From Margin to Center: Initiation and Development of an Environmental School from the Ground Up', in *Ecological Education in Action*, 1999, 79-102, here 99.

Another fascinating project, along similar lines as EMS, but rather than focusing on the entire curriculum just attempting to "reconnect public elementary schools with the traditional knowledge and natural heritage of their local communities", is the *Common Roots* project.[539]

- *Worcester College of Technology*: Environmental Education for Sustainability (environmental protection):

> Worcester College of Technology over the last three years [1997-1999] has required all HE students in all Dep[artmen]ts to take a Sustainability Unit. This was originally written by the Further Education Development Agency but rewritten by ourselves and accredited. It is delivered by subject tutor (as opposed to environmental professionals) and included personal and vocational aspects of sustainability. Unfortunately funding changes have led us to drop this programme this year which is a major shame. However we have had a superb experience and the tutors involved have fundamentally changed their approach to their subject.[540]

The unit was "designed to offer our students a basic environmental literacy" and the learning outcomes emphasise "the importance of natural processes and resources, the impact of human activity on the environment, the legislative framework and the economic and social benefits of good practice" and the experiential learning approach stresses personal responsibility.[541] The fate of this successful and interesting scheme just reminds us that any such projects are always under threat from the Master Science.

19. Create time and space for transdisciplinary EfS:
The Interfakultäre Koordinationsstelle für Allgemeine Ökologie (IKAÖ) at the university of Berne has a very similar structure and approach to MGU's as described (see point 16 above) in that it is a subsidiary scheme open to all students of the university attempting to add transdisciplinary knowledge and understanding to disciplinary education: "The aim of the scheme is to produce biologists, psychologists, economists etc. who are capable of an indepth critical analysis of ecological questions and interdisciplinary work." (Helmut Brandl *et al.*, 1999, 154) I therefore don't want to go into too much detail,[542] but would just like to mention two problems the IKAÖ ran into and one useful, transferable tool:

---

[539] See Joseph Kiefer and Martin Kemple' 'Stories from Our Common Roots: Strategies for Building an Ecologically Sustainable Way of Learning', in *Ecological Education in Action*, 1999, 21-45, here 21.

[540] Personal email from Iain Patton, Environmental Co-ordinator, Worcester College, 21.9.1999.

[541] Scheme of Work for "Environmental Education for Sustainability" (17.6.1997, [1-3]), as supplied by Iain Patton.

[542] For further information see Defila/Di Giulio, 1995; Yetergil *et al.*, 2000, 154; and the IKAÖ website [http://ikaoewww.unibe.ch].

- The timetabling of transdisciplinary teaching is a grave problem. Especially the natural sciences and law have very tight timetables, with set hours, and take no account of other disciplines. This tendency to squeeze out time will get worse with the proposed shortening of study times in Switzerland and Germany. The best strategy would be to free up one half day per week which is reserved for interdisciplinary modules.[543]
- In fact, if we take seriously the long-term necessity and value of EfS, the way forward would be at least an equal allocation of time and resources to disciplinary and transdisciplinary teaching which would necessitate a restructuring of the university as a whole (Defila/Di Giulio, 1995, 244).
- The concrete teaching experience at the IKAÖ has led tutors to the insight that in order to enable EfS the most important aspect is self-reflection of the sciences involved: "Our aim is to enable students to understand the specific perspective of their own discipline, to communicate this to others, especially lay people, and to contribute with their disciplinary knowledge and skills to an interdisciplinary working process." Based on this and the acknowledgement that templates for such courses are often missing they have developed a guide for a "general introductory course into the sciences for interdisciplinary-ecological degree schemes". It attempts to give a sound basis for interdisciplinary work, dealing with the most common problems and their solutions.[544]

20. If all else fails, make EfS core in your discipline
At many universities, rather than attempting to provide EfS for all their students, individual departments have started to offer core courses, often compulsory ones, with EfS content.

- At the University of Oxford (UK) the Archaeology and Anthropology department offers a "compulsory course on Environment and Culture" as does the Human Sciences department with a course on "human populations and their environment". Environmental economics is a core topic in the departments of Economics & Management and Philosophy, Politics & Economics. 80 per cent of Biological Sciences students take the option "Environmental Biology" whereas one of the main subject streams in Earth Sciences is "Environmental Geology".[545]
- At the economic university of St. Gallen, Switzerland, the IWÖ (Institut for economy and ecology) provides compulsory courses for all students which also form part of the exam requirements: students take "natural foundations of economic activity" in the first semester and "the enterprise

---

[543] Personal conversation with Prof. Ruth Kaufmann-Hayoz, head IKAÖ, Berne, 9.11.2000.
[544] Unfortunately this guide is only available in German so far (Defila *et al.*, 2000).
[545] Oxford Environmental Liaison Group (1999), *Oxford Environment. The University of Oxford's expertise and knowledge in environmental research and teaching* (University of Oxford), 20.

in the social and ecological environment" in the fourth semester. There is also an interdisciplinary Ph.D. programme called "Economy and Ecology".[546]

- *The US Experience:*

> In 1989, Tufts University became the first US university to make environmental literacy a goal for all graduates by creating the Tufts Environmental Literacy Institute. The Institute develops the capability of faculty from all disciplines to integrate environmental and sustainability concerns into their teaching. A consortium of seventeen colleges, based at Clark Atlanta University and that serve African American, Hispanic and Native American populations, has made significant changes in curriculum, operations and community outreach to promote environmental justice and sustainability. In the last three years, Northern Arizona University has revised eighty-eight courses from nearly every discipline to make environmental and sustainability concerns a central thrust in the curriculum. (Cortese, 1999a)

21. Indigenous Education
Kawagley and Barnhardt have developed educational principles out of the indigenous world view of Alaska Native people: they can serve very well as EfS parameters:

"Indigenous View	Educational Application
Long-term perspective	Education must be understood (and carried out) across generations
Interconnectedness of all things	Knowledge is bound to the context in which it is to be used (and learned), and all elements are interrelated
Adaptation to change	Education must continuously be adapted to fit the times and place
Commitment to the commons	The whole is greater than the sum of its parts"[547]

22. Cultivate the natural scientists and vice versa
The technical university of Zurich, Eidgenössische Technische Hochschule (ETH), does on a compulsory basis what its counterpart, the university of Zurich, does not (yet) do in the opposite direction. All students who want to graduate from the ETH have to pursue courses within the integrated faculty of social, legal and human sciences in order to broaden their horizons and to enable them to

---

[546] Thomas Dyllick, 'Universität St. Gallen – IWÖ-HSG: Vom Einmann-Betrieb zum Institut für Wirtschaft und Ökologie', in *GAIA*, 7 (1998) 2, 153-154, here 153.
[547] Angayuqaq Oscar Kawagley and Ray Barnhardt, 'Education Indigenous to Place: Western Science Meets Native Reality', in *Ecological Education in Action*, 1999, 117-140, here 134.

undertake interdisciplinary work. This is also part of their final exam.[548] When the university, on the other hand, will introduce compulsory natural science education for students of their social and human sciences faculties is anyone's guess.

23. A pot full of other ideas
- *Universitat Politècnica de Catalunya (UPC)*: This technical university in Barcelona has proven to be very innovative in terms of institutional and curriculum greening. It produced its first Environmental plan in 1996 and has published an environmental report on a yearly basis since. Here I just want to focus on the most interesting ideas developed there which could and should be replicated at other educational institutions:
  o The UPC Environmental Plan stresses institutional commitment to greening "as one of the main political commitments of the University". In policy terms this has led to the situation that *all vice-rectors* are specifically responsible for all the environmental issues in their area (Capdevila *et al.*, 1999, 3).
  o A compulsory part of every final thesis project is environmental impact. This was considered an "urgent priority, since this was the fastest and most effective way of reaching the general student body." (Capdevila *et al.*, 1999, 5)
  o There is a dedicated staff development course "to provide an incentive for curricular greening in all subject areas": "The course, which is 16 hours in duration, provides a basic overview of environmental issues and is given by the University's most qualified experts." (Capdevila *et al.*, 1999, 7)
  o In 1998 the university published the book "Environment and Technology. The UPC Environment Guide". 65 staff from 24 departments produced it and it was offered free of charge to 2,280 lecturers. By 1999 700 teaching staff (32 per cent of total) had taken up the offer (Capdevila *et al.*, 1999, 7).
  The authors of the report note that all this activity has so far "only achieved minor changes in the teaching and research being carried out, mainly because this is a large-scale project which requires faculty to change their mentality." (Capdevila *et al.*, 1999, 10)
- The "Studium Generale Groningen" offers a comprehensive series of lectures entitled "Duurzaam over de drempel: Mens en milieu in multidisciplinair perspectief" (Permanently over the threshold: humankind and environment in a multidisciplinary perspective). The topics covered are: sustainability in philosophy and religion; history and sustainability; ar-

---

[548] See Claudia Wirz, 'Geisteswissenschaften als Pflicht an der ETHZ', in *Neue Zürcher Zeitung*, No. 210, 10.9.1999, 13; homepage of the GESS faculty [http://www.gess.ethz.ch/].

chaeology and sustainability; sociology and sustainability; climate and sustainability; land use and sustainability; ecology and sustainability; literature and sustainability.[549]

- *Curriculum reform in Toronto*: The lesson from a recent reform of the curriculum by the Toronto Board of Education is that such reforms should always be done in consultation with the addressees and the wider community. It was based on a large consultation process within the community and "although the notion of 'sustainability' was not imposed, it emerged as an essential requirement in the course of the consultation" (UNESCO, 1997, 25, §75). Conclusion:

> Much of the success of the Toronto reform is due to the fact that it was not – and was not seen to be – an effort to change education to meet goals set by an elite or unduly influenced by outside pressures. The impetus to change came from within. The new curriculum had equal or greater academic rigour, but far greater relevance to life outside school walls. What it demonstrates is that education for sustainable development is simply good education, and that good education needs to make children aware of the growing interdependence of life on earth – interdependence among peoples and among natural systems – in order to prepare them for the future. (UNESCO, 1997, 26, §78)

- *Rensselaer: Leaving a legacy:* Students majoring in EEVP (Ecological Economics, Values and Policy) start a project for institutional sustainability ("Forge a partnership between local farms and Rensselaer's dining halls", "Establish a yard and food waste compost program", "Green Rensselaer procurement policy" etc.) "as a freshman and see the project through to completion (likely to take several years). Students will leave legacies upon graduation making Rensselaer a model of institutional sustainability".[550]

24. Don't reinvent the wheel

Contrary to the misconception that EfS is only just starting and that there are no decent programmes, templates and concrete ideas around (as some commented in response to my questionnaire), there is by now a whole host of help available. Here I am just picking out some of the internet resources which I found most helpful:

---

[549] Contact: Prof. Dr. L. Hacquebord, Arctic Centre, University of Groningen, PO Box 716, 9700 AS Groningen, The Netherlands, Tel. +31 50 3636834, FAX. +31 50 3634900, email: l.hacquebord@let.rug.nl; website: http://www.let.rug.nl/arctic/.

[550] Steve Breyman, 'Sustainability Through Incremental Steps? The Case of Campus Greening at Rensselaer', in *Sustainability and University Life*, 1999, 79-87, here 84-85.

- *Second Nature: Education for Sustainability*
  [http://www.secondnature.org/]:
  This is, without a doubt, the most comprehensive site on EfS. Under "Vision" you will find background information, documents and publications. The most useful part is the "Resource Center" with the following sections:
  - o   Calendar: Sustainability events of interest to higher education
  - o   Sustainability Contacts: Information about other relevant organizations and individuals
  - o   Bibliography: Books, articles, videos and other resources
  - o   Courses & Methods: Course syllabi and course projects pertaining to sustainability
  - o   Curriculum Greening Guide: A guide to incorporating sustainability concepts into your curriculum
  - o   EFS Profiles: Sustainability efforts at higher education institutions
  - o   EFS Writings: Recommended writings about Education for Sustainability
- *Blueprint for a green campus: The Campus Earth Summit Initiatives for Higher Education*
  [http://www.envirocitizen.org/cgv/blueprint/index.html]:
  The document gives very detailed, practical guidance for greening the curriculum and the institution in all its aspects. The "Executive Summary of Recommendations" is a good starting point for anybody who wants to initiate change towards more sustainable institutional practice.
- *Association of University Leaders for a Sustainable Future: Leadership for Global Environmental Literacy* [http://www.ulsf.org/index.html]:
  This is the home of the Talloires Declaration which set the ball rolling for EfS in the university sector by accepting the responsibility universities have for a sustainable future. There is a very helpful section on "Programs & Services", including the "Sustainability Assessment Questionnaire" for colleges and universities.
- *The Forum for the Future Higher Education 21 Project*
  The HE 21 Project was aimed at generating and promoting best practice for sustainability across the HE sector. It developed very useful hands-on advice on a whole range of sustainability issues relevant to universities. Unfortunately, the original HE21 website has been discontinued by the Forum for the Future, even though it is widely referenced. After some pressure from the EfS community in the UK, the follow-on project HEPS (Higher Education Partnership for Sustainability) which is also run by the Forum for the Future, has at least made the best practice bulletins on resource efficiency, waste, purchasing, environmental management systems, student initiatives, community learning and biodiversity available for download again at http://www.heps.org.uk/pages/download.asp.

- *Sustainable Development on Campus: Tools for Campus Decision Makers* [http://iisd1.iisd.ca/educate/default.htm]: This falls into five useful sections:
  - o Declarations: context and purpose for implementing EfS on campus.
  - o Learn here: learn about integrating EfS into the curriculum and then putting it into practice on campus.
  - o Policy Bank: Find campus policies on procurement, recycling, waste management, and related issues. Use the prototype policy evaluation tool. Submit your policies.
  - o Resource links to reports and guides for university leadership, green campus administration, curriculum issues and student action.
- *Greening the Campus: what can I do?* [http://www.swan.ac.uk/environment/greening.htm]: A practical guide, advising on implementation of various greening aspects (such as guiding principles, paper, PC usage, energy, purchasing, office pollution, recycling/waste, transport/car sharing, water, curriculum, staff development), including additional links.
- I have found the *Education for Sustainable Development Toolkit*, developed by Rosalyn McKeown (1999), to be a very useful tool for games to introduce the concept of sustainability [http://www.esdtoolkit.org/].

25. Don't forget to live

We have seen above that some of the strongest opposition to change doesn't stem from a lack of knowledge or awareness, but from the absence of political will, from entrenched habits and traditions, from structures which prevent, often on the level of practicalities, a sustainable, effective and sensible organisation of daily life in an educational institution. It is therefore worthwhile, when planning any institutional or curricular greening, to bear this in mind, to look very closely at work routines, habits and perceived needs of employees. The next step is to work out new routines and organisational structures which favour sustainable behaviour, and incentives to adopt these. Undesirable behaviour can be discouraged through financial disincentives and by linking it to a negative image or unfavourable career structures.[551]

26. Start from real life

We have already referred to institutions which more or less organise all their teaching from a sustainability core, as it were (see IIUEPS and EMS above as well as boxes 21 and 22 on p. 296f.). Yet there are examples of learning

---

[551] See Karl-Werner Brand, Angelika Poferl and Karin Schilling, 'Umweltmentalitäten. Wie wir die Umweltthematik in unser Alltagsleben integrieren', in *Umweltbildung und Umweltbewußtsein*, 1998, 39-68, here 49-50.

opportunities which are more limited, but possibly easier to implement. Stephanie Kaza has provided a whole list of interesting ideas for the adaptation of experiential, real-life learning to university courses, especially in the humanities.

- The first idea I found particularly important since this aspect is often forgotten. It focuses on the personal dimension of learning, trying to further reflection and integration of the studies into one's own life. Kaza attempts this with an ongoing journal of subjective responses to the class discussion.[552]

- Life cycle analysis (including "energy costs, groundwater pollution, shipping distances, volume consumed (...), labor policies in the production place, associated environmental hazards, and alternative, less environmentally destructive options") of products students hold dear, always wear and consume, in order to connect the learning experience to their daily lives: coffee, beer, soy milk, tampons, paper. Other options are complete campus analysis of food operations, investments by the relevant university, pesticide use, water management, waste, procurement, paper, energy etc.[553]

- Holding a model Earth summit or a Human Rights tribunal. The value of organising such "complex and highly choreographed events", publicly visible, lies in their mobilising capacity: students have to organise, communicate with each other and people from outside such as the media, they have to prepare statements, documentation, testimonials; they are engaged "at the feeling level" and are encouraged to "speak out and give voice to their concerns"; in other words it is a thoroughly empowering process, but one which only can happen in a community/group.[554]

27. Don't wait for any leaders to get round to acting: do it yourself
The history of environmental education and now EfS, the process of implementation of *Agenda 21*, but also the numerous international conferences on sustainability have shown that it is both illusory and unnecessary to wait for some global action from government or business leaders. Don't look for any excuses, there is only one way: get up and get going. The real potential for change lies in local communities, not states, global bodies or the like:

> If sustainable development/creating a sustainable future is to be implemented, and there is a sense of urgency, it will happen at the local level, not at the national and international level. These levels must support local initiatives – thus national and international programs should be defined after you define what you would

---

[552] Stephanie Kaza, 'Liberation and Compassion in Environmental Studies', in *Ecological Education in Action*, 1999, 143-160, here 149.
[553] Ibid. 151.
[554] Ibid. 158.

like to do at the local level. Otherwise, it will never happen! (Jean Perras in *ESDebate*, 2000, 40)

28. EfS means learning for all

Stephen Sterling has pointed out that whatever we do in the context of EfS, and wherever we do it, we need to be continuously open for self-reflection on what we are doing and what we are claiming: We need "some humility and readiness to learn – we are all learners in the changes we are interested in. And that includes 'experts'!" (in *ESDebate*, 2000, 27)

Bibliography

Structure:
1. Sustainability Education Resources
2. Sustainability Resources
3. Other Resources

[*Note*: URLs are correct as at 25 January 2002 (this applies to the whole text).]

Sustainability Education Resources

Shirley Ali Khan (1995), *Taking Responsibility: Overview* (London, Pluto Press; WWF UK [The Environmental Agenda: Taking Responsibility: Promoting Sustainable Practice through Higher Education Curricula]).

Shirley Ali Khan (1996), *Environmental Responsibility: A Review of the 1993 TOYNE REPORT*, commissioned by Welsh Office, Department of the Environment, Department for Education and Employment (London, HMSO).

Aaron S. Allan (1998), *Greening the Campus: Institutional Environmental Change at Tulane University* (Honors Thesis, Tulane University) [http://www.tulane.edu/~greenclb/thesis].

Anna Ashmole (1996), *Curriculum Greening: A resource pack for integrating environmental perspectives into courses* (University of Edinburgh's Environmental Agenda) [http://www.cecs.ed.ac.uk/greeninfo/gcpack/index2.htm].

Bedingungen umweltverantwortlichen Handelns von Individuen. Proceedings des Symposiums "Umweltverantwortliches Handeln" vom 4.-6./7. September 1996 in Bern (1997), ed. by Ruth Kaufmann-Hayoz (Universität Bern, Interfakultäre Koordinationsstelle für Allgemeine Ökologie [IKAÖ]).

Blueprint for a Green Campus: The Campus Earth Summit Initiatives for Higher Education (January 1995), ed. by Campus Earth Summit [http://www.envirocitizen.org/cgv/blueprint/index.html].

Dietmar Bolscho and Hansjörg Seybold (1996), *Umweltbildung und ökologisches Lernen: ein Studien- und Praxisbuch* (Berlin, Cornelsen Scriptor).

C. A. Bowers (1995), *Educating for an Ecologically Sustainable Culture: Rethinking Moral Education, Creativity, Intelligence, and Other Modern Orthodoxies* (New York, State University of New York Press [SUNY Series in Environmental Public Policy]).

C. A. Bowers (1997), *The Culture of Denial: Why the Environment Movement Needs a Strategy for Reforming Universities and Public Schools* (New York, State University of New York Press [SUNY Series in Environmental Public Policy]).

C. A. Bowers (2000), *Let Them Eat Data: How Computers Affect Education, Cultural Diversity, and the Prospects of Ecological Sustainability* (Athens; London, University of Georgia Press).

Helmut Brandl, Ruth Förster, Peter M. Frischknecht, Marianne Klug Arter, Christine Künzli, Gerhard Schneider and Susanne Ulbrich (1999), 'Die Evaluation von Lehrveranstaltungen in Umweltwissenschaften und Ökologie – Sichtweisen der Hochschulen', in *GAIA. Ecological Perspectives in Science, Humanities, and Economics*, 8 (1999) 2, pp. 150-158.

Ali Brownlie (1998), *Learning for a Global Society* (London, Development Education Association).

Paul Burger (1998), 'Ein engeres Band zwischen Natur- und Sozialwissenschaften knüpfen', in *GAIA*, 7 (1998) 3, pp. 234-237.

Ivan Capdevila, Jordi Corominas, Antonio Torres, Jordi Bruno, Pere Botella, Lluís Jofre (1999), 'Curriculum Greening at the Technical University of Catalonia', paper given at: *International Conference on Industrial Ecology and Sustainability*, Troyes, France, 20-21 September 1999 [unpublished].

Centre for Educational Research and Innovation (1995), *Environmental Learning for the 21st Century* (Paris, OECD).

The Concept of Sustainability in Higher Education (1999), ed. by W. Bor, P. van den Holen and A. Wals (Rome, FAO).

Anthony D. Cortese (1999), 'Education for Sustainability: The University as a Model of Sustainability' (Second Nature) [http://resources.secondnature.org/pdf/pres/univmodel.pdf]

Anthony D. Cortese (1999a), 'Education for Sustainability: The Need for a New Human Perspective' (Second Nature) [http://www.secondnature.org/pdf/pres/humanpersp.pdf].

A Curriculum for Global Citizenship: Oxfam's Development Education Programme (1997) (London, Oxfam).

CVCP/NUS/SCOP (1999), *Statement of Environmental Intent* (London, Committee of Vice-Chancellors and Principals of the Universities of the United Kingdom (CVCP) [document I/99/58 (a) and (b)]).

Rico Defila and Antonietta Di Giulio (1995), 'Ein übergreifendes Orientierungsangebot für alle Fächer? Die Studien in Allgemeiner Ökologie an der Universität Bern', in *Das Hochschulwesen*, 43 (1995) 4, pp. 240-246.

Rico Defila, Antonietta Di Giulio and Matthias Drilling (2000), *Leitfaden Allgemeine Wissenschaftspropädeutik für interdisziplinär-ökologische Studiengänge* (Bern, IKAÖ [=Allgemeine Ökologie zur Diskussion gestellt; No. 4]).

Distance Education and Environmental Education (1998), ed. by Walter Leal Filho and Farrukh Tahir (Frankfurt/M., Peter Lang [=*Environmental Education, Communication and Sustainability*; Bd. 2]).

Ecological Education in Action. On Weaving Education, Culture, and the Environment (1999), ed. by Gregory A. Smith and Dilafruz R. Williams (Albany, NY, State University of New York Press).

Education and Learning for Sustainable Consumption (1999), ed. by OECD Environment Directorate and OECD Centre for Educational Research and Innovation (CERI) (Paris, OECD [document: COM/ENV/CERI(99)64] [final report of Workshop held 14-15 September 1998, Paris]).

Education for Sustainability (1996), ed. by John Huckle and Stephen Sterling, foreword by Jonathon Porritt (London, Earthscan).

Education for Sustainable Development. *The Development Education Journal.* 5 (1999) 2.

John Elliott (1995), 'Das dynamische Curriculum: ein Ergebnis des ENSI-Projekts', in *Politik und Praxis der Umwelterziehung: Beiträge der internationalen OECD-Konferenz vom 6.-11.3.1994 in Braunschweig*, ed. by Norbert Reichel (Frankfurt/M.; Bern, Lang [=Bildungsforschung internationaler Organisationen; Bd. 12]), pp. 127-147.

Endogenisierung der Umweltwissenschaften im Bereiche der universitären Lehre: der Fall der Biodiversität (1997), Workshop 17/18 October 1996, organised by Kommission für Umweltwissenschaften der Schweizerischen Hochschulkonferenz (SHK) (Bern, KUW/SHK).

Environmental Education. A Pathway to Sustainability (1993), ed. by John Fien (Geelong, Deakin University Press).

Environmental Education for Sustainability: Good Environment, Good Life (1998), ed. by Walter Leal Filho and Mauri Ahlberg (Frankfurt/M., Peter Lang [=*Environmental Education, Communication and Sustainability*; Bd. 4]).

Environmental Education for the 21st Century: International and Interdisciplinary Perspectives (1997), ed. by Patricia J. Thompson (New York, Peter Lang).

Environmental Education in the European Union (1997), coordinated by Pierre Giolitto for the European Commission (Luxembourg, EUR-OP).

Environmental Responsibility [Toyne Report] (1993), ed. by Peter Toyne for the Department of Education/Department of the Environment (London, HMSO).

The Environmental Responsibility of Students in the UK (1995), Conference hosted by Community Environmental Education Developments, 2-5 July 1995, St. Peter's Campus, University of Sunderland [post conference information pack].

ESDebate: International Debate on Education for Sustainable Development (2000), ed. by Frits Hesselink, Peter Paul van Kempen and Arjen Wals (Gland; Cambridge, IUCN) [including CD-Rom].

The Essex Report: Workshop on the Principles of Sustainability in Higher Education (1995) [http://resources.secondnature.org/programs/starfish/biblio.nsf/667792b38551cf7a852563 d50075b64f/ad2b6c4c05e91d32852565db00575391?OpenDocument&ExpandSection=1# _Section1].

John Fien (1993), *Education for the Environment. Critical Curriculum Theorising and Environmental Education* (Geelong, Deakin University Press).

R. Warren Flint, William McCarter and Thomas Bonniwell (2000), 'Interdisciplinary Education in Sustainability: Links in Secondary and Higher Education. The Northampton Legacy Program', in *International Journal of Sustainability in Higher Education*, 1 (2000) 2, pp. 191-202.

Förderung umweltbezogener Lernprozesse in Schulen, Unternehmen und Branchen (1996), ed. by Michel Roux and Silvia Bürgin (Basel; Boston; Berlin, Birkhäuser [=Themenhefte Schwerpunktprogramm Umwelt]).

Fool's Gold: A Critical Look at Computers in Childhood (2000), ed. by Colleen Cordes and Edward Miller (College Park, MD, Alliance for Childhood) [http://www.allianceforchild hood.net/projects/computers/computers_reports.htm].

Greening the College Curriculum: A Guide to Environmental Teaching in the Liberal Arts (1996), ed. by Jonathan Collett and Stephen Karakashian (Washington, DC; Covelo, CA, Island Press).

John Holt (1991), *How Children Learn. Revised Edition* (London, Penguin Books).

C. Hopkins, J. Damlamian and G.L. Ospina (1996), 'Evolving towards education for sustainable development: An international perspective', in *Nature & Resources*, 32 (1996) 3, pp. 2-11.

Jonathan Howard, David Mitchell, Dirk Spennemann and Marci Webster-Mannison (2000), 'Is today shaping tomorrow for tertiary education in Australia? A comparison of policy and practice', in *International Journal of Sustainability in Higher Education*, 1 (2000) 1, pp. 83-96.

Ivan Illich (1971), *Deschooling Society* (New York, Harper & Row). [http://www.la.psu.edu/philo/ illich/deschool/index.html]

Ivan Illich (1971a), 'The Alternative to Schooling', in *Saturday Review*, LIV (1971) 25 [June 19], pp. 44-48 & 59-60.

Implementing Sustainable Development at University Level: A Manual of Good Practice (1996), ed. by Walter Leal Filho, Frances MacDermott and Jenny Padgham (Geneva, CRE-Copernicus).

Interdisziplinäre Themen in Fachstudiengängen: Umweltwissenschaften, Ethik und Gender Studies (2000), Workshop 14/15 September 2000 in Thun, organised by Kommission für

Umweltwissenschaften der Schweizerischen Hochschulkonferenz (SHK) (Bern, KUW/SHK).

Bob Jickling (1992), 'Why I Don't Want My Children to Be Educated for Sustainable Development', in *The Journal of Environmental Education*, 23 (1992) 4, pp. 5-8.

Bob Jickling (1999), 'Beyond Sustainability: Should We Expect More From Education?', paper presented at: *Environmental Education Association of South Africa*, Grahamstown, 7-9 September 1999 [unpublished].

Bob Jickling and Helen Spork (1998), 'Education for the Environment: a critique', in *Environmental Education Research*, 4 (1998) 3, pp. 309-327.

Regula Kyburz-Graber (1999), 'Environmental Education as Critical Education: how teachers and students handle the challenge', in *Cambridge Journal of Education*, 29 (1999) 3, pp. 415-432.

Regula Kyburz-Graber, Dominique Högger and Arnold Wyrsch (2000), *Sozio-ökologische Umweltbildung in der Praxis. Hindernisse. Bedingungen. Potentiale. Schlussbericht zum Forschungsprojekt "Bildung für eine nachhaltige Schweiz" SPP-Umwelt des Schweizerischen Nationalfonds* (Zurich, Höheres Lehramt Mittelschulen).

Lifelong Learning and Environmental Education (1997), ed. by Walter Leal Filho (Frankfurt/M., Peter Lang).

Rosalyn McKeown (1999), *Education for Sustainable Development Toolkit*. With assistance from Charles A. Hopkins and Regina Rizzi (University of Tennesse, Center for Geography and Environmental Education) [http://www.esdtoolkit.org/].

Martha C. Nussbaum (1998), *Cultivating Humanity. A Classical Defense of Reform in Liberal Education* (Cambridge, CA; London, Harvard University Press).

David W. Orr (1994), *Earth in Mind: On Education, Environment, and the Human Prospect* (Washington, DC, Island Press).

David W. Orr (1996), 'Reinventing Higher Education', in *Greening the College Curriculum* (1996), pp. 8-23.

David W. Orr (1999), 'Reassembling the Pieces. Ecological Design and the Liberal Arts', in *Ecological Education in Action* (1999), pp. 229-236.

David W. Orr (1999a), 'Transformation or Irrelevance: The Challenge of Academic Planning for Environmental Education in the 21st Century', in *Sustainability and University Life* (1999), pp. 219-233.

Edmund O'Sullivan (1999), *Transformative Learning. Educational Vision for the 21st Century* (London, Zed Books).

Malcolm Plant (1998), *Education for the Environment: Stimulating Practice* (Dereham, Peter Francis Publishers).

Neil Postman (1999), *Building a Bridge to the Eighteenth Century. How the Past Can Improve Our Future* (New York, Knopf).

Benjamin H. Strauss (1996), *The Class of 2000 Report: Environmental Education, Practices and Activism on Campus* (Nathan Cummings Foundation [http://www.ncf.org/reports/program/rpt_campus2000/toc.html]).

Sustainability and University Life (1999), ed. by Walter Leal Filho (Frankfurt/M., Peter Lang [=*Environmental Education, Communication and Sustainability*; Bd. 5]).

Sustainability, Development and Environmental Education: potentials and pitfalls (1994), *The Development Education Journal*. Issue 2. December 1994.

Sustainable Development Education Panel (1999), *First Annual Report 1998* (London, DETR [Department of the Environment, Transport and the Regions]).

The Talloires Declaration of the University Presidents for a Sustainable Future (1995), 'Report and Declaration of The Presidents Conference', Tufts University European Center, Talloires,

France, October 4-7, 1990, quoted in *The Environmental Responsibility of Students in the UK* (1995) [declaration online at http://www.ulsf.org/programs_talloires.html].

Thompson Island Summit Proceedings: Education for a Sustainable Future, Promoting Sustainability Education in Higher Education (1996) [http://resources.secondnature.org/ programs/starfish/biblio.nsf/667792b38551cf7a852563d50075b64f/f9902adf4a10852d85 2565db005752f9?OpenDocument].

Umweltbildung im 20. Jahrhundert. Anfänge, Gegenwartsprobleme, Perspektiven (2001), ed. by Regula Kyburz-Graber, Ulrich Halder, Anton Hügli and Markus Ritter in collaboration with Kirsten Schlüter (Münster, Waxmann [=Umwelt – Bildung – Forschung; Bd. 7]).

Umweltbildung in Schule und Hochschule. Proceedings des Symposiums "Umweltverantwortliches Handeln" vom 4.-6./7. September 1996 in Bern (1997), ed. by Ruth Kaufmann-Hayoz (Universität Bern, IKAÖ).

Umweltbildung in Theorie und Praxis (1998), ed. by Adelheid Stipproweit, Günther Seeber and Fritz Marz (Landau, Knecht [=Landauer Universitätsschriften: Umweltwissenschaft und Umweltbildung; Bd. 3]).

Umweltbildung und Umweltbewußtsein. Forschungsperspektiven im Kontext nachhaltiger Entwicklung (1998), ed. by Gerhard de Haan and Udo Kuckartz (Opladen, Leske + Budrich [=Schriftenreihe 'Ökologie und Erziehungswissenschaft' der Arbeitsgruppe 'Umweltbildung' der Deutschen Gesellschaft für Erziehungswissenschaft; Bd. 1]).

Umweltproblem Mensch. Humanwissenschaftliche Zugänge zu umweltverantwortlichem Handeln (1996), ed. by Ruth Kaufmann-Hayoz and Antonietta Di Giulio (Bern, Haupt).

Umweltschutz und Nachhaltigkeit an Hochschulen. Konzepte – Umsetzung (1998), ed. by Walter Leal Filho (Frankfurt/M., Peter Lang [=*Umweltbildung, Umweltkommunikation und Nachhaltigkeit*; Bd. 1]).

UNESCO (1997), *Educating for a Sustainable Future – a Transdiciplinary Vision for Concerted Action* (Paris, UNESCO) [UNESCO document EPD.97/CONF.401/CLD.1].

Lienhard Valentin (2000), *Mit Kindern neue Wege gehen. Erziehung für eine Welt von morgen* (Reinbek bei Hamburg, Rowohlt [=rororo 60826]).

Hans van Weenen (2000), 'Towards a vision of a sustainable university', in *International Journal of Sustainability in Higher Education*, 1 (2000) 1, pp. 20-34.

Luis E. Velazquez, Nora E. Munguia and Miguel A. Romo (1999), 'Education for Sustainable Development: The engineer of the 21st Century', in *European Journal of Engineering Education*, 24 (1999) 4, pp. 359-370.

Frederic Vester (1998), *Denken, Lernen, Vergessen. Was geht in unserem Kopf vor, wie lernt das Gehirn, und wann läßt es uns im Stich?* (Munich, dtv [=dtv 33045]).

Vivian Wylie (1995), *Taking Responsibility: Humanities and Social Sciences* (London, Pluto Press; WWF UK [The Environmental Agenda: Taking Responsibility: Promoting Sustainable Practice through Higher Education Curricula]).

Devrim Yetergil, Helmut Brandl, Paul Burger, Rico Defila and Antonietta Di Giulio (2000), 'Chancen inter- und transdisziplinärer Wissenschaftspraxis an Schweizer Hochschulen', in *GAIA*, 9 (2000) 2, pp. 149-158.

Die Zukunft denken – die Gegenwart gestalten. Handbuch für Schule, Unterricht und Lehrerbildung zur Studie "Zukunftsfähiges Deutschland" (1997), ed. by Landesinstitut für Schule und Weiterbildung des Landes Nordrhein-Westfalen (Weinheim; Basel, Beltz).

Online resource collections:

Second Nature: Education for Sustainability [http://www.secondnature.org/].

Association of University Leaders for a Sustainable Future: Leadership for Global Environmental Literacy [http://www.ulsf.org/index.html].

ESSENCE Reports on higher environmental education programmes and environmental labour market in countries of the European Union [http://www.vsnu.nl/servlet/nl.gx.vsnu.client. http.ShowObject?id=6769].

Sustainability Resources

ActionAid, Berne Declaration, GeneWatch uk and the Swedish Society for Nature Conservation (2000), *Syngenta: Switching off farmers' rights?* (Bern, EvB) [http://www.evb.ch/cm_data/syngenta_e_0.pdf].

Agenda 21: Earth Summit - The United Nations Programme of Action from Rio (1993) (New York, United Nations) [http://www.un.org/esa/sustdev/agenda21text.htm].

Gregory Bateson (2000), *Steps to an Ecology of Mind*. With a new Foreword by Mary Catherine Bateson (Chicago; London, University of Chicago Press).

Sharon Beder (1997), *Global Spin: the Corporate Assault on Environmentalism* (Dartington, Resurgence).

Veronika Bennholdt-Thomsen and Maria Mies (1999), *The Subsistence Perspective. Beyond the Globalised Economy* (London, New York, Zed Books; Australia, Spinifex Press).

Murray Bookchin (1980), *Toward an Ecological Society* (Montréal; Buffalo, Black Rose Books).

Bringing the Food Economy Home (2000), ed. by Helena Norberg-Hodge, Todd Merrifield and Steven Gorelick (Dartington, ISEC [International Society for Ecology and Culture]).

Rachel Carson (1991), *Silent Spring,* introduction by Lord Shackleton, preface by Julian Huxley (London, Penguin [Original 1962]).

Alan Carter (1999), *A Radical Green Political Theory* (London, Routledge [=Routledge Innovations in Political Theory 1]).

dekila chungyalpa, Andrea Durbin, Dawn Montanye and Jon Sohn (2000), *Dubios Development: How the World Bank's Private Arm Is Failing the Poor and the Environment* (Washington, DC, Friends of the Earth).

Clifford Cobb, Gary Sue Goodman and Mathis Wackernagel (1999), *Why bigger isn't better: The Genuine Progress Indicator – 1999 update* (Oakland, Redefining Progress) [http://www.rprogress.org/pubs/gpi1999/gpi1999.html].

The Cornerhouse (1998), *Briefing 10 - Food? Health? Hope?: Genetic Engineering and World Hunger* [http://cornerhouse.icaap.org/briefings/10.html].

Robert Costanza et al. (1997), 'The value of the world's ecosystem services and natural capital', in *Nature*, 387 (1997) 6630 [15 May], pp. 253-260.

H. E. Daly and J.B. Cobb (1989), *For the Common Good* (Boston, Beacon Press).

Vine Deloria jr. (1970), *We Talk – You Listen* (Golden, CO, Fulcrum Publishing) [(1978), *Nur Stämme werden überleben. Indianische Vorschläge für eine Radikalkur des wildgewordenen Westens* (Munich, Trikont)].

The Development Dictionary: A Guide to Knowledge As Power (1992), ed. by Wolfgang Sachs (London, Zed Books).

Hans-Peter Dürr (2000), *Für eine zivile Gesellschaft. Beiträge zu unserer Zukunftsfähigkeit,* ed. by Frauke Liesenborghs (Munich, dtv).

Factor Four: Doubling Wealth – Halving Resource Use. A report to the Club of Rome (1997), by Ernst Ulrich von Weizsäcker, Amory B. Lovins and L. Hunter Lovins (London, Earthscan).

Terry Fenge (1994), *Toward Sustainable Development in the Circumpolar North* (CARC [Canadian Arctic Resources Committee] briefings) [http://www.carc.org/pubs/briefs/brief1.htm].

Jack D. Forbes (1981), *Die Wétiko-Seuche. Eine indianische Philosophie von Aggression und Gewalt* (Wuppertal, Peter Hammer) [(1992), *Columbus and Other Cannibals: The Wetiko Disease of Exploitation, Imperialism, and Terrorism* (Brooklyn, Autonomedia)].

Richard W. Franke and Barbara H. Chasin (1994), *Kerala. Radical Reform as Development in an Indian State* (Oakland, Food First Books).

Getting Down to Earth: Practical Applications of Ecological Economics (1996), ed. by Robert Costanza, Olman Segura, and Juan Martinez-Alier [International Society for Ecological Economics] (Washington, DC, Island Press).

Globalising Poverty: The World Bank, IMF and WTO – their policies exposed. The Ecologistreport, September 2000.

Greening the North: A Post-industrial Blueprint for Ecology and Equity (1998), ed. by Wolfgang Sachs, Reinhard Loske, Manfred Linz *et al.* in association with Wuppertal Institute for Climate, Environment and Energy (London, Zed Books).

Jed Greer and Kenny Bruno (1997), *Greenwash: The Reality behind Corporate Environmentalism* (New York, Apex Press).

The Group of Lisbon (1995), *Limits to competition* (Cambridge, MA, MIT Press).

A Guide to World Resources 2000-2001: People and Ecosystems: The Fraying Web of Life (2000), ed. by UNDP, UNEP, World Bank and WRI (Washington, DC, World Resources Institute) [http://www.wri.org/wr2000/index.html].

Gertrude Hirsch Hadorn (1995), 'Beziehungen zwischen Umweltforschung und disziplinärer Forschung', in *GAIA*, 4 (1995) 5-6, pp. 302-314.

Gertrude Hirsch Hadorn (1999), 'Nachhaltige Entwicklung und der Wert der Natur', in *GAIA*, 8 (1999) 4, pp. 269-274.

Vittorio Hösle (1994), *Philosophie der ökologischen Krise. Moskauer Vorträge* (Munich, Beck [=Beck'sche Reihe 432]).

Human Development Report 1996 (1996), ed. by the United Nations Development Programme (UNDP) (New York: Oxford, Oxford University Press).

Human Development Report 1998 (1998), ed. by the United Nations Development Programme (UNDP) (New York: Oxford, Oxford University Press).

Human Development Report 1999 (1999), ed. by the United Nations Development Programme (UNDP) (New York: Oxford, Oxford University Press).

Ivan Illich (1973), *Tools for Conviviality* (New York, Harper & Row) [http://www.la.psu.edu/philo/illich/tools/intro.html].

Michael Jacobs (1996), *The Politics of the Real World. Meeting the New Century. Written and edited for the* Real World *coalition* (London, Earthscan).

Hans Jonas (1984), *The imperative of responsibility: in search of an ethics for the technological age*, translated by Hans Jonas, with the collaboration of David Herr (Chicago, University of Chicago Press).

Rolf Jucker (1997), 'Zur Kritik der real existierenden Utopie des Status quo', in *Zeitgenössische Utopieentwürfe in Literatur und Gesellschaft. Zur Kontroverse seit den achtziger Jahren*, ed. by Rolf Jucker (Amsterdam; Atlanta, Rodopi [=Amsterdamer Beiträge zur neueren Germanistik; Bd. 41]), pp. 13-78.

Rolf Jucker (1998), 'Toward Dematerialization: The Path of Ethical and Ecological Consumption', in *ON...*, 12.1.1998 [http://www.onweb.org/features/new/dematerial/dematerial.html].

Joshua Karliner (1997), *The Corporate Planet. Ecology and Politics in the Age of Globalization* (San Francisco, Sierra Club Books).

Kerala: The Development Experience. Reflections on Sustainability and Replicability (2000), ed. by Govindan Parayil (London, Zed Books).

Paul Kingsnorth (2000), 'Shadows in the Kingdom of Light', in *The Ecologist*, 30 (2000) 8, pp. 34-39.

Manfred Linz (2000), *Wie kann geschehen, was geschehen muß? Ökologische Ethik am Beginn dieses Jahrhunderts* (Wuppertal, Wuppertal Institut für Klima, Umwelt, Energie [Arbeitsgruppe neue Wohlstandsmodelle]) [=Wuppertal Papers No. 111; pdf version] [http://www.wupperinst.org/Publikationen/WP/WP111.pdf].

The Living Planet Report 2000 (2000), ed. by WWF, UNEP World Conservation Monitoring Centre, Redifining Progress and the Centre for Sustainability Studies [http://panda.org/livingplanet/lpr00/].

The Lugano Report: On Preserving Capitalism in the Twenty-first Century (1999), with an annexe and afterword by Susan George (London; Sterling, VA, Pluto Press).

C. Douglas Lummis (1996), *Radical Democracy* (Ithaca; London, Cornell University Press).

John Madeley (1999), *Big Business, Poor Peoples* (London, Zed Books).

Hans-Peter Martin and Harald Schumann (1997), *The Global Trap: Globalization and the assault on prosperity and democracy*, translated by Patrick Camiller (London; New York, Zed Books).

Bill McKibben (1997), *Hope, Human and Wild. True Stories of living lightly on Earth* (St. Paul, Minnesota, Hungry Mind Press).

Donella H. Meadows et al. (1972), *The limits to growth: a report for the Club of Rome's Project on the Predicament of Mankind* (London, Earth Island).

Donella H. Meadows, Dennis L. Meadows and Jørgen Randers (1992), *Beyond the limits: global collapse or a sustainable future* (London, Earthscan).

Norman Myers, with Jennifer Kent (1998), *Perverse Subsidies. Tax $s Undercutting Our Economies and Environments Alike* (Winnipeg, International Institute for Sustainable Development).

Natural Capitalism: the Next Industrial Revolution (2000), by Paul Hawken, Hunter L. Lovins and Amory B. Lovins (London, Earthscan) [http://www.natcap.org].

Bernard J. Nebel and Richard T. Wright (2000), *Environmental Science: The Way the World Works*, 7th edition (Upper Saddle River, NJ, Prentice Hall).

Helena Norberg-Hodge (2000), *Ancient Futures: Learning from Ladakh* (London, Rider Books) [see also video with the same title, http://www.isec.org.uk/ISEC/av.html].

Helga Nowotny (2000), 'Sozial robustes Wissen und nachhaltige Entwicklung', in *GAIA*, 9 (2000) 1, pp. 1-2.

Öko-intelligentes Produzieren und Konsumieren (1997), ed. by Friedrich Schmidt-Bleek, Ursula Tischner and Thomas Merten (Basel; Boston, Birkhäuser [=Wuppertal Texte]).

Our Common Future (1987), ed. by the World Commission on Environment and Development (Oxford, Oxford University Press) [=Brundtland report].

Der patentierte Hunger: Patente auf Leben und ihre Auswirkungen auf die Ernährungssicherheit (2000) (Zurich, Erklärung von Bern [=EvB-Dokumentation; No. 5]).

Jonathon Porritt and James Wilsdon (1998), *Making Sense of Sustainability. An internal Forum for the Future paper* [http://www.forumforthefuture.org.uk/].

Jonathon Porritt (2000), *Playing Safe: Science and the Environment* (London, Thames & Hudson [=Prospects for Tomorrow Series]).

Carl-Ludwig Reichert (1974), *Red Power. Indianisches Sein und Bewußtsein heute* (Munich, Piper [=Serie Piper 80]).

Hans Ruh (1995), *Störfall Mensch: Wege aus der ökologischen Krise* (Gütersloh, Kaiser; Gütersloher Verlags-Haus [=Kaiser-Taschenbücher 141]).

Wolfgang Sachs (1999), *Planet Dialectics. Explorations in Environment and Development* (London, Zed Books).

Friedrich Schmidt-Bleek (1997), *Wieviel Umwelt braucht der Mensch? Faktor 10 – das Maß für ökologisches Wirtschaften* (Munich, dtv [=dtv 30580]).

E.F. Schumacher (1993), *Small is Beautiful. A Study of Economics as if People Mattered* (London, Vintage [Original 1973]).

Walter and Dorothy Schwarz (1999), *Living Lightly* (Chipping Norton, Jon Carpenter Publishing).

Sharing Nature's Interest. Ecological Footprints as an indicator of sustainability (2000), by Nicky Chambers, Craig Simmons and Mathis Wackernagel (London; Sterling, VA, Earthscan).

Vandana Shiva (1991), *The violence of the green revolution. Third World agriculture, ecology and politics* (London, Zed Books).

Vandana Shiva (1992), 'Recovering the real meaning of sustainability', in *The environment in question: Ethics and global issues*, ed. by David E. Cooper and Joy A. Palmer (London, Routledge), pp. 187-193.

Vandana Shiva, Afsar H. Jafri, Gitanjali Bedi and Radha Holla Bhar (1997), *The Enclosure and Recovery of the Commons: Biodiversity, Indigenous Knowledge and Intellectual Property Rights* (New Delhi, The Research Foundation for Science, Technology and Ecology).

Vandana Shiva (1997), *Biopiracy: The Plunder of Nature and Knowledge* (Cambridge, MA, South End Press).

Vandana Shiva (2000), 'Globalization and Poverty', in *Resurgence*, No. 202, September/October, pp. 15-19.

State of the World 1998. A Worldwatch Institute Report on Progress Toward a Sustainable Society (1998), ed. by Lester R. Brown, Christopher Flavin, Hilary French *et al.* (New York, Norton).

State of the World 2000 (2000), ed. by Lester R. Brown, Christopher Flavin, Hilary French *et al.* (New York, Norton).

Dieter Steiner (1998), 'Wie nachhaltig ist das Konzept der Nachhaltigkeitsforschung?', in *GAIA*, 7 (1998) 4, pp. 308-312.

Immanuel Stieß and Peter Wehling (1997), 'Nachhaltige Entwicklung der Sozialwissenschaften?', in *GAIA*, 6 (1997) 2, pp. 120-124.

Sustainability: Searching for Solutions, New Internationalist, No. 329, November 2000. [http://www.oneworld.org/ni/issue329/contents.htm]

Taking Nature into Account: Toward a Sustainable National Income. A Report to the Club of Rome (1995), ed. by Wouter van Dieren (New York, Springer).

Paul W. Taylor (1986), *Respect for Nature* (Princeton, Princeton University Press).

Frederic Vester (1997), *Neuland des Denkens. Vom technokratischen zum kybernetischen Zeitalter* (Munich, dtv [=dtv 33001]).

Mathis Wackernagel and William Rees (1996), *Our ecological footprint: Reducing human impact on the Earth*, illustrated by Phil Testemale (Gabriola Island, BC, New Society Publishers [=The new catalyst bioregional series; 9]).

Hugh Warwick (2000), 'Guilty as Charged. Report from India on a unique "Citizens' Jury" Project, set up to decide the fate of GM crops', in *The Ecologist*, 30 (2000) 7, pp. 52-53.

Ernst Ulrich von Weizsäcker (1992), *Ecological tax reform: a policy proposal for sustainable development* (London, Zed Books).

Ernst Ulrich von Weizsäcker (1997), *Grenzen-los? Globalisierung zwischen Bedrohung und Chance* (Basel; Boston, Birkhäuser).

Ernst Ulrich von Weizsäcker (2000), 'Efficiency and Justice', in *Resurgence*, No. 202, September/October, pp. 20-22.

Edward O. Wilson (1992), *The Diversity of Life* (Cambridge, MA, Belknap Press of Harvard University Press).

Sybille Wölfling Kast (1999), 'Was hindert uns daran, umweltfreundlich zu handeln? Eine psychologische Perspektive', in *GAIA*, 8 (1999) 4, pp. 279-287.
WWF (1998), *The Living Planet Report 1998* [http://www.panda.org/livingplanet/lpr/].
Zukunftsfähiges Deutschland. Ein Beitrag zu einer global nachhaltigen Entwicklung (1996), ed. by BUND and MISEREOR (Basel, Birkhäuser).

[see also: http://www.swan.ac.uk/german/modulepages/green.htm]

Other Resources

Theodor W. Adorno and Max Horkheimer (1973), *Dialectic of enlightenment*, translated by John Cumming (London, Allen Lane) [Original 1944].
James Agee and Walker Evans (1960), *Now Let Us Praise Famous Men. Three Tenant Families* (Cambridge, MA, The Riverside Press [Original Boston, Houghton, Mifflin Company, 1941]).
Alas Poor Darwin. Arguments Against Evolutionary Psychology (2000), ed. by Hilary Rose and Steven Rose (London, Jonathan Cape).
Günther Anders (1988), *Die Antiquiertheit des Menschen. Bd. 1: Über die Seele im Zeitalter der zweiten industriellen Revolution* (Munich, C.H. Beck [=Beck'sche Reihe 319]).
Hannah Arendt (1958), *The Human Condition* (Chicago, University of Chicago Press).
Hannah Arendt (1970), *On Violence* (New York, Harcourt Brace).
Lothar Baier (1993), *Die verleugnete Utopie. Zeitkritische Texte* (Berlin, Aufbau).
Pierre Bourdieu (1992), *Die feinen Unterschiede. Kritik der gesellschaftlichen Urteilskraft* (Frankfurt/M., Suhrkamp [=stw 658]).
Volker Braun (1998), *Wir befinden uns soweit wohl. Wir sind erst einmal am Ende. Äußerungen* (Frankfurt/M., Suhrkamp [=edition suhrkamp 2088]).
Volker Braun (2000), *Das Wirklichgewollte* (Frankfurt/M., Suhrkamp).
Elias Canetti (1998), *Crowds and Power* (New York, Noonday Press).
Alex Carey (1997), *Taking the Risk out of Democracy. Corporate Propaganda versus Freedom and Liberty*, ed. by Andrew Lohrey (Urbana; Chicago, Chicago University Press).
Noam Chomsky (1989), *Necessary Illusions. Thought Control in Democratic Societies* (London, Pluto Press).
Noam Chomsky (1992), *The Chomsky Reader*, ed. by James Peck (London, Serpent's Tail).
Noam Chomsky (1992a), *Deterring Democracy* (London, Vintage).
Noam Chomsky and Edward S. Herman (1994), *Manufacturing Consent. The Political Economy of the Mass Media* (London, Vintage).
Demokratie radikal (1992), *Widerspruch*, 12 (1992) 24.
'Discomfort and Joy. An interview with Bill Joy by Zac Goldsmith' (2000), in *The Ecologist*, 30 (2000) 7, pp. 35-39.
Umberto Eco (1987), *Lector in fabula. Die Mitarbeit der Interpretation in erzählenden Texten*, translated from Italian by Heinz-Georg Held (Munich; Wien, Hanser [=Edition Akzente]).
The ecocriticism reader: landmarks in literary ecology (1996), ed. by Cheryll Glotfelty and Harold Fromm (Athens; London, University of Georgia Press).
Europe, Inc. – Regional & Global Restructuring and the Rise of Corporate Power (2000), by Belen Balanya, Ann Doherty, Olivier Hoedeman, Adam M'anit and Erik Wesselius from CEO (Corporate Europe Observatory) with a foreword by George Monbiot (London, Pluto Press).
Frantz Fanon (1967), *The Wretched of the Earth* (Harmondsworth, Penguin).

Thomas Ferguson (1995), *Golden Rule. The Investment Theory of Party Competition and the Logic of Money-Driven Political Systems* (Chicago; London, University of Chicago Press).

Paul Feyerabend (1975), *Against Method: Outline of an Anarchistic Theory of Knowledge* (London, Verso).

Erich Fromm (1982), *Über den Ungehorsam und andere Essays* (Stuttgart, Deutsche Verlagsanstalt).

Donna J. Haraway (1997), *Modest_Witness@Second_Millennium. FemaleMan©_Meets_Onco Mouse™. Feminism and Technoscience* (New York; London, Routledge).

Terence M. Holmes (1995), *The Rehearsal of Revolution: Georg Büchner's Politics and his Drama 'Dantons Tod'* (Bern, Peter Lang).

Will Hutton (1996), *The State We're In* (London, Vintage).

Aldous Huxley (1977), *Brave New World* (London, Grafton).

Rolf Jucker (1995), 'Zeitgenössische Literatur und die Forderung nach Neuheit. Überlegungen im Ausgang von Christoph Heins Text *Lorbeerwald und Kartoffelacker*', in *Literaturkritik und erzählerische Praxis. Deutschsprachige Erzähler der Gegenwart*, ed. by Herbert Herzmann (Tübingen, Stauffenburg [=Stauffenburg Colloquium; Bd. 34]), pp. 49-59.

Rolf Jucker (1999), '"Mitdenken mit der Welt" – Volker Braun als Produktivkraft', in *Volker Braun Arbeitsbuch*, ed. by Frank Hörnigk (Berlin, Theater der Zeit, Literaturforum im Brecht-Haus), pp. 110-112.

Rolf Jucker (2000), 'Stefan Schütz', in *Deutsche Dramatiker des 20. Jahrhunderts*, ed. by Alo Allkemper and Norbert Otto Eke (Berlin, Erich Schmidt Verlag), pp. 692-710.

Naomi Klein (2001), *No Logo* (London, Flamingo).

Robert Kurz (1999), *Schwarzbuch Kapitalismus. Ein Abgesang auf die Marktwirtschaft* (Frankfurt/M., Eichborn).

Remo H. Largo (1999), *Kinderjahre. Die Individualität des Kindes als erzieherische Herausforderung* (Zurich; Munich, Piper).

Heinz Lippuner and Heinrich Mettler (1997), 'Literatur als Ort des Nirgendwo', in *Zeitgenössische Utopieentwürfe in Literatur und Gesellschaft. Zur Kontroverse seit den achtziger Jahren*, ed. by Rolf Jucker (Amsterdam; Atlanta, Rodopi [=Amsterdamer Beiträge zur neueren Germanistik; Bd. 41]), pp. 139-158.

Richard Llewellyn (1991), *How Green Was My Valley* (London, Penguin) [Original 1939].

C.B. Macpherson (1962), *The Political Theory of Possessive Individualism. Hobbes to Locke* (Oxford, Oxford University Press).

Peter Marshall (1993), *Demanding the Impossible: A History of Anarchism* (London, Fontana Press).

Heiner Müller (1994), 'Es gibt ein Menschenrecht auf Feigheit', in Heiner Müller, *Gesammelte Irrtümer 3. Texte und Gespräche* (Frankfurt/M., Verlag der Autoren [Theaterbibliothek]).

Heiner Müller (1998), 'Mommsens Block', in Heiner Müller, *Gedichte. Werke 1*, ed. by Frank Hörnigk (Frankfurt/M., Suhrkamp).

Nineteen Eighty-Four: Science between Utopia and Dystopia (1984), ed. by Everett Mendelsohn and Helga Novotny (Dordrecht, D Reidel Publisher [=Sociology of the Sciences: A Yearbook 8]).

Helga Nowotny (2000a), 'Re-thinking Science: From Reliable Knowledge to Socially Robust Knowledge', in *Jahrbuch 2000 des Collegium Helveticum der ETH Zurich*, ed. by Helga Nowotny, Martina Weiss and Karin Hänni (Zurich, vdf Hochschulverlag), pp. 221-244.

Christian Nürnberger (1999), *Die Machtwirtschaft. Ist die Demokratie noch zu retten?* (Munich, dtv).

Ian Pearson and Chris Winter (2000), *Where's IT Going?* (London, Thames & Hudson [Prospects for Tomorrow]).

Christian Pohl (2000), 'Wie transdisziplinäre Forschung mit Grenzen umgeht', in *Jahrbuch 2000 des Collegium Helveticum der ETH Zurich*, ed. by Helga Nowotny, Martina Weiss and Karin Hänni (Zurich, vdf Hochschulverlag), pp. 379-401.

Neil Postman (1987), *Amusing Ourselves to Death: Public discourse in the age of showbusiness* (London, Methuen).

Steven Rose (1998), *Lifelines: Biology Beyond Determinism* (Oxford; New York, Oxford University Press).

Arundhati Roy (2000), '"I wish I had the guts to shut up". Interview by Paul Kingsnorth', in *The Ecologist*, 30 (2000) 6, pp. 29-33.

Wilhelm Schmid (2000), 'Was ist und zu welchem Zweck betreibt man Askese? Kleine Geschichte eines missverstandenen Begriffs', in *Neue Rundschau*, 111 (2000) 4, pp. 9-14.

Sie bewegt sich doch. Ein Weltbilder-Lesebuch (1993), ed. by Martin Bauer and Otto Kallscheuer (Berlin, Rotbuch).

Alan Sokal and Jean Bricmont (1998), *Intellectual Impostures. Postmodern philosophers' abuse of science* (London, Profile Books).

John Stauber and Sheldon Rampton (1995), *Toxic Sludge is Good For You! Lies, Damn Lies and the Public Relations Industry* (Monroe, ME, Common Courage Press).

Clifford Stoll (2000), *High-Tech Heretic. Reflections of a Computer Contrarian* (New York, Anchor Books).

Utopien – Die Möglichkeit des Unmöglichen (1989), ed. by Hans-Jürg Braun (Zurich, vdf Hochschulverlag).

Granville Williams (1994), *Britain's Media: How They Are Related. Media Ownership & Democracy* (London, CPBF).

Christa Wolf (1989), *Accident: A Day's News*. Translated by Heike Schwarzbauer and Rick Takvorian (New York, Farrar Straus & Giroux).

Winfried Wolf (1996), *Car Mania. A Critical History of Transport* (London, Pluto Press).

George Woodcock (1983), *Der gewaltlose Revolutionär. Leben und Wirken Mahatma Gandhis* (Kassel-Bettenhausen, Zündhölzchen Verlag).

Yusuf Yeşilöz (1998), *Reise in die Abenddämmerung. Eine Erzählung* (Zurich, Rotpunktverlag).

Yusuf Yeşilöz (2000), *Steppenrutenpflanze. Eine kurdische Kindheit* (Zurich, Rotpunktverlag).

Howard Zinn (1996), *A People's History of the United States. From 1492 to the Present* (London; New York, Longman).

Abbreviations

CEO	Chief Executive Officer *or* Corporate Europe Observatory
CFC	Chlorofluorocarbons (gases causing depletion of stratospheric ozone layer)
DDT	Dichlorodiphenyltrichloroethane (highly toxic synthetic insecticide)
EfS	Education for Sustainability
ERT	European Roundtable of Industrialists
ESD	Education for Sustainable Development
EU	European Union
FAO	Food and Agriculture Organization of the United Nations
FDI	Foreign Direct Investment
FSC	Forest Stewardship Council
GM(O)	Genetically Modified (Organisms)
GNP/GDP	Gross National Product/ Gross Domestic Product
GPI	Genuine Progress Indicator
HDI	Human Development Index
I(C)T	Information (and Communication) Technology
ICLEI	International Council for Local Environmental Initiatives
IKAÖ	Interfakultäre Koordinationsstelle für Allgemeine Ökologie, Berne.
IMF	International Monetary Fund
ISEW	Index of Sustainable Economic Welfare
ISO	International Organization for Standardization
IUCN	The World Conservation Union
LCA	Life Cycle Analysis
MAI	Multilateral Agreement on Investment
MGU	Mensch – Gesellschaft – Umwelt, Basle
MIPS	Material Intensity Per Service Unit
NGO	Non-Governmental Organization
OECD	Organization for Economic Cooperation and Development
PC	Personal Computer
PR	Public Relations
PVC	Polyvinyl Chloride (industrial polymer, potential hormone disruptor)
SAP	Structural Adjustment Programme
SNI	Sustainable National Income
TABD	Transatlantic Business Dialogue
TINA	There Is No Alternative
TNC	Transnational Corporations
UNCED	United Nations Conference on Environment and Development
UNCTAD	United Nations Conference on Trade and Development
VNR	Video News Release
WTO	World Trade Organisation
WWF	World Wide Fund for Nature

Index

Walter Leal Filho (ed.)

Environmental Careers, Environmental Employment and Environmental Training

International Approaches and Contexts

Frankfurt/M., Berlin, Bern, Bruxelles, New York, Oxford, Wien, 2001.
199 pp., num. fig. and tab.
Environmental Education, Communication and Sustainability.
Edited by Walter Leal Filho. Vol. 9
ISBN 3-631-38686-9 · pb. € 30.20 /us-$31.95 / £ 20.00*
US-ISBN 0-8204-5449-4

This book presents an overview of experiences, projects and approaches related to employment in the environment sector and of trends related to sustainability. It also contains an article on career prospects for women in the field of engineering, of which environmental engineering is an important component. This publication, prepared as part of the project „Careers in Engineering" funded by the EU's LEONARDO Programme, documents a variety of experiences on environmental training useful to those involved with curriculum development, curriculum planning and other aspects of environmental education.

Contents: Experiences of Women Engineers in Ireland · Fostering Employment in the Environment Sector in Europe · Continuing Education in the Environmental Sciences · Environmental Education at Universities of Applied Sciences · Professionalization and Professional Activities in the Swiss Market for Environmental Services · The Environmental Protection Industry and Environmental Jobs in the U.S.A.

Frankfurt/M · Berlin · Bern · Bruxelles · New York · Oxford · Wien
Distribution: Verlag Peter Lang AG
Jupiterstr. 15, CH-3000 Bern 15
Telefax (004131) 9402131

*The €-price includes German tax rate
Prices are subject to change without notice
Homepage http://www.peterlang.de